IET CONTROL, ROBOTICS AND SENSORS SERIES 117

Data Fusion in Wireless Sensor Networks

The IET International Book Series on Sensors

IET International Book Series on Sensing–Call for Authors

The use of sensors has increased dramatically in all industries. They are fundamental in a wide range of applications from communication to monitoring, remote operation, process control, precision and safety, and robotics and automation. These developments have brought new challenges such as demands for robustness and reliability in networks, security in the communications interface, and close management of energy consumption. This book series covers the research and applications of sensor technologies in the fields of ICTs, security, tracking, detection, monitoring, control and automation, robotics, machine learning, smart technologies, production and manufacturing, photonics, environment, energy, and transport.

Proposals for coherently integrated international multiauthored edited or coauthored handbooks and research monographs will be considered for this book series. Each proposal will be reviewed by the IET Book Series Editorial Board members with additional external reviews from independent reviewers. Please email your book proposal to vmoliere@theiet.org or author_support@theiet.org.

Data Fusion in Wireless Sensor Networks

A statistical signal processing perspective

Edited by
Domenico Ciuonzo and Pierluigi Salvo Rossi

The Institution of Engineering and Technology

Published by The Institution of Engineering and Technology, London, United Kingdom

The Institution of Engineering and Technology is registered as a Charity in England & Wales (no. 211014) and Scotland (no. SC038698).

First published 2019

The Institution of Engineering and Technology
Michael Faraday House
Six Hills Way, Stevenage
Herts, SG1 2AY, United Kingdom

www.theiet.org

British Library Cataloguing in Publication Data
A catalogue record for this product is available from the British Library

ISBN 978-1-78561-584-9 (hardback)
ISBN 978-1-78561-585-6 (PDF)

Typeset in India by MPS Limited
Printed in the UK by CPI Group (UK) Ltd, Croydon

To my wife Rosaria and my family
— Domenico Ciuonzo

To my three little pirates: Federico, Daniele, and Claudio
— Pierluigi Salvo Rossi

Contents

About the editors

Domenico Ciuonzo was born in Aversa, Italy, on June 29, 1985. He received the B.Sc. and M.Sc. (both *summa cum laude*) degrees in computer engineering and the Ph.D. degree in electronic engineering from the University of Campania "L. Vanvitelli," Aversa, Italy, in 2007, 2009, and 2013, respectively. In 2011, he was involved in the Visiting Researcher Programme of NATO CMRE, La Spezia, Italy. In 2012, he was a visiting scholar with the Electrical and Computer Engineering Department, University of Connecticut, Storrs, CT, USA. In 2015 and 2016, he held visiting appointments at the Department of Electronics and Telecommunications, Norwegian University of Science and Technology, Trondheim, Norway. In 2018, he was a Visiting Researcher at Centre Tecnologic de Telecomunicacions de Catalunya (CTTC), Castelldefels, Spain. From 2013 to 2014, he was a Post-Doctoral Researcher with the Department of Industrial and Information Engineering, University of Campania, "L. Vanvitelli." From 2014 to 2016, he was a Post-Doctoral Researcher with the Department of Electrical Engineering and Information Technology, University of Naples "Federico II," Italy. Since 2017, he is a Researcher at NM-2 s.r.l., Naples, Italy. His reviewing activity was recognized by the IEEE Communications Letters, the IEEE Transactions on Communications, and the IEEE Transactions on Wireless Communications which nominated him "Exemplary Reviewer" in 2013/2017, 2014 and 2017, respectively. Since 2016, he is an IEEE Senior Member. In 2018, he achieved the national Italian qualification for Associate Professor in "Telecommunications" from the Italian Ministry of University and Research (MIUR). Since 2014, he is/has been member of the editorial board of different IEEE, IET and Elsevier journals. His research interests fall within the areas of data fusion, statistical signal processing, wireless sensor networks, machine learning and network analytics.

Pierluigi Salvo Rossi was born in Naples, Italy, on April 26, 1977. He received the Dr. Eng. degree in telecommunications engineering (*summa cum laude*) and the Ph.D. degree in computer engineering, in 2002 and 2005, respectively, both from the University of Naples "Federico II," Italy. From 2005 to 2008, he worked as a postdoc at the Department of Computer Science and Systems, University of Naples "Federico II," Italy, at the Department of Information Engineering, Second University of Naples, Italy, and at the Department of Electronics and Telecommunications, Norwegian University of Science and Technology, Trondheim, Norway. From 2008 to 2014, he was an assistant professor (tenured in 2011) in telecommunications at the Department of Industrial and Information Engineering, Second University of Naples, Italy. From

2014 to 2016, he was an associate professor in signal processing with the Department of Electronics and Telecommunications, Norwegian University of Science and Technology, Trondheim, Norway. From 2016 to 2017, he was a full professor in signal processing with the Department of Electronic Systems, Norwegian University of Science and Technology, Trondheim, Norway. Since 2017, he is a principal engineer with the Department of Advanced Analytics and Machine Learning, Kongsberg Digital AS, Norway. He held visiting appointments at the Department of Electrical and Computer Engineering, Drexel University, Philadelphia, PA, USA, at the Department of Electrical and Information Technology, Lund University, Sweden, at the Department of Electronics and Telecommunications, Norwegian University of Science and Technology, Trondheim, Norway, and at the Excellence Center for Wireless Sensor Networks, Uppsala University, Sweden. He is an IEEE senior member and serves as a senior editor for the IEEE Communications Letter (since 2016) and an associate editor for the IEEE Transactions on Wireless Communications (since 2015). He was an associate editor for the IEEE Communications Letter (from 2012 to 2016) and a Guest Editor for Elsevier Physical Communication (2012). His research interests fall within the areas of communications, machine learning, signal processing and sensor networks.

List of contributors

Paolo Braca
NATO Center for Maritime Research and Experimentation
La Spezia, Italy
paolo.braca@cmre.nato.int

Pradeep Chennakesavula
Department of Electrical Engineering
National Tsing Hua University
Hsinchu, Taiwan
pradeepch@erdos.ee.nthu.edu.tw

Domenico Ciuonzo
Network Measurement and Monitoring
Naples, Italy
domenico.ciuonzo@ieee.org

Edward J. Coyle
School of Electrical and Computer Engineering
Georgia Institute of Technology
Atlanta, GA, USA
ejc@gatech.edu

Subhrakanti Dey
School of Information Technology and Mathematical Sciences
University of South Australia
Adelaide, Australia
subhra.dey@unisa.edu.au

Xiaoxi Guo
Department of Electrical and Electronic Engineering
University of Melbourne
Melbourne, Australia
xiaoxig@student.unimelb.edu.au

Y.-W. Peter Hong
Department of Electrical Engineering
National Tsing Hua University
Hsinchu, Taiwan
ywhong@ee.nthu.edu.tw

Bhavya Kailkhura
Lawrence Livermore National Labs
Livermore, CA, USA
kailkhura1@llnl.gov

Soummya Kar
Department of Electrical and Computer Engineering
Carnegie Mellon University
Pittsburgh, PA, USA
soummyak@andrew.cmu.edu

Seksan Laitrakun
School of Information Technology
Mae Fah Luang University
Chiang Rai, Thailand
seksan.lai@mfu.ac.th

Alex S. Leong
Department of Electrical Engineering and Information Technology
Paderborn University
Paderborn, Germany
alex.leong@upb.de

Shaowei Lin
Pillar of Engineering Systems and Design
Singapore University of Technology
Singapore
shaowei_lin@sutd.edu.sg

Zuozhu Liu
Pillar of Information Systems Technology and Design
Singapore University of Technology
Singapore
zuozhu_liu@mymail.sutd.edu.sg

Roberto López-Valcarce
Department of Signal Theory and Communications
University of Vigo
Vigo, Spain
valcarce@gts.uvigo.es

Stefano Marano
Department of Information and Electrical Engineering and Applied Mathematics
University of Salerno
Salerno, Italy
marano@unisa.it

Vincenzo Matta
Department of Information and Electrical Engineering and Applied Mathematics
University of Salerno
Salerno, Italy
vmatta@unisa.it

Alba Pagès-Zamora
Department of Signal Theory and Communications
Polytechnic University of Barcelona
Barcelona, Spain
alba.pages@upc.es

Deepa Phanish
School of Electrical and Computer Engineering
Georgia Institute of Technology
Atlanta, GA, USA
deepa.phanish@gatech.edu

Tony Q.S. Quek
Pillar of Information Systems Technology and Design
Singapore University of Technology
Singapore
tonyquek@sutd.edu.sg

Anit Kumar Sahu
Department of Electrical and Computer Engineering
Carnegie Mellon University
Pittsburgh, PA, USA
anits@ece.cmu.edu

Pierluigi Salvo Rossi
Department of Advanced Analytics and Machine Learning
Kongsberg Digital
Trondheim, Norway
salvorossi@ieee.org

Pramod K. Varshney
Department of Electrical Engineering and Computer Science
Syracuse University
Syracuse, NY, USA
varshney@syr.edu

Aditya Vempaty
IBM Thomas J. Watson Research Center
Yorktown Heights, NY, USA
avempat@us.ibm.com

Peter Willett
Department of Electrical and Computer Engineering
University of Connecticut
Storrs, CT, USA
peter.willett@uconn.edu

Thakshila Wimalajeewa
Department of Electrical Engineering and Computer Science
Syracuse University
Syracuse, NY, USA
twwewelw@syr.edu

Wenyu Zhang
Department of Statistical Science
Cornell University
Ithaca, NY, USA
wz258@cornell.edu

Introduction

The Internet-of-Things revolution is coming to reality and most of real-life scenarios are experiencing the pervasive presence of network-enabled devices, in most cases sensors. The deployment of heterogeneous sensors at several locations enables collection of different kinds of (big amount of) data about the surrounding scenario. A wireless sensor network is the typical solution for data collection, data processing, and inference in most of Internet-of-Things applications.

Wireless sensor networks have been studied extensively in various research contexts with results explored many applications, such as surveillance, security, traffic control, health care, environmental monitoring, and industrial monitoring. These networks are often tailored for a specific application, thus their design and deployment account for specific purposes. However, severe bandwidth and energy limitations on the network make resource allocation a crucial issue for the effective system design.

A wireless sensor network consists of a number of small and inexpensive nodes dispersed over a geographic area to estimate/detect one (or more) parameter(s) of interest. Each node has sensing capabilities, limited computational and storage capabilities, limited energy availability, and is connected through low-power wireless communications to similar neighbor nodes and/or to a more sophisticated node, namely, the fusion center. Coordination of the phases in which the sensors collect the information, share the information, and process the information to achieve a common understanding of the surrounding environment is the key aspect of a wireless sensor network.

This edited book deals with data fusion in wireless sensor networks from a statistical signal-processing perspective. It develops through four parts, each made of three chapters. The first part focuses on problems related to the sensing phase (i.e., when the wireless sensor network collects the information), while the second and third parts concentrate on problems related to the reporting phase (i.e., when the wireless sensor network shares the information), in centralized and decentralized architectures, respectively. Finally, the last part emphasizes potential benefits from cross-layer approaches related to energy harvesting and security requirements.

Part I – Sensing model uncertainty

Optimal strategies for wireless sensor networks have been developed in many relevant applications. Unfortunately, in most cases, due to a lack of knowledge of the parameters of the phenomenon of interest to be inferred, it is not possible to employ optimal solutions. Alternative approaches to the generalized likelihood-ratio test which are less demanding from a computational perspective exist. Methodologies from compressive sensing and deep-learning frameworks show to be promising in acquiring signals to perform a variety of inference tasks such as detection, classification and estimation where complete signal reconstruction is not necessary.

Chapter 1 (by D. Ciuonzo, P. Salvo Rossi, and P. Willett) investigates distributed detection of a noncooperative target with a wireless sensor network. Both the cases of deterministic and random unknown signals are analyzed, resulting in one-sided and two-sided testing problems with nuisance parameters present only under the target-present hypothesis, respectively. Generalized versions of well-known score test are introduced within Davies' framework and compared to the generalized likelihood-ratio test in terms of performance and computational complexity. A convincing rationale for threshold-optimization is also provided.

Chapter 2 (by T. Wimalajeewa and P. K. Varshney) discusses how the compressive sensing theory is modified/extended so that a given inference task is performed with a desired performance criteria exploiting sparsity prior in temporal as well as spatial domains. The exploitation of joint sparse structures of multiple sensor observations to compute decision statistics is discussed. Additionally, this chapter discusses how to directly solve inference problems in the compressed domain when the signals of interest are not necessarily sparse. Sparse signal processing under practical considerations including quantization, imperfect channel conditions and physical layer secrecy is further reviewed.

Chapter 3 (by Z. Liu, W. Zhang, S. Lin, and T. Q. S. Quek) shows how neural networks and deep learning techniques can be employed in heterogeneous sensor data fusion. The basic idea of deep learning and training issues in deep neural networks are introduced with concrete examples. Then, deep learning models are adapted to solve practical sensor data fusion problems. Finally, a real-world case is studied in details, from data preprocessing and problem formulation to performance evaluation.

Part II – Reporting channel uncertainty

When used as a shared channel for distributed detection and estimation tasks, the wireless medium raises issues which differ in part from the analogous counterparts in communication systems. Despite being relatively simple, centralized architectures with a star topology are sufficient to introduce and discuss some of the main challenges related to wireless channels in sensor networks. Protocols like parallel access, multiple access and random access are discussed and results from communication theory and array processing related to the design of multiple-input–multiple-output systems are shown to be suitable in wireless sensor networks as well. Recent trends in communication systems, i.e., the use of large-size antenna arrays, are proved to provide new insights into design of future wireless sensor networks.

Chapter 4 (by S. Laitrakun, D. Phanish, and E. J. Coyle) explores applications for large, shared-channel, wireless sensor networks that have a limited time to collect and process each local sensor information to produce an improved decision/estimate.

Since the wireless sensor network has a time-delay constraint, this chapter investigates the design of a reliability-based splitting algorithm into the random access protocol. Performance trends and advantages of the proposed random-access scheme are underlined through theoretical findings and numerical results.

Chapter 5 (by D. Ciuonzo and P. Salvo Rossi) analyzes binary-decision fusion over a virtual multiple-input–multiple-output channel in presence of multipath small-scale fading. Both the cases of instantaneous and statistical channel-state information at the fusion center are considered. Optimal fusion rule is analyzed and different sub-optimal fusion rules are considered as appealing alternatives in terms of performance with improved numerical stability, reduced complexity, and lower system-knowledge requirements.

Chapter 6 (by D. Ciuonzo, P. Salvo Rossi, and S. Dey) explores channel-aware decentralized detection and estimation over a virtual (or distributed) multiple-input–multiple-output channel in the large-array regime at the fusion center. As for detection, the focus is the development and performance evaluation of (widely) linear fusion rules as opposed to the optimum-but-unsuitable log-likelihood ratio. As for estimation, amplify-and-forward mechanism from the sensors to the fusion center is assumed and a minimum mean squared error approach for estimating the source is considered in two different scenarios: (i) constrained total power minimization, and (ii) constrained mean squared error minimization.

Part III – Distributed inference over graphs

In dynamic scenarios, exploiting the graph structure of the sensor network may represent the only viable approach to distributed detection and estimation problems. Computations are carried out within the network in order to avoid raw data transmission to a fusion center, thus resulting in improved robustness and energy savings. Consensus by sensor gossiping is a popular paradigm for modern sensor networks designed for inference purposes, which ensures information retrieval from any subset of sensors at an arbitrary instant of time. The network continuously senses the surrounding dynamic environment, while consensus among its nodes is simultaneously enforced. Recursive online (in contrast to offline batch processing) detection algorithms are preferable in most multi-agent energy-constrained networked scenarios, where the priority is to obtain reasonable inference performance by expending fewer amount of resources. In the presence of incomplete data, the expectation-maximization algorithm is a popular means to iteratively compute the maximum likelihood estimate.

Chapter 7 (by P. Braca, S. Marano, V. Matta, and P. Willett) reviews the state of the art of consensus techniques, with special focus on the running consensus scheme. This chapter discusses their applications to detection and estimation problems, and finally presents several new direction of investigations.

Chapter 8 (by A. K. Sahu and S. Kar) considers the problem of recursive composite hypothesis testing in a network of sparsely connected agents. The focus of this chapter is on distributed composite hypothesis testing in multi-agent networks in which the goal is to estimate the state (possibly high dimensional) of the environment and detect which hypothesis is in force based on the sensed information across all the agents at all times. Distributed recursive generalized likelihood ratio test type algorithms are discussed and corresponding parameters characterized for a broad class of nonlinear observation models.

Chapter 9 (by R. López-Valcarce and A. Pagès-Zamora) reviews the formulation of the centralized expectation-maximization estimation algorithm and then discusses distributed versions well suited for implementation in sensor networks. The propagation of information throughout the network is embedded in the iterative parameter update by appropriately combining two terms for information diffusion and information averaging (consensus) in the update equations at each node. The chapter focuses on two relevant sensor network applications: (i) distributed estimation of the parameters of a Gaussian mixture and (ii) parameter estimation under a linear model with an unknown number of randomly malfunctioning sensors.

Part IV – Cross-layer issues

The list of problems which cannot be addressed at physical layer or benefit significantly from the interaction of multiple layers is large. The last part of this book focuses on two among the most relevant at the times of writing: energy and security. Energy harvesting techniques, which enable the conversion of ambient energy (e.g., solar, vibrational, or thermal energy) to electric energy, are being developed and used to prolong sensor lifetime with large impact in large-scale sensor networks. The broadcast nature of the wireless medium leaves itself vulnerable to eavesdropping by malicious users. Physical layer security is one of the framework which may support secure communication in sensor networks. The emergence of big and dirty data era demands new distributed solutions to tackle the problem of inference with corrupted data. While the impact of imperfect transmission channels or sensor faults on distributed inference systems is well studied, corrupted data models are philosophically different.

Chapter 10 (by Y.-W. P. Hong and P. Chennakesavula) examines distributed estimation for energy-harvesting wireless sensor networks, where the energy available at the sensors is converted entirely from ambient sources. This chapter considers both the cases where the energy arrival is random and where it is controlled by wireless charging stations. Furthermore, the sensor deployment problem is examined for the purpose of reconstructing the entire random field.

Chapter 11 (by X. Guo, A. Leong, and S. Dey) investigates decentralized remote estimation problems via multiple sensors in the presence of an eavesdropper. With the objective of minimizing average distortion at the remote estimator while keeping the average distortion at the eavesdropper above an acceptably large value, to render the eavesdropper estimate essentially meaningless. Various problem setups involving multiple sensors (with single or multiple antennas) with signal-processing techniques such as artificial noise, beamforming and optimum sensor power allocation are investigated. This chapter also studies the related problem of distortion outage minimization at the remote estimator while keeping the distortion outage at the eavesdropper at an acceptably high level. This latter problem is important in the context of delay-limited estimation.

Chapter 12 (by A. Vempaty, B. Kailkhura, and P. K. Varshney) discusses the effect of corrupted data on distributed inference networks and design strategies to ensure reliable inference performance for several practical and emerging network topologies (parallel, tree, and peer-to-peer topologies). It focuses on a generalization of the classical Byzantine generals problem in the context of distributed inference to different topologies. This chapter provides fundamental limits of distributed inference with Byzantines for several practical network topologies and presents optimal countermeasures from a network designer's perspective.

Acknowledgment

We would like to thank Val Moliere and Olivia Wilkins for their precious help and administrative support during the editorial preparation of the book.

Domenico Ciuonzo
Network Measurement and Monitoring (NM2), Naples, Italy

Pierluigi Salvo Rossi
Kongsberg Digital AS, Trondheim, Norway

Part I

Sensing model uncertainty

Chapter 1

Generalized score-tests for decision fusion with sensing model uncertainty

Domenico Ciuonzo[1], Pierluigi Salvo Rossi[2], and Peter Willett[3]

This chapter investigates distributed detection of a phenomenon of interest (POI) via decision fusion in wireless sensor networks (WSNs). The decisions are collected by a fusion center (FC), which is in charge of performing a more accurate global decision. So as to account for a realistic scenario, it is assumed that the POI presents a signature with limited spatial extent, and its exact location and emitted amplitude (or energy) are not known. More specifically, when the POI is present, the sensors observe a signal with an attenuation depending on the distance between the sensor and the (unknown) target position, embedded in Gaussian noise. The unavailability of a completely specified model defeats the applicability of the well-known (optimal) likelihood-ratio (LR) test (LRT). As a consequence, in the general case, the FC is usually in charge of solving a composite hypothesis test and the generalized LRT (GLRT) is commonly employed. Unfortunately, in these scenarios, its complexity is typically high. Accordingly, the present chapter discusses the development of generalized score tests as alternatives with reduced computational complexity. After a brief recall of the GLRT for the considered problems, fusion rules corresponding to generalized versions of well-known score tests are introduced, based on Davies' framework, since the resulting problems include nuisance parameters only under the POI-present hypothesis. The focus is on two relevant signal models, i.e., the cases of random and deterministic unknown signals, leading to one-sided and two-sided testing, respectively. Finally, a convincing (semi-theoretical) rationale for threshold-optimization is presented and analyzed.

1.1 Uncertainty in decision fusion sensing model

WSNs have attracted significant attention due to their potential improved capabilities in performing detection and estimation, with a wide range of applications, comprising battlefield surveillance, reconnaissance, security, traffic, and environmental

[1]Network Measurement and Monitoring (NM2), Italy
[2]Department of Advanced Analytics and Machine Learning, Kongsberg Digital AS, Norway
[3]ECE Department, University of Connecticut, USA

monitoring [1]. Distributed detection of a POI is among the fundamental tasks that a WSN needs to accomplish [2]. Relevant examples of a POI may be the presence of an uncooperative target in a surveillance scenario, the transmission of a primary user in a cognitive-radio system [3] or revealing a dangerous emission, such as an oil-spill source measured by an underwater sensor network or a radioactive source from a set of Geiger counters.

Due to stringent bandwidth and energy constraints, it is often assumed that each sensor quantizes its own observation with a single bit before transmission to the FC, which takes a global decision. In this context, the optimal test (under Bayesian/Neyman–Pearson frameworks) at each sensor is known to be a one-bit quantization of the local LR, i.e., a LRT.

Unfortunately in most cases, due to a lack of knowledge of the parameters of the POI to be detected, it is not possible to compute the LRT at each sensor. Also, even when the sensors *can* compute their local LRT, the search for local quantization thresholds is well-known to be exponentially complex [4,5]. Thus the bit of information being sent is usually the result of a "dumb" quantization [6,7] or represents the estimated decision regarding the detection event, based on the local measurement. In both cases, the bits from the sensors are collected by the FC and combined via a specifically designed fusion rule aiming at improved detection performance.

The optimal fusion rule in both Neyman–Pearson and Bayesian senses, which is derived from the LRT [8], is commonly referred to as *Chair–Varshney* (CV) rule. It amounts to a threshold detector on the weighted sum of binary sensor detections, with each weight depending on sensor activation probabilities (in most cases, the sensor detection and false-alarm rates). Unfortunately, also these weights typically depend on some relevant *unknown parameters* of the POI [2]. For example, the sensor detection probability is seldom known or difficult to estimate when the detection event relates to a POI described by a spatially limited signature. In fact, in the latter case, such activation probability depends on the (unknown) constituent parameters of the POI to be detected at the FC, such as its amplitude (or average power) and location (see Figure 1.1). These assumptions well fit a realistic scenario, i.e., there is no preliminary agreement between POI and sensors in order to exchange the information related to the (average) emitted power or make it possible to be estimated.

Some simple fusion approaches, based on the counting rule (CR) or channel-aware statistics, have been proposed in the literature to overcome such unavailability [9–12]. In the former case, the FC counts the number of local detections in the WSN and compares it with a threshold [13]. Unfortunately, though CR is a very reasonable approach arising from different rationales [2,14,15], it does not make any attempt to use information about the contiguity of sensors that declare (potential) POI presence. Therefore, the CR suffers from performance degradation when trying to detect spatial events.

Based on these considerations, several studies have focused on design of fusion rules filling the performance gap toward the ideal CV rule. On the other hand, in some particular scenarios, the UMP test is independent of the unknown parameters under the alternative hypothesis, so they do not need to be estimated [16,17]. Nonetheless, in the general case, the FC is usually in charge of solving a composite hypothesis test, and the GLRT is commonly employed [8].

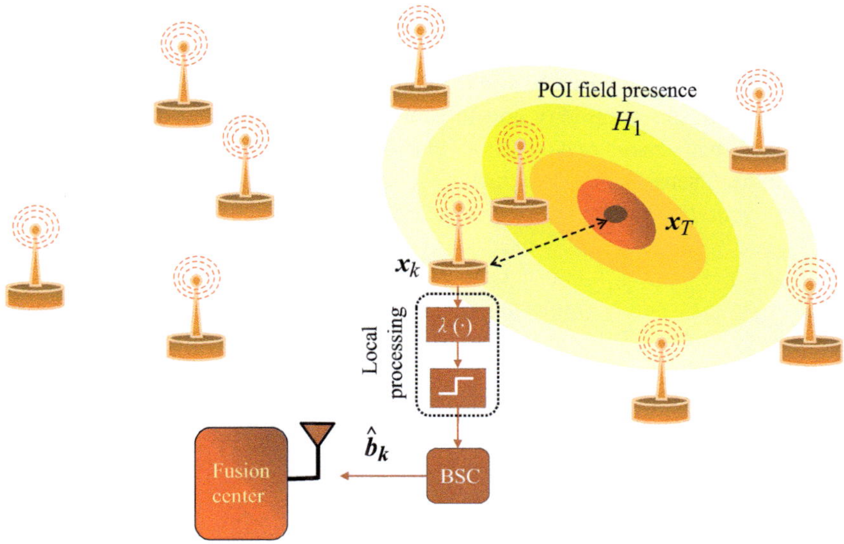

Figure 1.1 Distributed detection of a POI with a spatial signature

On the other hand, score tests represent simpler and appealing alternatives for dealing with two-sided (Rao test) and one-sided (locally optimum detection, LOD) tests. Unfortunately, their application in this context is precluded by the absence of POI location parameter information.

The aim of this chapter is then to provide a discussion on the design and optimization of generalized forms of score tests for decentralized detection of POIs whose emitted energy and location are generally both unknown. The technical difficulty related to the application of common score tests is obviated by resorting to the Davies rationale [18]. For the considered problem, two relevant scenarios pertaining to *non-fluctuating* and *fluctuating* signal cases are taken into consideration. Additionally, their performance improvement is investigated under the lens of local quantizers' optimization.

The rest of the chapter is organized as follows. Section 1.2 formulates the considered distributed detection problem, whereas Section 1.3 provides the founding rationale for the development of generalized score tests. Differently, in Section 1.4, sensors threshold optimization is approached via a semi-theoretical rationale. Final considerations and suggested further reading in Section 1.5 end the chapter.

1.2 Problem statement

1.2.1 Sensing model

In this chapter, a binary hypothesis test is considered where a collection of sensors $k \in \mathcal{K} \triangleq \{1, \ldots, K\}$ is deployed in a surveillance area to monitor the absence ($\mathcal{H}_0$) or presence ($\mathcal{H}_1$) of a POI having an isotropic partially specified spatial signature.

In other terms, when the POI is present (i.e., $\mathcal{H}_1$), it is assumed that its radiated (amplitude) signal is isotropic and experiences (distance dependent) path-loss and additive white Gaussian noise, before reaching individual sensors.

Additionally, for simplicity and to limit information exchange overhead, it is hypothesized that sensors make their local decisions *individually* without collaboration. Then, each sensor is faced to tackle the following *composite* hypothesis testing in the *deterministic (non-fluctuating)* signal case:

$$\begin{cases} \mathcal{H}_0 & : \; y_k = w_k \\ \mathcal{H}_1 & : \; y_k = \theta g(\boldsymbol{x}_T, \boldsymbol{x}_k) + w_k \end{cases} \qquad k \in \mathcal{K}; \tag{1.1}$$

whereas in the *random (fluctuating)* signal case:

$$\begin{cases} \mathcal{H}_0 : \; y_k = w_k \\ \mathcal{H}_1 : \; y_k = \xi_k g(\boldsymbol{x}_T, \boldsymbol{x}_k) + w_k \end{cases} \qquad k \in \mathcal{K}. \tag{1.2}$$

In both these scenarios, the scalar $y_k \in \mathbb{R}$ denotes the kth sensor measurement and $w_k \sim \mathcal{N}(0, \sigma_{w,k}^2)$ the corresponding measurement noise, where $\mathcal{N}(\mu, \sigma^2)$ denotes the usual Gaussian pdf. Here, due to spatial separation of the sensors, the noise contributions w_ks are assumed to be statistically independent. Furthermore, $\boldsymbol{x}_T \in \mathbb{R}^d$ denotes the *unknown* POI position, while $\boldsymbol{x}_k \in \mathbb{R}^d$ denotes the *known* kth sensor position, with the pair $(\boldsymbol{x}_T, \boldsymbol{x}_k)$ *uniquely* determining the value of $g(\boldsymbol{x}_T, \boldsymbol{x}_k)$, denoting the amplitude attenuation function (AAF). Indeed, although the sensor may accurately estimate its own position $\boldsymbol{x}_k$, the POI location $\boldsymbol{x}_T$ is *unknown* in the practice. With reference to the specific AAF, two common examples [19,20] are the *power law* and the *exponentially* attenuated models, defined respectively as follows:

$$g_{\mathrm{P}}(\boldsymbol{x}_T, \boldsymbol{x}_k) \triangleq \frac{1}{\sqrt{1 + ((\|\boldsymbol{x}_T - \boldsymbol{x}_k\|)/\eta)^{\alpha}}}, \qquad g_{\mathrm{E}}(\boldsymbol{x}_T, \boldsymbol{x}_k) \triangleq \sqrt{\exp\left[-\left(\frac{\|\boldsymbol{x}_T - \boldsymbol{x}_k\|}{\eta}\right)^2\right]}. \tag{1.3}$$

In (1.3), the parameter η controls the (approximate) spatial signature extent produced by both AAFs, while α is a positive coefficient that dictates the rapidity of signal decay as a function of the distance in the case of a power law AAF.

Additionally, in the *non-fluctuating* signal case, the emitted signal θ is assumed *unknown* and *deterministic*. On the other hand, in the *fluctuating* (random) signal case, the fading coefficient $\xi_k \sim \mathcal{N}(0, \theta)$ ($\theta > 0$ denotes the average power in this case) is adopted to model fluctuations in the received signal strength at kth sensor. In the present discussion, also the fading coefficients ξ_ks are assumed statistically independent.

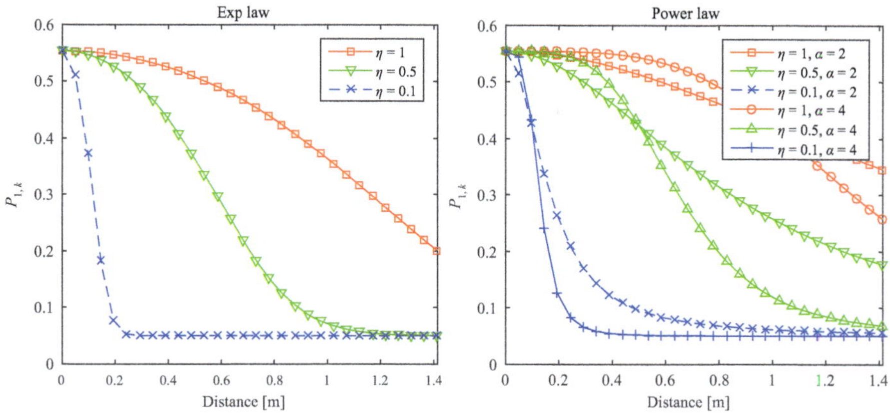

Figure 1.2 $P_{1,k}$ *vs. distance* $\|\boldsymbol{x}_T - \boldsymbol{x}\|$ *for different values of* η *(resp. of* η *and* α*) in the exponential (resp. power law) AAF case in the left (resp. right) figure. Setup parameters:* $P_{0,k} = 0.05$ *and* $(\theta/\sigma^2_{w,k})_{dB} = 10$

In view of the aforementioned assumptions, the measured signal y_k is distributed under both hypotheses as follows:

$$y_k \mid \mathscr{H}_0 \sim \mathscr{N}(0, \sigma^2_{w,k}) \tag{1.4}$$

$$y_k \mid \mathscr{H}_1 \sim \begin{cases} \mathscr{N}(\theta g(\boldsymbol{x}_T, \boldsymbol{x}_k), \sigma^2_{w,k}) & \text{(non-fluctuating case)} \\ \mathscr{N}(0,\, \theta g^2(\boldsymbol{x}_T, \boldsymbol{x}_k) + \sigma^2_{w,k}) & \text{(fluctuating case)} \end{cases} \tag{1.5}$$

1.2.2 Local processing and reporting

In the rest of the discussion, to cope with stringent bandwidth and energy budgets in WSNs, it is assumed that kth sensor quantizes y_k into only *one bit* of information, denoted with b_k.

More specifically, in the *deterministic* case, a simple approach is pursued, which avoids tackling the resulting (local) *composite hypothesis testing*, i.e., a naïve one-bit (deterministic) measurement quantization is performed, that is:

$$b_k \triangleq u\,(y_k - \tau_k), \quad k \in \mathscr{K}, \tag{1.6}$$

with $u(\cdot)$ denoting the Heaviside (unit) step function and τ_k the quantizer threshold. Then, in this case, the bit (viz. activation) probabilities have the following form:

$$P_{0,k} \triangleq \Pr\{y_k \geq \tau_k | \mathscr{H}_0\} = Q\left(\frac{\tau_k}{\sigma_{w,k}}\right); \tag{1.7}$$

$$P_{1,k}(\theta,\, \boldsymbol{x}_T) \triangleq \Pr\{y_k \geq \tau_k | \mathscr{H}_1\} = Q\left(\frac{\tau_k - \theta g(\boldsymbol{x}_T, \boldsymbol{x}_k)}{\sigma_{w,k}}\right); \tag{1.8}$$

where $Q(\cdot)$ denotes the usual complementary cumulative distribution function of a standard normal random variable. It should be noticed that in this case, since only a one-bit quantization of the measurement is applied (instead of performing a local decision statistic), these probabilities cannot be interpreted as the common local false-alarm and detection probabilities. Some relevant profiles of bit probability under $\mathcal{H}_1$ are depicted in Figure 1.2.

On the other hand, in the *random* case, although each sensor is still faced with a composite hypothesis test, it can be shown that such difficulty can be circumvented. Indeed, by considering a local decision procedure based on the Neyman–Pearson lemma [8], i.e., a test based on the *local* log-likelihood ratio (LLR) of kth sensor, i.e., $\lambda_k \triangleq \ln[(p(y_k|\mathcal{H}_1))/(p(y_k|\mathcal{H}_0))]$, it can be shown that the corresponding statistic is in the form:

$$\lambda_k = \alpha_k + \beta_k y_k^2, \tag{1.9}$$

where $\alpha_k \triangleq \frac{1}{2}\ln\{\sigma_{w,k}^2/[\sigma_{w,k}^2 + \theta g^2(\boldsymbol{x}_T, \boldsymbol{x}_k)]\}$ and $\beta_k \triangleq [\theta g^2(\boldsymbol{x}_T, \boldsymbol{x}_k)]/\{\sigma_{w,k}^2[\sigma_{w,k}^2 + \theta g^2(\boldsymbol{x}_T, \boldsymbol{x}_k)]\}$ are both *positive*. The result in (1.9) reveals that the LLR is increasing with y_k^2, *irrespective* of θ and $\boldsymbol{x}_T$. Thus, by Karlin–Rubin theorem [21, p. 65], the following *local processing*:

$$b_k \triangleq u(y_k^2 - \tau_k), \quad k \in \mathcal{K}, \tag{1.10}$$

which corresponds to an energy test deciding in favor of $\mathcal{H}_1$ (resp. $\mathcal{H}_0$) when $y_k^2 > \tau_k$ (resp. $y_k^2 \leq \tau_k$), is the *uniformly most powerful* (UMP) in a local sense, where τ_k is a suitable threshold.

Therefore, by assuming that each sensor implements processing in (1.10), the bit (viz. activation) probabilities are summarized as:

$$P_{0,k} \triangleq \Pr\{y_k^2 \geq \tau_k|\mathcal{H}_0\} = 2\,Q\left(\sqrt{\frac{\tau_k}{\sigma_{w,k}^2}}\right), \tag{1.11}$$

$$P_{1,k}(\theta, \boldsymbol{x}_T) \triangleq \Pr\{y_k^2 \geq \tau_k|\mathcal{H}_1\} = 2\,Q\left(\sqrt{\frac{\tau_k}{\sigma_{w,k}^2 + \theta g^2(\boldsymbol{x}_T, \boldsymbol{x}_k)}}\right), \tag{1.12}$$

where $P_{0,k}$ and $P_{1,k}$ collectively describe the local performance of kth sensor and assume the meaning of local false-alarm and detection probabilities, respectively [22]. It is worth noticing that a suitable (optimized) choice of τ_k in both deterministic and random signal cases is later investigated in Section 1.4.

Furthermore, to model energy-constrained reporting due to limited battery life, it is assumed that the bit b_k is sent over a binary symmetric channel (BSC) and the FC observes a (communication) error-prone $\hat{b}_k$. It is worth noticing that the BSC model arises when separation between sensing and communication layers is performed in the design phase, namely, a *decode-then-fuse* philosophy is pursued [11,23]. More specifically, the BSC model here considered is the following:

$$\hat{b}_k = \begin{cases} 1 - b_k & \text{with probability } P_{e,k} \\ b_k & \text{with probability } 1 - P_{e,k} \end{cases}. \tag{1.13}$$

Here, $P_{e,k}$ denotes the bit-error probability (BEP) of kth link, which is reasonably assumed to satisfy $P_{e,k} \leq 1/2$. Additionally, in the following discussion, it is hypothesized that $P_{e,k}$ values can be safely estimated by the FC (i.e., they are *known*). This is, for example, the case when coherent detection or noncoherent detection with orthogonal symbols is performed over a fading channel, as soon as the corresponding signal-to-noise ratio can be obtained, e.g., [15,24,25].

Finally, for sake of compactness, all the sensors' decisions are collected in the vector $\hat{\boldsymbol{b}} \triangleq [\hat{b}_1 \ \cdots \ \hat{b}_K]^T$.

1.2.3 Resulting hypothesis testing

By looking at both tests in (1.1) and (1.2), it can be observed that the *unknown* target position $\boldsymbol{x}_T$ is *observable* (i.e., can be estimated) at the FC *only* when the signal is present. This corresponds to the condition $\theta \neq \theta_0$ in the *deterministic* case, whereas for *random* case, this corresponds to the average power being $\theta > \theta_0$, where $\theta_0 \triangleq 0$.

As a consequence, in the *deterministic* case, the problem refers to a *two-sided* parameter test (i.e., $\{\mathcal{H}_0, \mathcal{H}_1\}$ corresponds to $\{\theta = \theta_0, \theta \neq \theta_0\}$) *with nuisance parameters* ($\boldsymbol{x}_T$) *present only under the hypothesis* $\mathcal{H}_1$ [18]. Similarly, in the *random* case, the problem can be cast as *a one-sided test with nuisance parameters* ($\boldsymbol{x}_T$) *present only under the hypothesis* $\mathcal{H}_1$ [18], where $\{\mathcal{H}_0, \mathcal{H}_1\}$ corresponds to $\{\theta = \theta_0, \theta > \theta_0\}$.

The problem here is the derivation of a (computationally) simple test (in the following discussion, it is used interchangeably the term "fusion rule") on the basis of vector $\hat{\boldsymbol{b}}$ in the form:

$$\Lambda(\hat{\boldsymbol{b}}) \underset{\hat{\mathcal{H}}=\mathcal{H}_0}{\overset{\hat{\mathcal{H}}=\mathcal{H}_1}{\gtrless}} \gamma_{\text{fc}}, \tag{1.14}$$

where γ_{fc} is the threshold chosen to ensure a certain *global* false-alarm rate at the sensor (in Neyman–Pearson approach) or to minimize the *global* fusion error–probability (in the Bayesian framework) [8]. The FC performance is evaluated in terms of its *false-alarm* and *detection* probabilities, defined as follows:

$$P_{\text{F}} \triangleq \Pr\{\Lambda > \gamma_{\text{fc}} | \mathcal{H}_0\}, \qquad P_{\text{D}} \triangleq \Pr\{\Lambda > \gamma_{\text{fc}} | \mathcal{H}_1\}. \tag{1.15}$$

The behavior of the global probability of detection (P_{D}) vs. the global probability of false alarm (P_{F}) is commonly denoted receiver operating characteristic (ROC) [2].

1.2.4 Background on clairvoyant LLR

It is apparent that, under hypothesis $\mathcal{H}_1$, the probability mass function (PMF) of $\hat{\boldsymbol{b}}$ assumes the explicit expression represented by the product of independent Bernoulli

PMFs (since the decisions $\hat{b}_k$ are conditionally *independent*, as an immediate consequence of mutual independence of w_ks and ξ_ks, decoupled quantization process and independent BSCs), that is,

$$P(\hat{\boldsymbol{b}}|\mathscr{H}_1) = \prod_{k=1}^{K} P(\hat{b}_k|\mathscr{H}_1) = \prod_{k=1}^{K} \left[\rho_{1,k}(\theta, \boldsymbol{x}_T)\right]^{\hat{b}_k} \left[1 - \rho_{1,k}(\theta, \boldsymbol{x}_T)\right]^{(1-\hat{b}_k)}, \quad (1.16)$$

where $\rho_{1,k}(\theta, \boldsymbol{x}_T)$ represents the *composite* activation probability under $\mathscr{H}_1$ which takes into account sensing and reporting uncertainty, that is,

$$\rho_{1,k}(\theta, \boldsymbol{x}_T) \triangleq \left[(1 - P_{e,k}) P_{1,k}(\theta, \boldsymbol{x}_T) + P_{e,k}(1 - P_{1,k}(\theta, \boldsymbol{x}_T))\right]. \quad (1.17)$$

A similar expression holds for $P(\hat{\boldsymbol{b}}|H_0)$ when replacing $\rho_{1,k}$ with $\rho_{0,k}$, i.e., the *composite* activation probability under $\mathscr{H}_0$, defined as:

$$\rho_{0,k} \triangleq \left[\left(1 - P_{e,k}\right) P_{0,k} + P_{e,k}(1 - P_{0,k})\right]. \quad (1.18)$$

The optimal decision statistic in both Neyman–Pearson and Bayesian senses is represented by the (global) LLR, given by:

$$\begin{aligned} \Lambda_{\mathrm{LLR}} &\triangleq \ln\left[\frac{P(\hat{\boldsymbol{b}}|\mathscr{H}_1)}{P(\hat{\boldsymbol{b}}|\mathscr{H}_0)}\right] = \sum_{k=1}^{K} \ln\left[\frac{P(\hat{b}_k|\mathscr{H}_1)}{P(\hat{b}_k|\mathscr{H}_0)}\right] \\ &= \sum_{k=1}^{K} \left\{\hat{b}_k \ln\left[\frac{\rho_{1,k}(\theta, \boldsymbol{x}_T)}{\rho_{0,k}}\right] + (1 - \hat{b}_k) \ln\left[\frac{1 - \rho_{1,k}(\theta, \boldsymbol{x}_T)}{1 - \rho_{0,k}}\right]\right\}. \end{aligned} \quad (1.19)$$

As anticipated in Section 1.1, the LLR is a weighted sum of sensor decisions, and it is commonly referred to as CV rule [26]. The computation of the latter weights requires knowledge of both $\boldsymbol{x}_T$ and θ, other than the BEPs $P_{e,k}$. Therefore, it is apparent that (1.19) should not be intended as a realistic element of comparison, but rather as an optimistic (upper) bound on the achievable performance, based on a *clairvoyant* assumption.

1.3 Design of generalized score tests

1.3.1 Counting rule (CR) and GLRT

In view of CV rule's infeasibility in the considered scenario, the need for alternative approaches arises.

A first simple approach can be pursued in the random case. Indeed, in the latter case, the condition $\rho_{1,k}(\theta, \boldsymbol{x}_T) \geq \rho_{0,k}\ \forall k \in \mathcal{K}$, holds. The aforementioned condition is satisfied as long as the reasonable conditions $P_{1,k}(\theta, \boldsymbol{x}_T) \geq P_{0,k}$ and $P_{e,k} \leq 1/2$ hold, respectively. The first condition implies that each reasonable local decision procedure would achieve an ROC—operation point that is more informative than an unbiased coin (i.e., above the chance line). Based on this observation, in the latter case, the well-known CR [2] may be applied, not requiring sensors' local performance for its implementation. This rule is widely used in DF due to its simplicity and lack of requirements about system knowledge. More specifically, CR is based on the following statistic:

$$\Lambda_{\mathrm{CR}} \triangleq \sum_{k=1}^{K} \hat{b}_k \,. \tag{1.20}$$

The above rule, despite its simplicity, has been obtained under different rationales in the literature [2,14,15].

In general, to take advantage of the spatial compactness of the POI, a common approach to detection in composite hypothesis testing problems is given by the GLRT, which has been derived and investigated in [7,27] for the model under investigation. Its implicit expression (in logarithmic form) is:

$$\Lambda_{\mathrm{GLR}} \triangleq 2 \cdot \ln\left[\frac{P(\hat{\boldsymbol{b}}; \hat{\theta}_1, \hat{\boldsymbol{x}}_T)}{P(\hat{\boldsymbol{b}}; \theta_0)}\right], \tag{1.21}$$

with

$$\ln P(\hat{\boldsymbol{b}}; \theta, \boldsymbol{x}_T) = \sum_{k=1}^{K} \left\{\hat{b}_k \ln\left[\rho_{1,k}(\theta, \boldsymbol{x}_T)\right] + (1 - \hat{b}_k) \ln\left[1 - \rho_{1,k}(\theta, \boldsymbol{x}_T)\right]\right\}, \tag{1.22}$$

being the log-likelihood function w.r.t. $(\theta, \boldsymbol{x}_T)$. A similar expression holds for $\ln P(\hat{\boldsymbol{b}}; \theta_0)$, when replacing $\rho_{1,k}(\theta, \boldsymbol{x}_T)$ with $\rho_{0,k}$. Also, the pair $(\hat{\theta}_1, \hat{\boldsymbol{x}}_T)$ represents the *maximum likelihood* (ML) *estimates* assuming that hypothesis $\mathcal{H}_1$ is true, that is,

$$(\hat{\theta}_1, \hat{\boldsymbol{x}}_T) \triangleq \arg\max_{(\theta, \boldsymbol{x}_T)} P(\hat{\boldsymbol{b}}; \theta, \boldsymbol{x}_T) \,. \tag{1.23}$$

The (log-)GLR statistic can be expressed equivalently as:

$$\Lambda_{\mathrm{GLR}} \triangleq 2 \cdot \ln\left[\frac{\max_{(\theta, \boldsymbol{x}_T)} P(\hat{\boldsymbol{b}}; \theta, \boldsymbol{x}_T)}{P(\hat{\boldsymbol{b}}; \theta_0)}\right] \tag{1.24}$$

$$= 2 \cdot \max_{(\theta, \boldsymbol{x}_T)} \ln\left[\frac{P(\hat{\boldsymbol{b}}; \theta, \boldsymbol{x}_T)}{P(\hat{\boldsymbol{b}}; \theta_0)}\right] = 2 \cdot \max_{(\theta, \boldsymbol{x}_T)} \Lambda_{\mathrm{LLR}}(\theta, \boldsymbol{x}_T), \tag{1.25}$$

where $\Lambda_{\mathrm{LLR}}(\theta, \boldsymbol{x}_T)$ is used to denote the LLR (cf. (1.19)) evaluated assuming the target position and amplitude (viz. average power), in the deterministic (resp. random) case corresponding to $\boldsymbol{x}_T$ and θ, respectively. Equation (1.21) reveals that Λ_{GLR} requires the solution to an optimization task, and the unavailability of the closed form of the ML estimate pair $(\hat{\theta}_1, \hat{\boldsymbol{x}}_T)$ hinders its practical implementation. This increases the computational complexity of its implementation and typically involves a grid approach on $(\theta, \boldsymbol{x}_T)$, see later discussion on computational complexity.

1.3.2 Generalized score tests

A different path for exploiting the two-sided and one-sided nature of the problem in the *deterministic* and *random* signal cases, respectively, consists in adopting the rationale proposed by Davies in [18]. The aforementioned approach allows the designer to extend score-based tests to the case of nuisance parameters present solely under $\mathcal{H}_1$. Indeed, score-based tests require the ML estimates of nuisances under $\mathcal{H}_0$ [8], which thus cannot be obtained, as they are not *observable*. The cornerstone of Davies' work is summarized as follows.

If $\boldsymbol{x}_T$ were known, a Rao (resp. LOD) statistic would represent the basis for a reasonable decision procedure for the corresponding two-sided (resp. one-sided) testing corresponding to the deterministic (resp. random) case [8]. However, since $\boldsymbol{x}_T$ is unknown in the considered setup, a *family of score statistics* is instead obtained by varying $\boldsymbol{x}_T$. Thus, to overcome this technical difficulty, Davies proposed the use of the *maximum* of the resulting family of the statistics, as rigorously formulated in what follows.

Davies approach [18]

Let $\Lambda(\hat{\boldsymbol{b}}, \boldsymbol{x}_T)$ be an appropriate statistic, i.e., having large values when $\mathcal{H}_1$ holds true, obtained by assuming $\boldsymbol{x}_T$ known. Then, the test statistic suggested for the case of unknown $\boldsymbol{x}_T$ is:

$$\Lambda(\hat{\boldsymbol{b}}) \triangleq \sup\left\{\Lambda(\hat{\boldsymbol{b}}, \boldsymbol{x}_T) : \boldsymbol{x}_T \in \mathrm{S}_{\boldsymbol{x}_T} \subset \mathbb{R}^d\right\} \tag{1.26}$$

where $\mathrm{S}_{\boldsymbol{x}_T}$ denotes the set of possible values of $\boldsymbol{x}_T$.

Davies' generalized score statistic can be interpreted as following a "GLRT-like" approach with respect to the unknown position $\boldsymbol{x}_T$. In what follows, the employed decision tests will be referred to as *Generalized Rao* (G-Rao) *and Generalized LOD (G-LOD)*, to underline the use of Rao and LOD as the inner statistics employed in Davies framework. In other terms, the aforementioned generalized score tests are defined as:

$$\Lambda_{\text{GRAO}}(\hat{\boldsymbol{b}}) \triangleq \max_{\boldsymbol{x}_T \in \mathrm{S}_{\boldsymbol{x}_T}} \Lambda_{\text{RAO}}(\hat{\boldsymbol{b}}, \boldsymbol{x}_T); \tag{1.27}$$

$$\Lambda_{\text{GLOD}}(\hat{\boldsymbol{b}}) \triangleq \max_{\boldsymbol{x}_T \in \mathrm{S}_{\boldsymbol{x}_T}} \Lambda_{\text{LOD}}(\hat{\boldsymbol{b}}, \boldsymbol{x}_T); \tag{1.28}$$

where $\Lambda_{\mathrm{RAO}}(\hat{\boldsymbol{b}}, \boldsymbol{x}_T)$ and $\Lambda_{\mathrm{LOD}}(\hat{\boldsymbol{b}}, \boldsymbol{x}_T)$ represent the Rao and LOD statistic obtained assuming known $\boldsymbol{x}_T$ in the corresponding composite hypothesis test. The appeal is motivated by a reduced complexity of test implementation, since for score statistics evaluation of $\hat{\theta}_1$ is not required, cf. (1.29), and thus a maximization w.r.t. the sole $\boldsymbol{x}_T$ is needed for their implementation (see (1.27) and (1.28)).

The Rao statistic, in the case of an unknown scalar parameter θ, follows the usual definition [8]:

$$\Lambda_{\mathrm{RAO}}(\hat{\boldsymbol{b}}, \boldsymbol{x}_T) \triangleq \frac{\left(\partial \ln P(\hat{\boldsymbol{b}}; \theta, \boldsymbol{x}_T)/\partial \theta\right)^2 \Big|_{\theta=\theta_0}}{\mathrm{I}(\theta_0, \boldsymbol{x}_T)}. \tag{1.29}$$

On the other hand, the LOD statistic is defined as follows [28]:

$$\Lambda_{\mathrm{LOD}}(\hat{\boldsymbol{b}}, \boldsymbol{x}_T) \triangleq \frac{\partial \ln\left[P(\hat{\boldsymbol{b}}; \theta, \boldsymbol{x}_T)\right]/\partial \theta \Big|_{\theta=\theta_0}}{\sqrt{I(\theta_0, \boldsymbol{x}_T)}}. \tag{1.30}$$

In the above definitions, $\partial \ln P(\hat{\boldsymbol{b}}; \theta, \boldsymbol{x}_T)/\partial \theta$ represents the *score function*, whereas the term $I(\boldsymbol{x}_T, \theta)$ denotes the *fisher information* (FI), defined as

$$I(\boldsymbol{x}_T, \theta) \triangleq \mathbb{E}\left\{\left(\frac{\partial \ln\left[P(\hat{\boldsymbol{b}}; \theta, \boldsymbol{x}_T)\right]}{\partial \theta}\right)^2\right\}. \tag{1.31}$$

It is worth highlighting that *both* the *score function* and the *FI* are derived *assuming* $\boldsymbol{x}_T$ *known,* which thus bypasses the nuisance problem. Additionally, these two quantities are required to be evaluated at θ_0 in (1.29) and (1.30). The explicit expression for G-Rao is:

$$\Lambda_{\mathrm{GRAO}}\left(\hat{\boldsymbol{b}}\right) = \max_{\boldsymbol{x}_T \in \mathrm{S}_{\boldsymbol{x}_T}} \frac{\left[\sum_{k=1}^{K} \nu_k(\hat{b}_k)\, g(\boldsymbol{x}_T, \boldsymbol{x}_k)\right]^2}{\sum_{k=1}^{K} \psi_{0,k}\, g^2(\boldsymbol{x}_T, \boldsymbol{x}_k)}, \tag{1.32}$$

whereas for G-LOD

$$\Lambda_{\mathrm{GLOD}}(\hat{\boldsymbol{b}}) = \max_{\boldsymbol{x}_T \in \mathrm{S}_{\boldsymbol{x}_T}} \frac{\sum_{k=1}^{K} \bar{\nu}_k(\hat{b}_k)\, g^2(\boldsymbol{x}_T, \boldsymbol{x}_k)}{\sqrt{\sum_{k=1}^{K} \bar{\psi}_{0,k}\, g^4(\boldsymbol{x}_T, \boldsymbol{x}_k)}}, \tag{1.33}$$

where the following *auxiliary definitions* have been employed:

$$\nu_k(\hat{b}_k) \triangleq \frac{\left[\hat{b}_k - \rho_{0,k}\right]}{\rho_{0,k}\,(1-\rho_{0,k})} \frac{(1-2\,P_{e,k})}{\sigma_{w,k}}\, p_{\mathcal{N}}\left(\frac{\tau_k}{\sigma_{w,k}}\right), \tag{1.34}$$

$$\bar{\nu}_k(\hat{b}_k) \triangleq \frac{\left[\hat{b}_k - \rho_{0,k}\right]}{\rho_{0,k}\,(1-\rho_{0,k})} \frac{(1-2P_{e,k})\sqrt{\tau_k}}{\sigma^3_{w,k}}\, p_{\mathcal{N}}\left(\sqrt{\frac{\tau_k}{\sigma^2_{w,k}}}\right), \tag{1.35}$$

and

$$\psi_{0,k} \triangleq \left\{\frac{(1-2\,P_{e,k})}{\sigma_{w,k}}\, p_{\mathcal{N}}\left(\frac{\tau_k}{\sigma_{w,k}}\right)\right\}^2 \frac{1}{\rho_{0,k}\,(1-\rho_{0,k})}, \tag{1.36}$$

$$\overline{\psi}_{0,k} \triangleq \left\{\frac{(1-2P_{e,k})\sqrt{\tau_k}}{\sigma^3_{w,k}}\, p_{\mathcal{N}}\left(\sqrt{\frac{\tau_k}{\sigma^2_{w,k}}}\right)\right\}^2 \frac{1}{\rho_{0,k}\,(1-\rho_{0,k})}. \tag{1.37}$$

A brief sketch of derivation is reported in the appendix at the end of this chapter.

1.3.3 Computational complexity

In this section, a summarizing comparison of GLRT and derived generalized score statistics (G-Rao and G-LOD) is provided, focusing on the computational complexity. To this end, in Table 1.1, the corresponding complexity required for their implementation is reported, by means of a *grid discretization.*

More specifically, to apply grid discretization, it is assumed that $\boldsymbol{x}_T$ and θ belong to limited sets $S_{\boldsymbol{x}_T} \subset \mathbb{R}^d$ and $S_\theta \subset \mathbb{R}$, respectively. The search space $(\boldsymbol{x}_T, \theta)$ is then discretized into:

- $N_{\boldsymbol{x}_T}$ position bins in $\mathbb{R}^d$, each one associated to a center bin position, say $\boldsymbol{x}_T[i]$, $i \in \{1, \ldots, N_{\boldsymbol{x}_T}\}$ and
- N_θ amplitude (power) bins in $\mathbb{R}$ (in $\mathbb{R}^+$) in the deterministic (resp. random) case, each one associated to a center bin amplitude (power), say $\theta[j], j \in \{1, \ldots, N_\theta\}$.

Then, starting from the alternative form of (1.25), the GLR statistic is approximated via the following grid search:

$$\Lambda_{\text{GLR}} \approx \max_{i=1,\ldots,N_{\boldsymbol{x}_T}} \max_{j=1,\ldots,N_\theta} 2 \cdot \Lambda_{\text{LLR}}(\boldsymbol{x}_T[i],\, \theta[j]); \tag{1.38}$$

where we recall that $\Lambda_{\text{LLR}}(\boldsymbol{x}_T[i],\, \theta[j])$ represents the expression of the clairvoyant LLR statistic obtained by evaluating the $\rho_{1,k}$s by replacing $\boldsymbol{x}_T$ and θ with $\boldsymbol{x}_T[i]$ and $\theta[j]$, respectively. Hence, for GLRT, a grid approach on $(\theta, \boldsymbol{x}_T)$ is typically involved [22,27,29], with complexity $\mathcal{O}(K\,N_{\boldsymbol{x}_T}\,N_\theta)$, where $N_{\boldsymbol{x}_T}$ (resp. N_θ) is the number of position (resp. power) bins employed.

A similar reasoning is applied for implementation of generalized score statistics, restricting, however, to the sole search space $\boldsymbol{x}_T$. Indeed, their *implementation* is

Table 1.1 Complexity of decision statistics when implemented with grid search

Fusion rule	**Search space**	**Computational complexity**
GLR	$(\boldsymbol{x}_T, \theta)$	$\mathcal{O}(K\,N_{\boldsymbol{x}_T}\,N_\theta)$
G-Rao / G-LOD	$(\boldsymbol{x}_T)$	$\mathcal{O}(K\,N_{\boldsymbol{x}_T})$

simpler since $\hat{\theta}_1$ is not needed and thus solely a grid w.r.t. $\boldsymbol{x}_T$ is required. More specifically, G-Rao is approximated as:

$$\Lambda_{\text{GRAO}} \approx \max_{i=1,\dots,N_{\boldsymbol{x}_T}} \frac{\left[\sum_{k=1}^{K} \nu_k(\hat{b}_k)\, g(\boldsymbol{x}_T[i], \boldsymbol{x}_k)\right]^2}{\sum_{k=1}^{K} \psi_{0,k}\, g^2(\boldsymbol{x}_T[i], \boldsymbol{x}_k)}. \tag{1.39}$$

whereas for G-LOD the following approximation holds:

$$\Lambda_{\text{GLOD}} \approx \max_{i=1,\dots,N_{\boldsymbol{x}_T}} \frac{\sum_{k=1}^{K} \bar{\nu}_k(\hat{b}_k)\, g^2(\boldsymbol{x}_T[i], \boldsymbol{x}_k)}{\sqrt{\sum_{k=1}^{K} \bar{\psi}_{0,k}\; g^4(\boldsymbol{x}_T[i], \boldsymbol{x}_k)}}. \tag{1.40}$$

Thus, in view of (1.39) and (1.40), their complexity is $\mathcal{O}(K\,N_{\boldsymbol{x}_T})$, thus providing a *significant complexity reduction* w.r.t. the GLR, as reported in Table 1.1.

1.4 Quantizer design

By looking at (1.39) and (1.40), it is apparent that both Λ_{GRAO} and Λ_{GLOD} (as well as Λ_{GLR}) are functions of the quantization thresholds τ_k. Indeed $\nu_k(\hat{b}_k)$ and $\psi_{0,k}$ *both depend* on τ_k, $k \in \mathcal{K}$, and the same applies to $\overline{\nu}_k(\hat{b}_k)$ and $\overline{\psi}_{0,k}$, $k \in \mathcal{K}$. Therefore, this section is devoted to investigating how these thresholds can be optimized to achieve improved performance. Before proceeding, the set of local thresholds is collected as $\boldsymbol{\tau} \triangleq [\tau_1 \;\cdots\; \tau_K]^T$.

It is worth noticing that optimal quantizers cannot be obtained as in [6,7], even in the asymptotic (i.e., large WSN) case, because no performance expressions are known in the literature for tests based on the Davies approach [18]. To this end, an intuitive rationale based on [6,7] is adopted in this chapter.

Specifically, it is known that the (position $\boldsymbol{x}_T$) clairvoyant Rao statistic $\Lambda_{\text{RAO}}(\cdot)$ is asymptotically (and assuming a weak-signal[1]) distributed as [8]:

$$\Lambda_{\text{RAO}}(\hat{\boldsymbol{b}}, \boldsymbol{x}_T, \boldsymbol{\tau}) \overset{a}{\sim} \begin{cases} \chi_1^2 & \text{under } \mathcal{H}_0 \\ \chi_1'^2\left(\lambda_Q(\boldsymbol{x}_T, \boldsymbol{\tau})\right) & \text{under } \mathcal{H}_1 \end{cases}, \tag{1.41}$$

[1]That is, $|\theta_1 - \theta_0| = c/\sqrt{K}$ for some constant $c > 0$ [8].

where $\chi_1^2(\cdot)$ denotes the central chi-square distribution with one degree of freedom. Similarly, $\chi_1'^2(\cdot)$ is used to denote its noncentral counterpart, with a noncentrality parameter defined as $\lambda_Q(\boldsymbol{x}_T, \boldsymbol{\tau}) \triangleq (\theta_1 - \theta_0)^2 \, \mathrm{I}(\theta_0, \boldsymbol{x}_T, \boldsymbol{\tau})$. With a slight abuse of notation, we have employed the terminology $\mathrm{I}(\theta, \boldsymbol{x}_T, \boldsymbol{\tau})$ to underline the FI dependence on $\boldsymbol{x}_T$ and $\boldsymbol{\gamma}$.

The noncentrality parameter is given explicitly as follows:

$$\lambda_Q(\boldsymbol{x}_T, \boldsymbol{\tau}) = \theta_1^2 \sum_{k=1}^{K} \psi_{0,k}(\tau_k) \, g^2(\boldsymbol{x}_T, \boldsymbol{x}_k), \tag{1.42}$$

with θ_1 being the true value under $\mathcal{H}_1$. Correspondingly, the (position $\boldsymbol{x}_T$) clairvoyant LOD statistic $\Lambda_{\mathrm{LOD}}(\cdot)$ has an analogous (weak signal) asymptotic distribution, reported in the following [8]:

$$\Lambda_{\mathrm{LOD}}(\hat{\boldsymbol{b}}, \boldsymbol{x}_T, \boldsymbol{\tau}) \overset{a}{\sim} \begin{cases} \mathcal{N}(0, 1) & \text{under} \quad \mathcal{H}_0 \\ \mathcal{N}\left(\delta_Q(\boldsymbol{x}_T, \boldsymbol{\tau}), 1\right) & \text{under} \quad \mathcal{H}_1 \end{cases}, \tag{1.43}$$

where the deflection measure is defined as $\delta_Q(\boldsymbol{x}_T, \boldsymbol{\tau}) \triangleq (\theta_1 - \theta_0)\sqrt{\mathrm{I}(\theta_0, \boldsymbol{x}_T, \boldsymbol{\tau})}$ and provided in closed form in the following:

$$\delta_Q(\boldsymbol{x}_T, \boldsymbol{\tau}) = \theta_1 \sqrt{\sum_{k=1}^{K} \bar{\psi}_{0,k}(\tau_k) \, g^4(\boldsymbol{x}_T, \boldsymbol{x}_k)}\,. \tag{1.44}$$

It is worth highlighting that we have used the notations $\psi_{0,k}(\tau_k)$ and $\overline{\psi}_{0,k}(\tau_k)$, as opposed to $\psi_{0,k}$ and $\overline{\psi}_{0,k}$, respectively, to *emphasize* the dependence of these terms on the local thresholds τ_ks *only*.

Clearly the larger $\lambda_Q(\boldsymbol{x}_T, \boldsymbol{\tau})$ and $\delta_Q(\boldsymbol{x}_T, \boldsymbol{\tau})$, the better the standard score tests will perform when the POI to be detected is *known to be located at* $\boldsymbol{x}_T$ (i.e., they have clairvoyant knowledge of the position). Also, it is apparent that $\lambda_Q(\boldsymbol{x}_T, \boldsymbol{\tau})$ and $\delta_Q(\boldsymbol{x}_T, \boldsymbol{\tau})$ are both functions of τ_k, $k \in \mathcal{K}$ (because of the $\psi_{0,k}$s and $\overline{\psi}_{0,k}$s, respectively).

For this reason, the threshold vector $\boldsymbol{\tau}$ is selected to maximize the noncentrality parameter $\lambda_Q(\boldsymbol{x}_T, \boldsymbol{\tau})$ in the *deterministic* case, that is,

$$\boldsymbol{\tau}^\star \triangleq \arg\max_{\boldsymbol{\tau}} \lambda_Q(\boldsymbol{x}_T, \boldsymbol{\tau}). \tag{1.45}$$

Similarly, in the *random* case, the threshold vector $\boldsymbol{\tau}$ is selected to maximize the deflection measure $\delta_Q(\boldsymbol{x}_T, \boldsymbol{\tau})$:

$$\boldsymbol{\tau}^\star \triangleq \arg\max_{\boldsymbol{\tau}} \delta_Q(\boldsymbol{x}_T, \boldsymbol{\tau}). \tag{1.46}$$

In general, both these maximizations would lead to optimized thresholds $\boldsymbol{\tau}^\star$ *depending on* $\boldsymbol{x}_T$, thus hindering their practical use.

However, for this specific task, these optimizations can be decoupled into the following set of K-independent threshold design problems, which are also *independent of* $\boldsymbol{x}_T$ (cf. (1.44)). Then, threshold design under the presented criterion appears to be feasible and with a complexity that scales *linearly* with K.

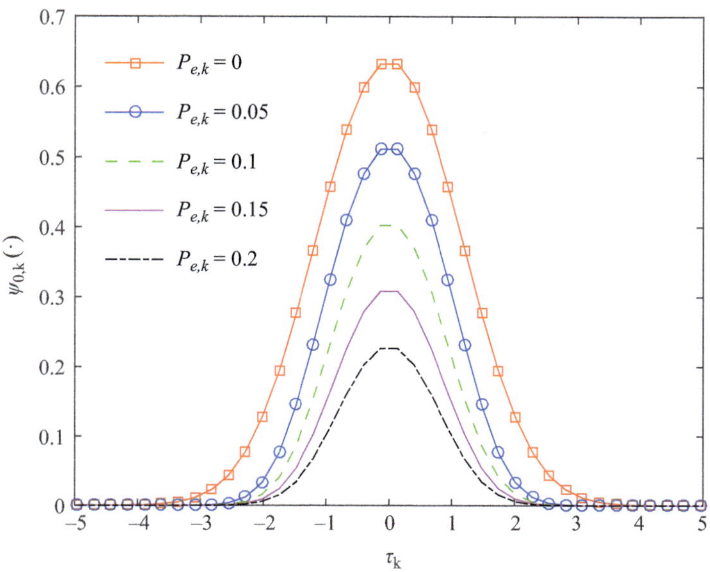

Figure 1.3 Threshold optimization in deterministic case for G-Rao: objective $\psi_{0,k}$ vs. threshold τ_k for different values of $P_{e,k}$

More specifically, in the case of G-Rao, each τ_k is selected as the solution of the following optimization:

$$\tau_k^\star = \arg\max_{\tau_k} \left\{ \psi_{0,k}(\tau_k) = \frac{p_{\mathcal{N}}^2(\tau_k/\sigma_{w,k})}{\Delta_k + Q(\tau_k/\sigma_{w,k})\left[1 - Q(\tau_k/\sigma_{w,k})\right]} \right\} \tag{1.47}$$

where $\Delta_k \triangleq [P_{e,k}(1 - P_{e,k})]/(1 - 2P_{e,k})^2$. It is known from the quantized estimation literature [30,31] that many unimodal and symmetric $p_{w_k}(\cdot)$s with $\mathbb{E}\{w_k\} = 0$ lead to $\tau_k^\star \triangleq \arg\max_{\tau_k} \psi_{0,k}(\tau_k) = 0$ (independent of Δ_k); one example is the considered Gaussian distribution. Other examples are the Laplace, Cauchy, and the widely used generalized normal distribution (only in the case $0 \leq \epsilon \leq 2$). Also, it has been shown in [6] that $\tau_k = 0$ is still a good (suboptimal) choice even when not corresponding to the optimizer for a specific noise pdf, especially in the case of noisy ($P_{e,k} \neq 0$) reporting channels. As a relevant example, in Figure 1.3, the objective $\psi_{0,k}(\tau_k)$ is illustrated in the case $\sigma_{w,k}^2 = 1$ for different BEP values.

On the other hand, in the case of G-LOD, an optimized τ_k is obtained by solving the following maximization:

$$\tau_k^\star = \arg\max_{\tau_k} \left\{ \overline{\psi}_{0,k}(\tau_k) = \frac{1}{(\sigma_{w,k}^2)^2} \frac{p_{\mathcal{N}}^2\left(\sqrt{\tau_k/\sigma_{w,k}^2}\right)(\tau_k/\sigma_{w,k}^2)}{\Delta_k + P_{0,k}(\tau_k)\left[1 - P_{0,k}(\tau_k)\right]} \right\}, \tag{1.48}$$

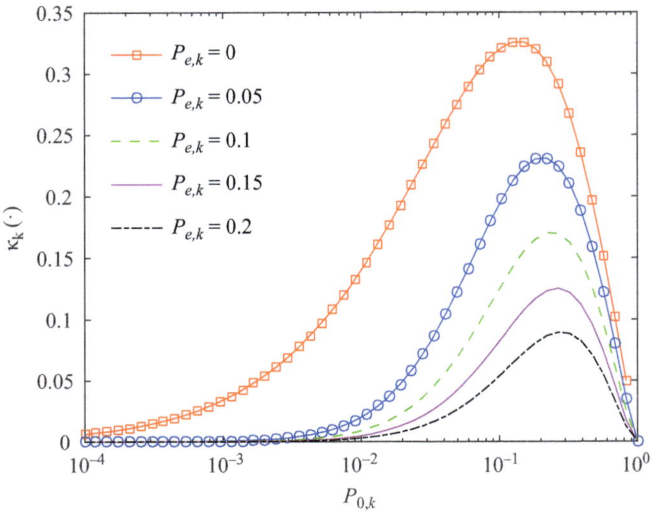

Figure 1.4 Threshold optimization in random case for G-LOD: Objective κ_k vs. local false-alarm probability $P_{0,k}$ for different values of $P_{e,k}$

The solution to (1.48) does not appear to possess a closed form. In this case, however, the maximization can be reformulated in terms of $P_{0,k}$, being in *bijective* correspondence with τ_k and having the meaning of a local false-alarm probability in the *random* signal case. The aforementioned reformulation leads to the following objective:

$$\kappa_k(P_{0,k}) = \frac{p^2_{\mathcal{N}}(Q^{-1}(P_{0,k}/2))[Q^{-1}(P_{0,k}/2)]^2}{\Delta_k + P_{0,k}(1 - P_{0,k})}. \tag{1.49}$$

Examples of the objective $\kappa_k(P_{0,k})$ corresponding to different values of $P_{e,k}$ (viz. Δ_k) are shown in Figure 1.4. The optimal $P^{\star}_{0,k}$ can be easily evaluated via a line search.

1.5 Conclusions and further reading

This chapter focused on the development of practical fusion rules for decentralized detection of a POI with a spatially limited extent and unknown emitting amplitude/power. Such case is typical when detecting noncooperative targets or leakage of hazardous material of unknown quantity. All the considered rules have accounted the case of imperfect reporting channels, modeled as BSCs with corresponding BEPs $P_{e,k}$ assumed known at the FC. The present problem, in the two signal scenarios considered, corresponds to one-sided and two-sided hypothesis tests, respectively, with nuisance parameters (i.e., $\boldsymbol{x}_T$) which are present only under $\mathcal{H}_1$ (viz. POI-present hypothesis). The GLRT, being the *de facto* approach, requires a grid search over the space $(\theta, \boldsymbol{x}_T)$, thus limiting its applicability in operational scenarios.

As a consequence, generalized score statistics, as an appealing alternative, have been discussed. The aforementioned rules arise from the maximization (w.r.t. $\boldsymbol{x}_T$) of a family of score decision statistics and present reduced complexity with respect to GLRT, requiring only a grid-based search with respect to $\boldsymbol{x}_T$.

The performances of these decision procedures strongly depend on the local quantization process. For this reason, a reasonable criterion for optimized sensor thresholds has also been developed; this is suggested by the asymptotic performance of position-clairvoyant score statistics. It has been shown that such criterion results in *a simple per-sensor line search*. This result was exploited to suggest how performance can be optimized for generalized score tests.

Finally, this section ends with suggestions for the interested reader. The first work concerning decentralized detection of POIs with a spatial extent is [29], where a GLRT which assumes a deterministic AAF and in the case of a known emitted power is developed, and its superiority is shown in comparison to the CR. In a similar setup, the (approximate) theoretical performance of CR are developed in [12]. In both works, the average emitted power of the target is however assumed *known*. An alternative approach for tackling POIs having a spatial extent with unknown location is represented by the scan statistics framework [32], which accounts also for random sensor positions. In [20], also a scan statistic and Bayesian-originated approaches were obtained and compared with existing alternatives. A stochastic AAF, to account for possible amplitude fluctuations subsuming the Rayleigh fading model, is assumed in both [20,32].

The rationale here presented for threshold optimization relies upon the works in [6,7], which tackle the development and threshold-optimization of fusion rules originated from composite hypothesis testing, in the case of known location. Specifically, the work in [7] investigates a GLRT in additive Gaussian noise, whereas [6] proposes a Rao test, as a low-complexity asymptotically optimal alternative, in the more general case of unimodal noise distributions. The numerical validation of the proposed rationale can be found instead in [22,33].

A remarkable study belonging to the same category of problems (unknown power, known location) is represented by [34], where an asymptotic locally optimum detector is obtained for a WSN with (random) sensors positions following a Poisson point process. Unfortunately, the deterministic AAF there employed implicitly assumed that the FC has available the target position. Thus, despite some presented numerical analysis investigating mismatched AAF performance, the above work has limited applicability.

An alternative approach to detect a POI with a spatially limited extent is represented by the work in [19], proposing a two-step DF algorithm in which sensors first correct their decisions on the basis of neighboring sensors and then make a collective decision as a network. For the proposed approach, an explicit formula for FC threshold choice (viz. false-alarm rate determination) is provided, based on normal approximation of the statistic under the target-absent hypothesis. Subsequently, a simple and more accurate threshold design based on the beta-binomial approximation is proposed in [35].

The GLRT for the (random) model discussed in this chapter has been first developed by [27], and compared to the CR, the CV rule and a GLRT based on the awareness of target-emitted power. It has been shown that the loss incurred by the proposed GLRT is marginal when compared to the "power-clairvoyant" GLRT. G-LOD (and other hybrid alternatives, based on Bayesian framework) is first proposed in [36], whereas G-Rao is first investigated in [22].

An alternative application of GLRT-based fusion of quantized data in WSN can be found in [37] for fusing conditionally dependent decisions.

A.1 Appendix: Sketch of generalized score tests derivation

The key step in obtaining generalized score statistics corresponds to deriving the position-clairvoyant score statistics $\Lambda_{\mathrm{RAO}}(\hat{\boldsymbol{b}}, \boldsymbol{x}_T)$ and $\Lambda_{\mathrm{LOD}}(\hat{\boldsymbol{b}}, \boldsymbol{x}_T)$ explicitly. In order to accomplish this task, the corresponding score function and FI are required to be obtained explicitly. First, exploiting the independence of sensors' measurements and reporting channels, $(\partial \ln P(\hat{\boldsymbol{b}}; \theta, \boldsymbol{x}_T))/\partial\theta$ can be simplified and expanded in the following form:

$$
\begin{aligned}
\frac{\partial \ln P(\hat{\boldsymbol{b}}; \theta, \boldsymbol{x}_T)}{\partial \theta} &= \sum_{k=1}^{K} \frac{\partial \ln P(\hat{b}_k; \theta, \boldsymbol{x}_T)}{\partial \theta} \\
&= \sum_{k=1}^{K} \frac{\partial \rho_{1,k}(\theta, \boldsymbol{x}_T)}{\partial \theta} \left[\frac{\hat{b}_k - \rho_{1,k}(\theta, \boldsymbol{x}_T)}{\rho_{1,k}(\theta, \boldsymbol{x}_T)\left[1 - \rho_{1,k}(\theta, \boldsymbol{x}_T)\right]} \right]. \qquad \text{(A.1)}
\end{aligned}
$$

It is worth noticing that the above expression holds in both *deterministic* and *random* signal cases. The sole difference is represented by the derivative of the composite activation probability under $\mathscr{H}_1$, given by:

$$
\frac{\partial \rho_{1,k}(\theta, \boldsymbol{x}_T)}{\partial \theta} = (1-2P_{e,k}) \times \begin{cases} p_{\mathcal{N}}\left(\dfrac{(\tau_k - \theta g(\boldsymbol{x}_T, \boldsymbol{x}_k))}{\sigma_{w,k}} \right) \dfrac{g(\boldsymbol{x}_T, \boldsymbol{x}_k)}{\sigma_{w,k}} & \text{(deterministic)} \\ p_{\mathcal{N}}\left(\sqrt{\dfrac{\tau_k}{\theta g^2(\boldsymbol{x}_T, \boldsymbol{x}_k) + \sigma_{w,k}^2}} \right) \dfrac{\sqrt{\tau_k}\, g^2(\boldsymbol{x}_T, \boldsymbol{x}_k)}{\left[\theta g^2(\boldsymbol{x}_T, \boldsymbol{x}_k) + \sigma_{w,k}^2\right]^{3/2}} & \text{(random)} \end{cases} \qquad \text{(A.2)}
$$

On the other hand, the closed form of $\mathrm{I}(\theta, \boldsymbol{x}_T)$ is [6,7]:

$$
\mathrm{I}(\theta, \boldsymbol{x}_T) = \sum_{k=1}^{K} \left\{ \frac{\partial \rho_{1,k}(\theta, \boldsymbol{x}_T)}{\partial \theta} \right\}^2 \frac{1}{\rho_{1,k}(\theta, \boldsymbol{x}_T)\left[1 - \rho_{1,k}(\theta, \boldsymbol{x}_T)\right]}. \qquad \text{(A.3)}
$$

Finally, evaluation at $\theta_0 = 0$ simply provides:

$$\left.\frac{\partial \ln\left[P\left(\hat{\boldsymbol{b}};\theta,\boldsymbol{x}_T\right)\right]}{\partial\theta}\right|_{\theta=\theta_0} = \sum_{k=1}^{K} \left.\frac{\partial \rho_{1,k}(\theta,\boldsymbol{x}_T)}{\partial\theta}\right|_{\theta=\theta_0} \left[\frac{\hat{b}_k - \rho_{0,k}}{\rho_{0,k}\left(1-\rho_{0,k}\right)}\right], \tag{A.4}$$

and

$$\mathrm{I}(\theta_0,\boldsymbol{x}_T) = \sum_{k=1}^{K} \left\{ \left.\frac{\partial \rho_{1,k}(\theta,\boldsymbol{x}_T)}{\partial\theta}\right|_{\theta=\theta_0} \right\}^2 \frac{1}{\rho_{0,k}\left(1-\rho_{0,k}\right)}, \tag{A.5}$$

for the score function and the FI, respectively. Finally, plugging (A.4) and (A.5), with the appropriate definition of $(\partial\rho_{1,k}(\theta,\boldsymbol{x}_T))/\partial\theta$ into (1.29) and (1.30), provides G-Rao and G-LOD statistics in (1.32) and (1.33), respectively.

References

[1] Chong CY, and Kumar SP. Sensor networks: evolution, opportunities, and challenges. Proceedings of the IEEE. 2003;91(8):1247–1256.

[2] Varshney PK. Distributed Detection and Data Fusion. 1st ed. Springer-Verlag New York, Inc.; 1996.

[3] Salvo Rossi P, Ciuonzo D, and Romano G. Orthogonality and cooperation in collaborative spectrum sensing through MIMO decision fusion. IEEE Transactions on Wireless Communications. 2013;12(11):5826–5836.

[4] Tsitsiklis JN. Decentralized detection. Advances in Statistical Signal Processing. 1993;2(2):297–344.

[5] Viswanathan R, and Varshney PK. Distributed detection with multiple sensors – Part I: Fundamentals. Proceedings of the IEEE. 1997;85(1):54–63.

[6] Ciuonzo D, Papa G, Romano G, *et al.* One-bit decentralized detection with a Rao test for multisensor fusion. IEEE Signal Processing Letters. 2013;20(9):861–864.

[7] Fang J, Liu Y, Li H, *et al.* One-bit quantizer design for multisensor GLRT fusion. IEEE Signal Processing Letters. 2013;20(3):257–260.

[8] Kay SM. Fundamentals of Statistical Signal Processing, Volume 2: Detection Theory. Upper Saddle River, NJ: Prentice Hall PTR; 1998.

[9] Aalo VA, and Viswanathan R. Multilevel quantisation and fusion scheme for the decentralised detection of an unknown signal. Proceedings of IEE – Radar, Sonar and Navigation. 1994;141(1):37–44.

[10] Chen B, Jiang R, Kasetkasem T, *et al.* Channel aware decision fusion in wireless sensor networks. IEEE Transactions on Signal Processing. 2004;52(12): 3454–3458.

[11] Ciuonzo D, Romano G, and Salvo Rossi P. Channel-aware decision fusion in distributed MIMO wireless sensor networks: decode-and-fuse vs. decode-then-fuse. IEEE Transactions on Wireless Communications. 2012;11(8): 2976–2985.

[12] Niu R, and Varshney PK. Performance analysis of distributed detection in a random sensor field. IEEE Transactions on Signal Processing. 2008;56(1):339–349.

[13] Niu R, Varshney PK, and Cheng Q. Distributed detection in a large wireless sensor network. Information Fusion. 2006;7(4):380–394. Special Issue on the 7th International Conference on Information Fusion-Part I.

[14] Ciuonzo D, De Maio A, and Salvo Rossi P. A systematic framework for composite hypothesis testing of independent Bernoulli Trials. IEEE Signal Processing Letters. 2015;22(9):1249–1253.

[15] Ciuonzo D, and Salvo Rossi P. Decision fusion with unknown sensor detection probability. IEEE Signal Processing Letters. 2014;21(2):208–212.

[16] Ciuonzo D, Romano G, and Salvo Rossi P. Optimality of received energy in decision fusion over Rayleigh fading diversity MAC with non-identical sensors. IEEE Transactions on Signal Processing. 2013;61(1):22–27.

[17] Salvo Rossi P, Ciuonzo D, Ekman T, *et al.* Energy detection for MIMO decision fusion in underwater sensor networks. IEEE Sensors Journal. 2015;15(3):1630–1640.

[18] Davies RD. Hypothesis testing when a nuisance parameter is present only under the alternative. Biometrika. 1987;74(1):33–43.

[19] Katenka N, Levina E, and Michailidis G. Local vote decision fusion for target detection in wireless sensor networks. IEEE Transactions on Signal Processing. 2008;56(1):329–338.

[20] Guerriero M, Svensson L, and Willett PK. Bayesian data fusion for distributed target detection in sensor networks. IEEE Transactions on Signal Processing. 2010;58(6):3417–3421.

[21] Lehmann EL, and Romano JP. Testing Statistical Hypotheses. New York: Springer Science & Business Media; 2006.

[22] Ciuonzo D, Salvo Rossi P, and Willett P. Generalized Rao test for decentralized detection of an uncooperative target. IEEE Signal Processing Letters. 2017;24(5):678–682.

[23] Ciuonzo D, Salvo Rossi P, and Dey S. Massive MIMO channel-aware decision fusion. IEEE Transactions on Signal Processing. 2015;63(3):604–619.

[24] Chaudhari S, Lundén J, Koivunen V, *et al.* Cooperative sensing with imperfect reporting channels: hard decisions or soft decisions? IEEE Transactions on Signal Processing. 2012;60(1):18–28.

[25] Proakis J. Digital Communications. 4th ed. New York: McGraw-Hill Science/Engineering/Math; 2000.

[26] Chair Z, and Varshney PK. Optimal data fusion in multiple sensor detection systems. IEEE Transactions on Aerospace and Electronic Systems. 1986; AES-22(1):98–101.

[27] Shoari A, and Seyedi A. Detection of a non-cooperative transmitter in Rayleigh fading with binary observations. In: IEEE Military Communications Conference (MILCOM); 2012. p. 1–5.

[28] Kassam SA, and Thomas JB. Signal detection in non-Gaussian noise. Springer-Verlag, New York; 1988.

[29] Niu R, and Varshney PK. Joint detection and localization in sensor networks based on local decisions. In: 40th Asilomar Conference on Signals, Systems and Computers; 2006. p. 525–529.

[30] Papadopoulos HC, Wornell GW, and Oppenheim AV. Sequential signal encoding from noisy measurements using quantizers with dynamic bias control. IEEE Transactions on Information Theory. 2001;47(3):978–1002.

[31] Rousseau D, Anand GV, and Chapeau-Blondeau F. Nonlinear estimation from quantized signals: quantizer optimization and stochastic resonance. In: Proc. 3rd Int. Symp. Physics in Signal and Image Processing; 2003. p. 89–92.

[32] Guerriero M, Willett P, and Glaz J. Distributed target detection in sensor networks using scan statistics. IEEE Transactions on Signal Processing. 2009;57(7):2629–2639.

[33] Ciuonzo D, and Salvo Rossi P. Quantizer design for generalized locally-optimum detectors in wireless sensor networks. IEEE Wireless Communications Letters. 2018;7(2):162–165.

[34] Sung Y, Tong L, and Swami A. Asymptotic locally optimal detector for large-scale sensor networks under the Poisson regime. IEEE Transactions on Signal Processing. 2005;53(6):2005–2017.

[35] Ridout MS. An improved threshold approximation for local vote decision fusion. IEEE Transactions on Signal Processing. 2013;61(5):1104–1106.

[36] Ciuonzo D, and Salvo Rossi P. Distributed detection of a non-cooperative target via generalized locally-optimum approaches. Information Fusion. 2017;36: 261–274.

[37] Iyengar SG, Niu R, and Varshney PK. Fusing dependent decisions for hypothesis testing with heterogeneous sensors. IEEE Transactions on Signal Processing. 2012;60(9):4888–4897.

Chapter 2

Compressed distributed detection and estimation

Thakshila Wimalajeewa[1] and Pramod K. Varshney[2]

Detection and estimation are two fundamental tasks that are performed by distributed sensor networks. It is a challenging problem to design efficient protocols and algorithms to perform these tasks taking the inherent scarce network resources, such as limited node power and the limited communication bandwidth, into account. Despite there being quite a rich literature related to energy efficiency in distributed sensor networks, there is still much ongoing research on investigating how to optimize power and communication bandwidth in processing high-dimensional (multimodal) data generated at emerging sensors with high fidelity and resolution. Recent advances in compressive sensing (CS) have led to novel ways of thinking about energy efficient signal processing in sensor networks. CS is well motivated for distributed sensor network applications since data compression at the sensors prior to transmission to the fusion center (FC) is vital to minimize the energy and communication requirements. The universal and agnostic nature of the CS measurement scheme is promising in acquiring compressed data for a variety of inference tasks. In many distributed sensor network applications, sparsity is a common characteristic that can be observed in various forms. While the CS theory widely focuses on sparse signal reconstruction, further research beyond the standard CS framework is needed to understand its applicability in solving a variety of inference problems. In this book chapter, our goal is to provide an up-to-date review on CS-based detection and estimation as applicable to sensor networks. After an introduction, we provide a brief overview of the theory of CS. Then, we discuss CS-based detection and parameter estimation problems considering different signal and noise models. The impact of compression via CS on detection and estimation is quantified in terms of different performance metrics.

2.1 Introduction

The wireless sensor network (WSN) technology has gained increasing attention by both the research community and actual users over the last two decades [1–4]. Applications of WSNs span a wide range including environmental monitoring [5], detection

[1]BAE Systems, Burlington, USA
[2]Department of Electrical Engineering and Computer Science, Syracuse University, USA

and classification [6–8], target/object tracking [9,10], industrial applications [11], and health care [12] to name a few. The WSN research mainly focuses on designing efficient protocols and algorithms to perform a given task for such applications taking the inherent resource constraints into account. There is abundant research performed in the last several years focusing on energy savings in WSNs. However, the emergence of new sensors with the capability of generating huge amounts of data makes real-time fusion of data in performing different inference tasks increasingly challenging.

CS has emerged as a promising paradigm for efficient acquisition of high dimensional sparse signals. In the CS framework, a sparse signal can be reliably recovered with a small number of random projections under certain conditions [13–16]. Since its inception over a decade ago, the CS theory has been successfully applied to a wide range of applications including compressive imaging [17], video processing [18], wireless channel estimation [19,20], sensor networks [21–25], cognitive radio networks [26], and radar signal processing [27] to name a few. Further, there exist several books, surveys, and tutorial papers covering different aspects of CS [14,28,29].

CS is a promising candidate for data acquisition in WSN applications which reduces energy and communication bandwidth requirements. Due to limited energy and communication resources in WSNs, data compression prior to transmission is vital. The universal and agnostic nature of the data-compression approach with CS enables the design of simple and flexible hardware for acquiring high-dimensional data for a variety of inference tasks. In many distributed sensor network applications, sparsity is a common characteristic that can be observed in various forms. Sparsity can be exploited in temporal data as well as in spatial data. A direct application of CS in WSNs is compressive data gathering exploiting temporal or spatial sparsity. The use of CS in data gathering directly relates to signal reconstruction as widely discussed in CS theory. However, in order to solve a given inference problem such as detection and classification, reconstruction of data observed at individual nodes may not be necessary. The use of CS for solving detection and estimation problems without signal reconstruction has gained considerable attention lately. The importance of understanding the role of CS theory in solving such inference problems is the motivation behind this book chapter. In detection problems, the most important task is to design a decision statistic from compressed data so that reliable detection can be performed with a desired performance level. In contrast to finding the number of compressed measurements to enable complete signal recovery as in the standard CS framework, here the goal is to control the number of CS measurements so that a desired detection performance is achieved. This requires the investigation of different metrics for performance analysis and quantification of the amount of information preserved under compression to construct decision statistics. Further, in contrast to complete signal reconstruction, detection can be performed with compressed data even when the signals are not necessarily sparse. Thus, CS-based detection requires design principles and performance evaluations that are different from those for complete signal reconstruction. In parameter estimation with compressed data, the goal is to estimate a continuous valued parameter or a set of parameters that govern the signal being observed. Similar to compressed detection, for compressed parameter estimation, sparsity prior of the signal of interest is not necessary. Here, the goal is to quantify the amount of information retained in the compressed domain so that a reliable estimator can be obtained. In both detection and estimation problems with compressed data, the performance and the specific design principles depend on how the signals or the parameters are modeled. Before discussing CS-based detection and estimation in detail, in the following, we provide a brief overview of CS.

2.2 Compressive sensing: background

The CS theory was motivated by the observation that many real world signals of interest are sparse. A length-N signal $\mathbf{x}$ is said to be sparse if it contains only a few nonzero elements, i.e., $||\mathbf{x}||_0 \ll N$ where $||\cdot||_p$ denotes the l_p norm. In the CS framework with a single measurement vector (SMV), one aims at solving for sparse $\mathbf{x}$ from the following underdetermined linear system:

$$\mathbf{y} = \mathbf{A}\mathbf{x} \tag{2.1}$$

with no noise and

$$\mathbf{y} = \mathbf{A}\mathbf{x} + \mathbf{v} \tag{2.2}$$

with noise, where $\mathbf{y}$ is the length-M compressed measurement vector, $\mathbf{A}$ is the $M \times N$ measurement matrix with $M < N$, and $\mathbf{v}$ is the $M \times 1$ additive noise vector. While recovering $\mathbf{x}$ from its compressed form $\mathbf{y}$ in (2.1) (or (2.2)) is ill-conditioned in general when $M < N$, it is possible to reconstruct $\mathbf{x}$ exactly under certain conditions on the measurement matrix $\mathbf{A}$ if $\mathbf{x}$ is sufficiently sparse [13,15,16]. Necessary and sufficient conditions satisfied by the matrix $\mathbf{A}$ so that $\mathbf{x}$ can be recovered from $\mathbf{y}$ have been discussed focusing on several properties [13,15,16,28]. One of the commonly used properties is the *restricted isometry property (RIP)* property.

Definition 2.1 (RIP [28]). *The matrix* $\mathbf{A}$ *is said to satisfy RIP of order* k *if there exists a* $\delta_k \in (0, 1)$ *such that*

$$(1 - \delta_k) \leq \frac{||\mathbf{A}\mathbf{x}||_2^2}{||\mathbf{x}||_2^2} \leq (1 + \delta_k) \tag{2.3}$$

for all $\mathbf{x} \in \Sigma_k$ *where*

$$\Sigma_k = \{\mathbf{x} : ||\mathbf{x}||_0 \leq k\}. \tag{2.4}$$

When the entries of $\mathbf{A}$ are chosen according to a Gaussian, Bernoulli or in general from a sub-Gaussian distribution, $\mathbf{A}$ satisfies RIP with high probability when $M = \mathcal{O}(k \log (N/k))$ [28]. In order to recover $\mathbf{x}$ from $\mathbf{y}$, the natural choice is to solve the following optimization problem (with no noise) [13,15,16,28]:

$$\min_{\mathbf{x}} ||\mathbf{x}||_0 \text{ such that } \mathbf{y} = \mathbf{A}\mathbf{x}. \tag{2.5}$$

Unfortunately, this l_0 norm minimization problem is generally computationally intractable. In order to approximately solve (2.5), several approaches have been proposed. A popular approach for signal reconstruction proposed in the CS theory is the so-called basis pursuit (BP) [30] in which the l_0 term in (2.5) is replaced by l_1 norm to get:

$$\min_{\mathbf{x}} ||\mathbf{x}||_1 \text{ such that } \mathbf{y} = \mathbf{A}\mathbf{x}. \tag{2.6}$$

Under some favorable conditions, the solution to (2.5) coincides with that in (2.6). In the presence of noise, BP denoising [30] aims at solving:

$$\min_{\mathbf{x}} ||\mathbf{x}||_1 \text{ such that } ||\mathbf{y} - \mathbf{A}\mathbf{x}||_2 \leq \varepsilon_1 \tag{2.7}$$

and least absolute shrinkage and selection operator (LASSO) [31–33] solves $\min_{\mathbf{x}} ||\mathbf{y} - \mathbf{A}\mathbf{x}||_2$ such that $||\mathbf{x}||_1 \leq \varepsilon_2$ or equivalently:

$$\min_{\mathbf{x}} \lambda ||\mathbf{x}||_1 + \frac{1}{2} ||\mathbf{y} - \mathbf{A}\mathbf{x}||_2 \tag{2.8}$$

where λ is the penalty parameter, and $\varepsilon_1, \varepsilon_2 > 0$.

While convex optimization–based techniques are promising in providing optimal and/or near optimal solutions to the sparse signal recovery problem, their computational complexity is relatively high. Greedy and iterative algorithms to solve the sparse signal recovery with less computational complexity than the optimization-based techniques have been proposed lately. In these approaches, the solution to (2.5) (or its noise-resistant extension) is computed in greedy or iterative manner. Examples of such algorithms include orthogonal matching pursuit (OMP) [34], compressive sampling matching pursuit [35], subspace pursuit (SP) [36], and iterative hard thresholding (IHT) [37]. Another class of sparse-recovery algorithms falls under the Bayesian formulation. In the Bayesian framework, the sparse signal reconstruction problem is formulated as a random signal estimation problem after imposing a sparsity promoting probability density function (pdf) on $\mathbf{x}$ in (2.2) [38–40].

Most of these recovery techniques with SMV have been extended for the multiple sensor setting by various authors. A review on CS recovery algorithms with single as well as multiple sensor observations can be found in [25,29,41].

In the next section, we introduce the CS-based detection problem.

2.3 Compressed detection

Theories and concepts developed in CS have been exploited in the recent literature to solve detection problems [42–51]. Before discussing CS-based detection approaches in detail, in the following, we briefly review detection with uncompressed data.

Consider a distributed sensor network with L number of sensor nodes observing a phenomenon of interest (PoI) where under hypothesis $\mathcal{H}_1$, the signal is present and under $\mathcal{H}_0$, there is only noise. The observation vector collected at the jth sensor node can be expressed as:

$$\begin{aligned} \mathcal{H}_1 &: \mathbf{x}_j = \mathbf{s}_j + \mathbf{v}_j \\ \mathcal{H}_0 &: \mathbf{x}_j = \mathbf{v}_j \end{aligned} \tag{2.9}$$

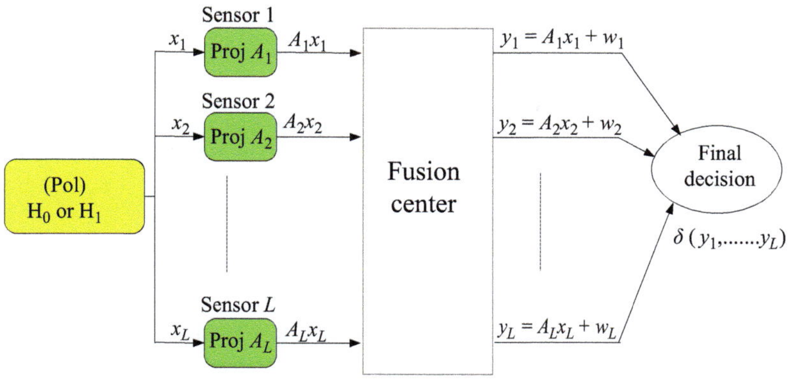

Figure 2.1 Collaborative compressed detection in a distributed sensor network

for $j = 1, \ldots, L$ where $\mathbf{s}_j \in \mathbb{R}^N$ is the signal observed at the jth sensor and $\mathbf{v}_j \in \mathbb{R}^N$ is the additive noise vector at the jth sensor. In order to solve the detection problem (2.9), observations collected at each node (or a summary) need to be sent to a central FC. If each sensor transmits its observations (without compression or local processing) to the FC via AWGN channels, the observations at the FC can be represented by

$$\begin{aligned} \mathcal{H}_1 : \tilde{\mathbf{x}}_j &= \mathbf{s}_j + \mathbf{v}_j + \mathbf{w}_j = \mathbf{s}_j + \tilde{\mathbf{w}}_j \\ \mathcal{H}_0 : \tilde{\mathbf{x}}_j &= \mathbf{v}_j + \mathbf{w}_j = \tilde{\mathbf{w}}_j \end{aligned} \tag{2.10}$$

for $j = 1, \ldots, L$ where $\mathbf{w}_j$ is the additive noise vector at the FC and $\tilde{\mathbf{w}}_j = \mathbf{v}_j + \mathbf{w}_j$. Suppose that the joint pdfs of $\tilde{\mathbf{x}} = [\tilde{\mathbf{x}}_1^T, \ldots, \tilde{\mathbf{x}}_L^T]^T$, where $\mathbf{x}^T$ denotes the transpose of $\mathbf{x}$, under both hypotheses are available, denoted by $f_1(\tilde{\mathbf{x}})$ and $f_0(\tilde{\mathbf{x}})$, respectively. The optimal detector (that minimizes the average probability of error in the Bayesian setting or maximizes the probability of detection keeping the probability of false alarm at a given value in the Neyman–Pearson (NP) setting) is given by the likelihood ratio test (LRT):

$$\mathrm{LR}(\tilde{\mathbf{x}}) \underset{\mathcal{H}_0}{\overset{\mathcal{H}_1}{\gtrless}} \tau \tag{2.11}$$

where $\mathrm{LR}(\tilde{\mathbf{x}}) = f_1(\tilde{\mathbf{x}})/f_0(\tilde{\mathbf{x}})$ and τ is the threshold. The optimality of (2.11) is violated when there are unknown parameters or when the exact knowledge of pdfs is not available. There are several suboptimal detectors proposed in the literature to account for such scenarios [52].

Now, assume that each sensor compresses its observations using random projections prior to transmission. The CS-based detection framework is illustrated in Figure 2.1.

The detection problem with compressed data at the FC is given by:

$$\begin{aligned}\mathcal{H}_1 : \mathbf{y}_j &= \mathbf{A}_j(\mathbf{s}_j + \mathbf{v}_j) + \mathbf{w}_j = \mathbf{A}_j\mathbf{s}_j + \tilde{\mathbf{w}}_j \\ \mathcal{H}_0 : \mathbf{y}_j &= \mathbf{A}_j\mathbf{v}_j + \mathbf{w}_j = \tilde{\mathbf{w}}_j\end{aligned} \tag{2.12}$$

for $j = 1, \ldots, L$ where $\tilde{\mathbf{w}}_j = \mathbf{A}_j\mathbf{v}_j + \mathbf{w}_j$. In the following sections, we discuss the solutions for (2.12) considering different models for $\mathbf{s}_j$ for $j = 1, \ldots, L$.

2.3.1 CS-based detection of known deterministic signals in the presence of iid noise

Suppose that the signals $\mathbf{s}_j$s for $j = 1, \ldots, L$ are deterministic and known. Further, the noise vectors at the nodes (before compression) and at the FC are assumed to be iid Gaussian so that $\tilde{\mathbf{w}}_j \sim \mathcal{N}(\mathbf{0}, \sigma_v^2\mathbf{A}_j\mathbf{A}_j^T + \sigma_w^2\mathbf{I})$ where $\mathbf{0}$ denotes a vector (with appropriate dimension) with all zeros and $\mathbf{I}$ denotes the identity matrix. The notation $\mathbf{x} \sim \mathcal{N}(\boldsymbol{\mu}, \boldsymbol{\Sigma})$ denotes that the random vector $\mathbf{x}$ is multivariate Gaussian with mean vector $\boldsymbol{\mu}$ and covariance matrix $\boldsymbol{\Sigma}$. The pdfs of $\mathbf{y} = [\mathbf{y}_1^T, \ldots, \mathbf{y}_L^T]^T$ under $\mathcal{H}_0$ and $\mathcal{H}_1$ are given by:

$$f_0(\mathbf{y}) = \prod_{j=1}^{L} \frac{e^{-(1/2)\mathbf{y}_j^T(\sigma_v^2\mathbf{A}_j\mathbf{A}_j^T + \sigma_w^2\mathbf{I})^{-1}\mathbf{y}_j}}{(2\pi)^{M/2}|\sigma_v^2\mathbf{A}_j\mathbf{A}_j^T + \sigma_w^2\mathbf{I}|^{1/2}} \quad \text{and} \tag{2.13}$$

$$f_1(\mathbf{y}) = \prod_{j=1}^{L} \frac{e^{-(1/2)(\mathbf{y}_j - \mathbf{A}_j\mathbf{s}_j)^T(\sigma_v^2\mathbf{A}_j\mathbf{A}_j^T + \sigma_w^2\mathbf{I})^{-1}(\mathbf{y}_j - \mathbf{A}_j\mathbf{s}_j)}}{(2\pi)^{M/2}|\sigma_v^2\mathbf{A}_j\mathbf{A}_j^T + \sigma_w^2\mathbf{I}|^{1/2}} \tag{2.14}$$

respectively, where $|\cdot|$ denotes the determinant of a matrix.

The LR test becomes:

$$\text{LR}(\mathbf{y}) = \frac{f_1(\mathbf{y})}{f_0(\mathbf{y})} \underset{\mathcal{H}_0}{\overset{\mathcal{H}_1}{\gtrless}} \tau$$

where $f_0(\mathbf{y})$ and $f_1(\mathbf{y})$ are as given in (2.13) and (2.14), respectively. The decision statistic, $\Lambda(\mathbf{y})$, of the log LR (LLR) test simplifies to $\Lambda(\mathbf{y}) = \sum_{j=1}^{L} \mathbf{y}_j^T(\sigma_v^2\mathbf{A}_j\mathbf{A}_j^T + \sigma_w^2\mathbf{I})^{-1}\mathbf{A}_j\mathbf{s}_j$ so that the LLR test decides between $\mathcal{H}_1$ and $\mathcal{H}_0$ using:

$$\Lambda(\mathbf{y}) \underset{\mathcal{H}_0}{\overset{\mathcal{H}_1}{\gtrless}} \tau_0 \tag{2.15}$$

where $\tau_0 = \sum_{j=1}^{L} \mathbf{s}_j^T\mathbf{A}_j^T(\sigma_v^2\mathbf{A}_j\mathbf{A}_j^T + \sigma_w^2\mathbf{I})^{-1}\mathbf{A}_j\mathbf{s}_j + \log\tau$. In order to compute the probabilities of false alarm and detection of the detector (2.15), the pdf of $\Lambda(\mathbf{y})$ under the two hypotheses needs to be computed. Since $\mathbf{y}_j$s are independent for $j = 1, \ldots, L$, we have:

$$\Lambda(\mathbf{y}) \sim \begin{cases} \mathcal{N}(0, \sum_{j=1}^{L} \mathbf{s}_j^T\mathbf{U}_j\mathbf{s}_j) & \text{under } \mathcal{H}_0 \\ \mathcal{N}(\sum_{j=1}^{L} \mathbf{s}_j^T\mathbf{U}_j\mathbf{s}_j, \sum_{j=1}^{L} \mathbf{s}_j^T\mathbf{U}_j\mathbf{s}_j) & \text{under } \mathcal{H}_1 \end{cases}$$

where $\mathbf{U}_j = \mathbf{A}_j^T(\sigma_v^2\mathbf{A}_j\mathbf{A}_j^T + \sigma_w^2\mathbf{I})^{-1}\mathbf{A}_j$. The probabilities of false alarm and detection are given by $P_f = \Pr(\Lambda(\mathbf{y}) \geq \tau_0|\mathcal{H}_0) = Q\left(\tau_0/\sqrt{\sum_{j=1}^{L} \mathbf{s}_j^T\mathbf{U}_j\mathbf{s}_j}\right)$ and

$P_d = \Pr(\Lambda(\mathbf{y}) \geq \tau_0 | \mathscr{H}_1) = Q\left(\frac{\tau_0 - \sum_{j=1}^{L} \mathbf{s}_j^T \mathbf{U}_j \mathbf{s}_j}{\sqrt{\sum_{j=1}^{L} \mathbf{s}_j^T \mathbf{U}_j \mathbf{s}_j}}\right)$ respectively, where $Q(x) = (1/\sqrt{2\pi}) \int_x^{\infty} e^{-(t^2/2)} dt$ is the Gaussian Q function. In order to ensure $P_f \leq \alpha_0$ in the NP setting, the threshold τ_0 is computed as $\tau_0 \geq Q^{-1}(\alpha_0)\sqrt{\sum_{j=1}^{L} \mathbf{s}_j^T \mathbf{U}_j \mathbf{s}_j}$ resulting in

$$P_d = Q\left(Q^{-1}(\alpha_0) - \sqrt{\sum_{j=1}^{L} \mathbf{s}_j^T \mathbf{U}_j \mathbf{s}_j}\right) \tag{2.16}$$

where $Q^{-1}(\cdot)$ denotes the inverse operation of the Q function.

2.3.1.1 Analysis when $\sigma_w^2 \to 0$

When $\sigma_w^2 \to 0$, we have $\mathbf{U}_j = (1/\sigma_v^2)\mathbf{P}_{\mathbf{A}_j}$ where $\mathbf{P}_{\mathbf{A}_j} = \mathbf{A}_j^T(\mathbf{A}_j\mathbf{A}_j^T)^{-1}\mathbf{A}_j$ is the orthogonal projection operator onto the range of $\mathbf{A}_j^T$ for $j = 1, \ldots, L$. Then, (2.16) reduces to:

$$P_d = Q\left(Q^{-1}(\alpha_0) - \frac{1}{\sigma_v}\sqrt{\sum_{j=1}^{L} ||\mathbf{P}_{\mathbf{A}_j}\mathbf{s}_j||_2^2}\right). \tag{2.17}$$

From (2.17), we can see that the impact of the random projection scheme on the detection performance is reflected in the quantity $||\mathbf{P}_{\mathbf{A}_j}\mathbf{s}_j||_2^2$ for $j = 1, \ldots, L$ which measures the sum of the powers of the signals $\mathbf{s}_j$ projected onto the row space of $\mathbf{A}_j$. We have $0 \leq ||\mathbf{P}_{\mathbf{A}_j}\mathbf{s}_j||_2^2 \leq ||\mathbf{s}_j||_2^2$ for $j = 1, \ldots, L$. Thus, in general, depending on how $\mathbf{A}_j$ is selected, the performance of the compressed detector can be significantly different from the uncompressed detector. For example, when $\mathbf{s}_j$ lies in the row space of $\mathbf{A}_j$, or equivalently $\mathbf{A}_j$ is selected as $\mathbf{s}_j^T$ (a row vector), we have $||\mathbf{P}_{\mathbf{A}_j}\mathbf{s}_j||_2^2 = ||\mathbf{s}_j||_2^2$ for $j = 1, \ldots, L$ yielding the same performance as that of the uncompressed detector. On the other hand, if $\mathbf{s}_j$ lies in the null space of $\mathbf{A}_j$, we have $||\mathbf{P}_{\mathbf{A}_j}\mathbf{s}_j||_2^2 = 0$ for $j = 1, \ldots, L$. However, if the elements of $\mathbf{A}_j$ are selected to be iid random and independent of $\mathbf{s}_j$ for $j = 1, \ldots, L$, so that it provides a stable embedding as defined below, we would expect that $||\mathbf{P}_{\mathbf{A}_j}\mathbf{s}_j||_2^2$ to be concentrated around $\frac{M}{N}||\mathbf{s}_j||_2^2$.

Definition 2.2 (ε-stable embedding [46]). *Let $\varepsilon \in (0, 1)$ and $\mathscr{S}$ and $\mathscr{X} \subset \mathbb{R}^N$ be given. The mapping $\mathbf{A}$ is a ε-stable embedding of $(\mathscr{S}, \mathscr{X})$ if $(1-\varepsilon)||\mathbf{s}-\mathbf{x}||_2^2 \leq ||\mathbf{As}-\mathbf{Ax}||_2^2 \leq (1+\varepsilon)||\mathbf{s}-\mathbf{x}||_2^2$ for all $\mathbf{s} \in \mathscr{S}$ and $\mathbf{x} \in \mathscr{X}$.*

It is noted that a matrix $\mathbf{A}$ satisfying RIP of order $2k$ is equivalent to being a ε-embedding of (Σ_k, Σ_k) or of $(\Sigma_{2k}, \{0\})$ where Σ_k is as defined in (2.4) [46]. In particular, if $\mathbf{A}_j$ is selected to be an orthoprojector so that $\mathbf{A}_j\mathbf{A}_j^T = \mathbf{I}$ or the elements of $\mathbf{A}_j$ are selected from $\mathscr{N}(0, (1/N))$, $\sqrt{(N/M)}\mathbf{P}_{\mathbf{A}_j}$ provides an ε-embedding of

$(\mathcal{S}, \{0\})$ with high probability. Then, for any $\mathbf{s}_j \in \mathcal{S}$ for $j = 1, \ldots, L$, the probability of detection is bounded by:

$$P_d \leq Q\left(Q^{-1}(\alpha_0) - \sqrt{\frac{M}{N}}\sqrt{(1+\varepsilon)}\frac{\sqrt{\sum_{j=1}^{L}||\mathbf{s}_j||_2^2}}{\sigma_v}\right) \quad \text{and}$$

$$P_d > Q\left(Q^{-1}(\alpha_0) - \sqrt{\frac{M}{N}}\sqrt{(1-\varepsilon)}\frac{\sqrt{\sum_{j=1}^{L}||\mathbf{s}_j||_2^2}}{\sigma_v}\right) \tag{2.18}$$

for $0 < \varepsilon < 1$. As $\varepsilon \to 0$, we have,

$$P_d \approx Q\left(Q^{-1}(\alpha_0) - \sqrt{\frac{M}{N}\gamma_{\text{av}}}\right) \tag{2.19}$$

where $\gamma_{\text{av}} = (\sum_{j=1}^{L}||\mathbf{s}_j||_2^2)/\sigma_v^2$ denotes the average signal-to-noise ratio (SNR) of the uncompressed signals at all the nodes. Bounding the Q-function using $Q(x) \leq \frac{1}{2}e^{-(x^2/2)}$, (2.18) can be written as:

$$P_d \geq Q\left(Q^{-1}(\alpha_0) - \sqrt{(1-\varepsilon)\frac{M}{N}\gamma_{\text{av}}}\right) = 1 - Q\left(\sqrt{(1-\varepsilon)\frac{M}{N}\gamma_{\text{av}}} - Q^{-1}(\alpha_0)\right)$$

$$\geq 1 - \frac{1}{2}e^{-(1/2)\left(\sqrt{\frac{M\eta_1}{N}} - Q^{-1}(\alpha_0)\right)^2}$$

where $\eta_1 = (1-\varepsilon)\gamma_{av}$. Thus, given that $(M/N) \geq ([Q^{-1}(\alpha_0)]^2)/\eta_1$, $P_d \to 1$ exponentially fast as M/N increases for given SNR.

2.3.1.2 Performance comparison with uncompressed data

In the case where detection is performed with uncompressed data as given in (2.10), the probability of detection of the NP detector (keeping the probability of false alarm under α_0) is given by [52]:

$$P_d^u = Q\left(Q^{-1}(\alpha_0) - \sqrt{\gamma_{av}}\right). \tag{2.20}$$

From (2.19) and (2.20), it is observed that the impact of the random projection scheme on the probability of detection is reflected in the SNR term which is reduced by a factor of M/N with compressed data compared to its uncompressed counterpart. When the average SNR of the uncompressed signals is large, the performance of the compressed detector coincides with that of the uncompressed detector although there is a distortion in the output SNR. This is illustrated in Figure 2.2. In Figure 2.2, the receiver operating characteristic (ROC) curve of the NP detector with compressed data is plotted for different values of the compression ratio $c_r = M/N$. For given SNR and L, the performance improves as c_r increases and coincides with the uncompressed detector as c_r exceeds a certain value. The compressed detector is capable of providing similar performance as the uncompressed detector for a certain values of parameters.

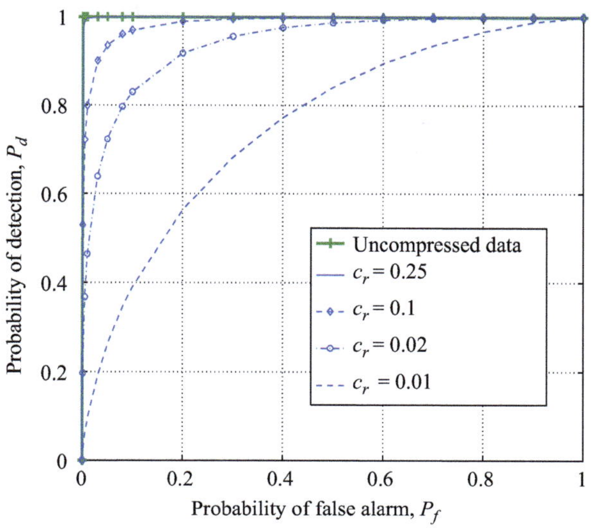

Figure 2.2 ROC for the NP detector with compressed data; $L = 10$, $N = 1{,}000$, $\gamma_{av} = 10\,dB$

Thus, when the problem is detection (but not exact signal recovery), compression via random projections can still be beneficial even without having the sparsity prior.

2.3.2 CS-based detection of unknown sparse signals in the presence of iid noise

Now suppose that the signals of interest $\mathbf{s}_j$s in (2.12) are deterministic and unknown but can be assumed to be sparse in a certain basis. More specifically, let $\mathbf{s}_j = \boldsymbol{\Psi}_j\boldsymbol{\theta}_j$ where $\boldsymbol{\Psi}_j \in \mathbb{R}^{N\times N}$ is an orthonormal basis and $\boldsymbol{\theta}_j$ is the $N \times 1$ coefficient vector with only $k \ll N$ nonzero elements. Then, the detection problem in (2.12) can be equivalently expressed as:

$$\begin{aligned}\mathscr{H}_1 : \mathbf{y}_j &= \mathbf{B}_j\boldsymbol{\theta}_j + \tilde{\mathbf{w}}_j \\ \mathscr{H}_0 : \mathbf{y}_j &= \tilde{\mathbf{w}}_j\end{aligned} \tag{2.21}$$

where $\mathbf{B}_j \triangleq \mathbf{A}_j\boldsymbol{\Psi}_j$ for $j = 1, \ldots, L$.

Let $\mathscr{U}_j$ be the set which contains the indices of the locations of nonzero coefficients in $\boldsymbol{\theta}_j$ which is defined by $\mathscr{U}_j := \{i \in \{1, \ldots, N\} \mid \boldsymbol{\theta}_j[i] \neq 0\}$ for $i = 1, \ldots, N$ and $j = 1, \ldots, L$ where $\boldsymbol{\theta}_j[i]$ denotes the ith element of $\boldsymbol{\theta}_j$. If $\mathscr{U}_j$ is exactly known, the detection problem in (2.21) reduces to:

$$\begin{aligned}\mathscr{H}_1 : \mathbf{y}_j &= \mathbf{B}_j(\mathscr{U}_j)\boldsymbol{\theta}_j(\mathscr{U}_j) + \tilde{\mathbf{w}}_j \\ \mathscr{H}_0 : \mathbf{y}_j &= \tilde{\mathbf{w}}_j\end{aligned} \tag{2.22}$$

for $j = 1, \ldots, L$ where $\mathbf{B}_j(\mathscr{U}_j)$ denotes the $M \times k$ submatrix of $\mathbf{B}_j$ in which columns are indexed by the ones in $\mathscr{U}_j$, and $\boldsymbol{\theta}_j(\mathscr{U}_j)$ is a $k \times 1$ vector containing nonzero elements in

$\boldsymbol{\theta}_j$ indexed by $\mathscr{U}_j$ for $j = 1, \ldots, L$. When $\mathbf{B}_j(\mathscr{U}_j)$ is known for $j = 1, \ldots, L$, (2.22) is the subspace detection problem which has been addressed previously [53–55]. Depending on how the unknown coefficient vector $\boldsymbol{\theta}_j(\mathscr{U}_j)$ is modeled, different detectors have been proposed. When $\boldsymbol{\theta}_j(\mathscr{U}_j)$ for $j = 1, \ldots, L$ is unknown, the optimal LR test cannot be performed since the exact knowledge of the pdf of $\mathbf{y}$ under $\mathscr{H}_1$ is not known. While there is no optimality guarantee associated, the generalized LRT (GLRT) is a widely used approach to solve detection problems in the presence of deterministic unknown parameters. In [53], a GLRT-based detector is proposed when $\boldsymbol{\theta}_j(\mathscr{U}_j)$ is assumed to be deterministic and $L = 1$. In [54], the analysis has been extended to the case when $\boldsymbol{\theta}_j(\mathscr{U}_j)$ is modeled as random. The problem with $L > 1$ is addressed in [55] where the authors have proposed adaptive subspace detectors when the coefficients $\{\boldsymbol{\theta}_j(\mathscr{U}_j)\}_{j=1}^{L}$ follow first and second order Gaussian models.

In the case of sparse signal detection, it is unlikely that the exact knowledge of $\mathscr{U}_j$ is available *a priori*. In other words, sparse signal detection needs to be performed when $\mathscr{U}_j$ is unknown for $j = 1, \ldots, L$. If the sparsity prior is neglected, the detection problem (2.21) can be solved using GLRT in which the unknown coefficient vectors $\boldsymbol{\theta}_j$s are estimated so that the likelihood function of $\mathbf{y}$ under $\mathscr{H}_1$ is maximized. With the maximum likelihood (ML) estimates, the decision statistic of the GLRT is given by $\Lambda(\mathbf{y}) = (\prod_{j=1}^{L} f_1(\mathbf{y}_j; \hat{\boldsymbol{\theta}}_{j,ML}))/(\prod_{j=1}^{L} f_0(\mathbf{y}_j))$ where $\hat{\boldsymbol{\theta}}_{j,ML}$ denotes the ML estimate of $\boldsymbol{\theta}_j$ based on $\mathbf{y}$ which can be obtained as [52]:

$$\hat{\boldsymbol{\theta}}_{j,ML} = (\mathbf{B}_j^T \mathbf{B}_j)^{-1} \mathbf{B}_j^T \mathbf{y}_j \tag{2.23}$$

for $j = 1, \ldots, L$. However, when $\boldsymbol{\theta}_j$ is sparse, the ML estimate in (2.23) is not capable of finding the sparse solution; thus, it is not accurate [56]. In order to perform the LR test with (2.21), $\hat{\boldsymbol{\theta}}_{j,ML}$ needs to be replaced by a better estimate.

2.3.2.1 Solving (2.21) by estimating the signals via sparse reconstruction

Let $\hat{\boldsymbol{\theta}}_{j,SR}$ denote an estimate of $\boldsymbol{\theta}_j$ for $j = 1, \ldots, L$ obtained using a sparse signal reconstruction technique. Then, the decision statistic for the LRT can be expressed as:

$$\Lambda(\mathbf{y}) = \frac{\prod_{j=1}^{L} f_1(\mathbf{y}_j; \hat{\boldsymbol{\theta}}_{j,SR})}{\prod_{j=1}^{L} f_0(\mathbf{y}_j)}. \tag{2.24}$$

In order to find $[\hat{\boldsymbol{\theta}}_{1,SR}, \ldots, \hat{\boldsymbol{\theta}}_{L,SR}]$ based on $\mathbf{y}$, one straight-forward way is to obtain each $\hat{\boldsymbol{\theta}}_{j,SR}$ based on $\mathbf{y}_j$ independently. However, in many practical scenarios, the sparse signals $[\boldsymbol{\theta}_1, \ldots, \boldsymbol{\theta}_L]$ share joint sparse structures that can be exploited to get better estimates for $[\boldsymbol{\theta}_1, \ldots, \boldsymbol{\theta}_L]$ jointly instead of estimating them independently. Simultaneous sparse signal reconstruction has been widely addressed in CS literature with different joint sparse structures of the multiple sparse signals [57,58].

2.3.2.2 Joint sparsity models

There are several joint sparsity models (JSMs) proposed in [57,58] for sensor network data. In the JSM-1 model discussed in [57], multiple sparse signals are assumed to have a common sparse component plus an innovation sparse component. In particular, each $\mathbf{s}_j$ can be represented as $\mathbf{s}_j = \boldsymbol{\Psi}\boldsymbol{\theta}_j + \tilde{\boldsymbol{\Psi}}_j\tilde{\boldsymbol{\theta}}_j$ where both $\boldsymbol{\theta}_j$ and $\tilde{\boldsymbol{\theta}}_j$ are sparse. $\boldsymbol{\theta}_j$s have the same support $\mathscr{U}$ for $j = 1, \ldots, L$ and the support of $\tilde{\boldsymbol{\theta}}_j$ can be different for different j. In the JSM-2 model in [57], it is assumed that all $\mathbf{s}_j$s are sparse in the same basis with the same sparsity pattern. More specifically, with JSM-2, we have $\mathbf{s}_j = \boldsymbol{\Psi}\boldsymbol{\theta}_j$ with $\mathscr{U}_j = \mathscr{U}$ for all j. This is the widely considered multiple measurement vector model in array signal processing and the sensor network literature. In the JSM-3 model [57], each $\mathbf{s}_j$ is assumed to have a nonsparse common component and a sparse innovation component. In the rest of the chapter, we restrict our analysis to the JSM-2 model due its versatility.

With the JSM-2 model, there are several algorithms proposed to estimate $[\boldsymbol{\theta}_1, \ldots, \boldsymbol{\theta}_L]$ simultaneously such as group LASSO [59], M-Focuss [60], S-OMP [61], SIHT [62], generalized SP [63], and MSBL [64]. A comprehensive list of such algorithms is provided in [25]. Using such an estimate $[\hat{\boldsymbol{\theta}}_{1,SR}, \ldots, \hat{\boldsymbol{\theta}}_{L,SR}]$, (2.24) can be computed. It is worth noting that, in CS, the most challenging task is signal reconstruction. Further, to construct a reliable decision statistic for detection, it is not necessary to reconstruct the signal completely. Sparse signal detection without complete signal reconstruction is discussed below.

2.3.2.3 Detection of sparse signals via partial estimation of signals/support set

In greedy type of sparse signal recovery algorithms, the common support set of the multiple sparse signals is found in a greedy manner. After estimating only a fraction of the support with less computational complexity than that needed for complete support set estimation, a decision statistic can be computed. In the following, we discuss the construction of a decision statistic for the detection problem (2.21) with only a partial estimate of $[\boldsymbol{\theta}_1, \ldots, \boldsymbol{\theta}_L]$. Suppose that we have estimated only $T_0 \leq k$ number of locations of the common support set, $\mathscr{U}$, of $[\boldsymbol{\theta}_1, \ldots, \boldsymbol{\theta}_L]$, using a sparse signal reconstruction technique. We further assume that the measurement matrix $\mathbf{A}_j$ is an orthoprojector so that $\mathbf{A}_j\mathbf{A}_j^T = \mathbf{I}$ for $j = 1, \ldots, L$. Then, $\tilde{\mathbf{w}}$ in (2.21) is distributed as $\tilde{\mathbf{w}} \sim \mathscr{N}(\mathbf{0}, \sigma_{\tilde{w}}^2\mathbf{I})$ where $\sigma_{\tilde{w}}^2 = \sigma_v^2 + \sigma_w^2$. Assuming that $\boldsymbol{\theta}_j$s have only T_0 number of nonzero locations, the GLRT decision statistic can be computed as [52]:

$$\Lambda(\mathbf{y}) = \sum_{j=1}^{L} ||\mathbf{P}_{j,T_0}\mathbf{y}_j||_2^2 \underset{\mathscr{H}_0}{\overset{\mathscr{H}_1}{\gtrless}} \tau_0, \tag{2.25}$$

where $\mathbf{P}_{j,T_0} = \mathbf{B}_j(\mathscr{U}(T_0))\left(\mathbf{B}_j(\mathscr{U}(T_0))^T\mathbf{B}_j(\mathscr{U}(T_0))\right)^{-1}\mathbf{B}_j(\mathscr{U}(T_0))^T$ is the projection operator to the space spanned by $\mathbf{B}_j(\mathscr{U}(T_0))$ and $\mathscr{U}(T_0) \subset \mathscr{U}$ denotes the set containing known T_0 indices of the support set. When $T_0 = k$, the decision statistic in (2.25) is the same as the GLRT decision statistic in [54] for the subspace detection problem.

In order to evaluate the performance of the detector in (2.25), computation of the pdf of $\Lambda(\mathbf{y})$ under the two hypotheses is required. It can be easily shown that $\Lambda(\mathbf{y})$ in (2.25) is distributed under $\mathscr{H}_0$ and $\mathscr{H}_1$ as $(\Lambda(\mathbf{y})/\sigma_{\tilde{w}}^2)|\mathscr{H}_0 \sim \mathscr{X}_{T_0L}^2$ and $(\Lambda(\mathbf{y})/\sigma_{\tilde{w}}^2)|\mathscr{H}_1 \sim \mathscr{X}_{T_0L}^2(\lambda_{T_0})$, respectively, where $\lambda_{T_0} = \sum_{j=1}^{L}(||\mathbf{P}_{j,T_0}\mathbf{B}_j\boldsymbol{\theta}_j||_2^2)/\sigma_{\tilde{w}}^2$. It is noted that $x \sim \mathscr{X}_n^2$

denotes that the random variable x has a central chi-squared distribution with degrees of freedom n while $x \sim \mathscr{X}_n^2(\lambda)$ denotes that x has a noncentral chi-squared distribution with degrees of freedom n and the noncentrality parameter λ. The probabilities of false alarm and detection for the test (2.25) are given by [65]:

$$P_f = 1 - F\left(\frac{T_0 L}{2}, \frac{\tau_0}{2\sigma_{\tilde{w}}^2}\right) \text{ and} \tag{2.26}$$

$$P_d = Q_{\frac{T_0 L}{2}}\left(\sqrt{\lambda_{T_0}}, \sqrt{\frac{\tau_0}{\sigma_{\tilde{w}}^2}}\right), \tag{2.27}$$

respectively, where $F(a,x) = \gamma(a,x)/\Gamma(a)$ is the regularized Gamma function, $\Gamma(\cdot)$ is the Gamma function, $\gamma(a,x)$ is the lower incomplete Gamma function given by $\gamma(a,x) = \int_0^x t^{a-1}e^{-t}dt$, and $Q_M(a,b) = \int_b^\infty x\left(\frac{x}{a}\right)^{M-1} e^{(-(x^2+a^2)/2)} I_{M-1}(ax)dx$ is the Marcum Q function with I_{M-1} being the modified Bessel function of order $M-1$.

Interpretation of λ_{T_0}

We can write λ_{T_0} as $\lambda_{T_0} = \sum_{j=1}^L (||\mathbf{B}_j\boldsymbol{\theta}_j||_2^2/\sigma_{\tilde{w}}^2)\kappa_{j,T_0}$ where $\kappa_{j,T_0} = ||\mathbf{P}_{j,T_0}\mathbf{B}_j\boldsymbol{\theta}_j||_2^2/||\mathbf{B}_j\boldsymbol{\theta}_j||_2^2$. Defining $\gamma_j = ||\boldsymbol{\theta}_j||_2^2/\sigma_{\tilde{w}}^2$, λ_{T_0} can be approximated by

$$\lambda_{T_0} \approx \frac{M}{N}\sum_{j=1}^{L}\gamma_j\kappa_{j,T_0} \tag{2.28}$$

since we have $||\mathbf{B}_j\boldsymbol{\theta}_j||_2^2 \approx (M/N)||\boldsymbol{\theta}_j||_2^2$ when $\mathbf{A}_j\mathbf{A}_j^T = \mathbf{I}$. Note that, $||\mathbf{P}_{j,T_0}\mathbf{B}_j\boldsymbol{\theta}_j||_2^2 = ||\mathbf{B}_j\boldsymbol{\theta}_j||_2^2 - ||\mathbf{P}_{j,T_0}^\perp\mathbf{B}_j\boldsymbol{\theta}_j||_2^2$ where $\mathbf{P}_{j,T_0}^\perp = \mathbf{I} - \mathbf{P}_{j,T_0}$. Thus, κ_{j,T_0} can be equivalently written as $\kappa_{j,T_0} = 1 - (||\mathbf{P}_{j,T_0}^\perp\mathbf{B}_j\boldsymbol{\theta}_j||_2^2)/(||\mathbf{B}_j\boldsymbol{\theta}_j||_2^2)$. The quantity $||\mathbf{P}_{j,T_0}^\perp\mathbf{B}_j\boldsymbol{\theta}_j||_2^2$ reflects the power of the sampled signal unaccounted for by the subspace spanned by $\mathbf{B}_j(\mathscr{U}(T_0))$. In other words, the impact of not knowing the complete support set of $\boldsymbol{\theta}_j$ on the detection performance is reflected by $||\mathbf{P}_{j,T_0}^\perp\mathbf{B}_j\boldsymbol{\theta}_j||_2^2$. When $T_0 = k$, we have $||\mathbf{P}_{j,T_0}^\perp\mathbf{B}_j\boldsymbol{\theta}_j||_2^2 = 0$ and $\kappa_{j,T_0} = 1$ and, therefore, $\lambda_k \approx (M/N)\sum_{j=1}^L \gamma_j$. If the desired probability of detection is τ_d and if $Q_{(kL/2)}\left(\sqrt{\lambda_k}, \sqrt{\tau_0/\sigma_v^2}\right) < \tau_d$, the desired P_d with the decision statistic (2.25) cannot be achieved even if all the indices of the common support are known correctly. Thus, detection by estimating only a fraction of the common support is sufficient only if $\tau_d < Q_{kL/2}\left(\sqrt{\lambda_k}, \sqrt{\tau_0/\sigma_v^2}\right)$. In that case, it is of interest to determine the minimum fraction of the common support that should be known to achieve the desired detection performance which will be discussed next.

Partial support set estimation

The goal is to find the minimum value of T_0 so that $P_d \geq \tau_d$ and $P_f \leq \alpha$. Finding a closed-form solution for T_0 is in general difficult. However, under the assumption that the nonzero coefficients in each θ_j are not significantly different from each other, and using approximations to P_d and P_f, we can find a closed-form solution for T_0 as discussed below. Letting $t = T_0$, the desired optimization problem can be cast as,

$$\min\ t \text{ such that}$$

$$P_d(t) \geq \tau_d, P_f \leq \alpha \text{ and } 1 \leq t < k, t \sim \text{integers} \tag{2.29}$$

with

$$P_d(t) = Q_{\frac{tL}{2}}\left(\sqrt{\lambda_t}, \sqrt{\frac{\tau_0}{\sigma_v^2}}\right) \tag{2.30}$$

where λ_t is as given in (2.28). The term t is related to λ_t via $\kappa_{j,t} = 1 - ((||\mathbf{P}_{j,t}^{\perp}\mathbf{B}_j\boldsymbol{\theta}_j||_2^2)/(||\mathbf{B}_j\boldsymbol{\theta}_j||_2^2)$. Let $\mathbf{P}_{j,t}^{\perp} = \mathbf{W}^T\boldsymbol{\Pi}\mathbf{W}$ where $\mathbf{W}$ is unitary and $\boldsymbol{\Pi}$ is a diagonal matrix with t zero elements and $(M-t)$ ones. Then, we can write:

$$||\mathbf{P}_{j,t}^{\perp}\mathbf{B}_j\boldsymbol{\theta}_j||_2^2 = ||\mathbf{P}_{j,t}^{\perp}\mathbf{B}_{j,k\setminus t}\boldsymbol{\theta}_{j,k\setminus t}||_2^2 = ||\boldsymbol{\Pi}\mathbf{W}\mathbf{B}_{j,k\setminus t}\boldsymbol{\theta}_{j,k\setminus t}||_2^2 \tag{2.31}$$

where $\mathbf{B}_{j,k\setminus t} = \mathbf{B}_j(\mathscr{U} \setminus \mathscr{U}(t))$ with cardinality of $\mathscr{U}$ denoted by $|\mathscr{U}|$ is k (similar definition holds for $\boldsymbol{\theta}_{j,k\setminus t}$). It is noted that $k \setminus t$ is a notation while $\mathscr{U} \setminus \mathscr{U}(t)$ denotes the set of indices in $\mathscr{U}$ that are not in $\mathscr{U}(t)$. Whenever $\mathbf{B}_j^T\mathbf{B}_j \approx (M/N)\mathbf{I}$, we have $\breve{\mathbf{B}}_j^T\breve{\mathbf{B}}_j \approx (M/N)\mathbf{I}$ since $\mathbf{W}$ is unitary where $\breve{\mathbf{B}}_j = \mathbf{W}\mathbf{B}_j$. Since $\boldsymbol{\Pi}$ has only $M-t$ ones, we may approximate (2.31) by:

$$||\boldsymbol{\Pi}\mathbf{W}\mathbf{B}_{j,k\setminus t}\boldsymbol{\theta}_{j,k\setminus t}||_2^2 = ||\boldsymbol{\Pi}\breve{\mathbf{B}}_{j,k\setminus t}\boldsymbol{\theta}_{j,k\setminus t}||_2^2 \approx \frac{M-t}{N}||\boldsymbol{\theta}_{j,k\setminus t}||_2^2.$$

Then, $\kappa_{j,t}$ can be approximated by $\kappa_{j,t} \approx 1 - ((M-t)/M)((||\boldsymbol{\theta}_{j,k\setminus t}||_2^2)/(||\boldsymbol{\theta}_j||^2))$. It is noted that $||\boldsymbol{\theta}_{j,k\setminus t}||_2^2$ depends on t and the nonzero coefficients of the signal indexed by $\mathscr{U} \setminus \mathscr{U}(t)$. Since the nonzero coefficients of $\boldsymbol{\theta}_j$ are unknown, the knowledge of $||\boldsymbol{\theta}_{j,k\setminus t}||_2^2$ for given t is not available. In the case where the nonzero coefficients of $\boldsymbol{\theta}_j$ do not deviate much from each other, we may approximate $(||\boldsymbol{\theta}_{j,k\setminus t}||_2^2/||\boldsymbol{\theta}_j||_2^2) \approx (k-t)/k$. In that case, $\kappa_{j,t}$ can be approximated by $\kappa_{j,t} \approx \frac{t}{k}(1 + ((k-t)/M))$. Then we have:

$$\lambda_t \approx \frac{Mt}{Nk}\left(1 + \frac{(k-t)}{M}\right)\sum_{j=1}^{L}\gamma_j. \tag{2.32}$$

In order to simplify the nonlinear constraint $P_d(t) \geq \tau_d$, we use the following approximations for P_d and P_f in (2.26) and (2.27), respectively, exploiting approximations of the tail probabilities of chi-squared random variables. The cumulative distribution function (cdf), $G(x, k)$, of a chi-squared random variable, $x \sim \mathscr{X}_k^2$, can be approximated by

$$G(x,k) \approx 1 - Q\left(\frac{\left(\frac{x}{k}\right)^{1/3} - \left(1 - \frac{2}{9k}\right)}{\sqrt{\frac{2}{9k}}}\right)$$

[66]. Using this approximation for P_f, we have,

$$P_f \approx Q\left(\frac{\left(\frac{\tau_0/\sigma_v^2}{tL}\right)^{1/3} - \left(1 - \frac{2}{9tL}\right)}{\sqrt{\frac{2}{9tL}}}\right)$$

and the threshold should be selected as:

$$\tau_0(t,\alpha) \approx \sigma_v^2 tL \left(\left(1 - \frac{2}{9tL}\right) + \sqrt{\frac{2}{9tL}} Q^{-1}(\alpha) \right)^3 \tag{2.33}$$

to maintain $P_f \leq \alpha$. Similarly, the cdf of a noncentral chi-squared random variable, $x \sim \mathscr{X}_k^2(\lambda)$, can be approximated by

$$G(x;k,\lambda) \approx 1 - Q\left(\frac{\left(\frac{x}{k+\lambda}\right)^h - (1 + hp(h-1-0.5(2-h)mp))}{h\sqrt{2p}(1+0.5mp)} \right)$$

where $h = 1 - (2/3)((k+\lambda)(k+3\lambda))/(k+2\lambda)^2, p = ((k+2\lambda)/(k+\lambda)^2)$, and $m = (h-1)(1-3h)$ [66]. With this approximation, P_d can be written as,

$$P_d \approx Q(g(t,\alpha,\gamma)) \tag{2.34}$$

with

$$g(t,\alpha,\gamma) = \left(\frac{\left(\frac{\tau_0(t,\alpha)}{\sigma_v^2(tL+\lambda_t)}\right)^h - (1 + hp(h-1-0.5(2-h)mp))}{h\sqrt{2p}(1+0.5mp)} \right) \tag{2.35}$$

where $h = 1 - (2/3)(((tL+\lambda_t)(tL+3\lambda_t))/(tL+2\lambda_t)^2), p = ((tL+2\lambda_t)/(tL+\lambda_t)^2)$, $m = (h-1)(1-3h)$, $\gamma = [\gamma_1, \ldots, \gamma_L]$ and $\tau_0(t,\alpha)$ is approximated as in (2.33). It is noted that λ_t on the right-hand side of (2.35) is given by (2.32) which is a function of γ.

With the approximations (2.32), (2.33), and (2.34), the optimization problem in (2.29) can be rewritten as:

$$\min \ t \text{ such that}$$

$$\tau_d - Q(g(t,\alpha,\boldsymbol{\gamma})) \leq 0,\ t - k < 0 \text{ and } 1 - t \leq 0, t \sim \text{integers}. \tag{2.36}$$

The theory developed in the area of integer nonlinear programming is much less mature than integer linear programming [67]. We aim to solve (2.36) based on rounding by relaxing the integer restriction. This is a heuristic method which is shown to be faster than most of the other algorithms developed for integer nonlinear programming [68]. After relaxing the integer restriction, the solution for t which satisfies the Karush–Kuhn–Tucker (KKT) conditions is summarized in the following proposition.

Proposition 2.1 ([49]). *Let $k > 1$. The solution for t in (2.36) which satisfies the KKT conditions is given by:*

$$\hat{t}_{cont} = \begin{cases} 1 & \text{if } \tau_d \leq Q(g(1)) \\ g^{-1}(Q^{-1}(\tau_d)) & \text{if } g'(t)|_{t=g^{-1}(Q^{-1}(\tau_d))} \leq 0 \ \& \ Q(g(1)) < \tau_d < Q(g(k)) \\ \text{infeasible} & \text{otherwise} \end{cases} \tag{2.37}$$

where $\hat{t}_{cont}$ is the continuous valued solution for t, $g(t)$ (we use $g(t)$ for $g(t,\alpha,\boldsymbol{\gamma})$ for simplicity) is as given in (2.35) and $g'(t)$ denotes the derivative of $g(t)$.

The integer-valued solution, $\hat{t}$, is obtained by rounding $\hat{t}_{cont}$ to the nearest integer. From Proposition 2.1, it is observed that, under certain conditions, i.e., when the parameters $\boldsymbol{\gamma}$, α, and τ_d are such that $\tau_d \leq Q(g(1, \alpha, \boldsymbol{\gamma}))$, it is sufficient to know only one support location of the sparse signal correctly to reach the desired detection performance.

Algorithm 1: S-OMP based sparse signal detection [49]

Inputs: $\mathbf{y} = [\mathbf{y}_1, \ldots, \mathbf{y}_L]^T$, $\{\mathbf{B}_j\}_{j=1}^L$, T_0

Outputs: Partial support set estimate $\hat{\mathscr{U}}(T_0)$, Decision statistic $\Lambda(\mathbf{y})$, Detection decision

1. Initialize $t = 1$, $\hat{\mathscr{U}}(0) = \emptyset$, residual vector $\mathbf{r}_{j,0} = \mathbf{y}_j$ for $j = 1, \ldots, L$
2. *For* $t = 1$ to $t = T_0$
3. Find the index $\beta(t)$ such that $\beta(t) = \arg \max\limits_{\omega=1,\ldots,N} \sum_{j=1}^L |\langle \mathbf{r}_{j,t-1}, \mathbf{B}_j(\omega)\rangle|$
4. Set $\hat{\mathscr{U}}(t) = \hat{\mathscr{U}}(t-1) \cup \{\beta(t)\}$
5. Compute the orthogonal projection operator:
$$\mathbf{P}_{j,t} = \mathbf{B}_j(\hat{\mathscr{U}}(t)) \left(\mathbf{B}_j(\hat{\mathscr{U}}(t))^T \mathbf{B}_j(\hat{\mathscr{U}}(t))\right)^{-1} \mathbf{B}_j(\hat{\mathscr{U}}(t))^T \text{ for } j = 1, \ldots, L$$
Update the residual: $\mathbf{r}_{j,t} = (\mathbf{I} - \mathbf{P}_{j,t})\mathbf{y}_j$ for $j = 1, \ldots, L$
6. End *For*
7. Set $\hat{\mathscr{U}}(T_0) = \hat{\mathscr{U}}(t)$
8. Detection decision: If $\Lambda(\mathbf{y}) = \sum_{j=1}^L ||\mathbf{P}_{j,T_0}\mathbf{y}_j||_2^2 \underset{\mathscr{H}_0}{\overset{\mathscr{H}_1}{\gtrless}} \tau_0$.

S-OMP based sparse signal detection via partial support set estimation

Proposition 2.1 provides insights on determining the number of elements to be estimated in the sparse signals to make a detection decision with the statistics given in (2.25). In the following, a practical algorithm based on S-OMP is discussed that performs detection via partial support set estimation. S-OMP proposed in [61] is a greedy algorithm which estimates the locations of the common support in a greedy manner. In Algorithm 1, the S-OMP algorithm is modified to construct the decision statistic (2.25) after estimating only a $T_0 \leq k$ number of elements in the common support set.

In Figure 2.3, the detection performance is illustrated when the partial support set is exactly known and when it is estimated via S-OMP in Algorithm 1. In the case where the partial support set is known, T_0 indices are selected randomly from $\mathscr{U}$ and the results are averaged over 20 trials. We let $T_0 = 1$ and the values of the parameters k, L, and N are provided in the figure caption. In Figure 2.3, we assume that $\mathbf{s}_j$s are sparse in the standard canonical basis so that $\boldsymbol{\Psi} = \mathbf{I}$. The nonzero coefficients of $\boldsymbol{\theta}_j$ are generated uniformly and randomly from $[-b, -a] \bigcup [a, b]$ with $a = 3$ and $b = 4$. The average SNR is defined as $\sum_{j=1}^L ||\boldsymbol{\theta}_j||_2^2/(LN\sigma_{\tilde{w}}^2)$. In Figure 2.3, we further plot the detection performance when the sparse signals are estimated via ML estimation ignoring sparsity, i.e. when the decision statistic is $\Lambda_{ML} = \sum_{j=1}^L ||\mathbf{P}_j\mathbf{y}_j||_2^2$ with $\mathbf{P}_j = \mathbf{B}_j\left(\mathbf{B}_j^T\mathbf{B}_j\right)^{-1}\mathbf{B}_j^T$. It is observed that exploitation of sparsity even with $T_0 = 1$

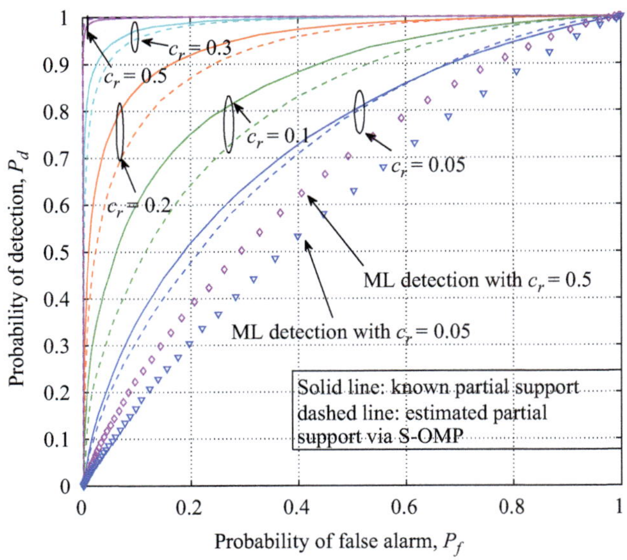

Figure 2.3 ROC curves for detection based on partial support set estimation, $N = 256$, $k = 5$, $T_0 = 1$, $L = 5$, Avg. SNR $= -6$ dB [49]

outperforms the ML-based detection approach. The performance of the S-OMP-based detection is close to the performance with known partial support for very small and relatively large $c_r = M/N$ values. When c_r is very small (e.g., ≈ 0.05 in Figure 2.3), the SNR of the compressed observation vector is small, thus even with the known support set, the power of the compressed observations projected onto the known subspace is not significant compared to the noise power projected onto the subspace. Thus, performance that is close and poor is observed when the partial support is known or estimated. On the other hand, when c_r is large, the estimated support set of size T_0 via Algorithm 1 is more accurate, resulting in performance that is close to the case where the partial support is exactly known.

2.3.3 *CS-based detection of random Gaussian signals in the presence of iid noise*

In Section 2.3.1, we have discussed compressed detection when $\mathbf{s}_j$s are deterministic and known, while in Section 2.3.2, the case where $\mathbf{s}_j$s are deterministic, unknown and sparse was considered. Now suppose that the signals of interest are iid Gaussian so that $\mathbf{s}_j \sim \mathcal{N}(\boldsymbol{\mu}, \mathbf{C}_s)$ for $j = 1, \ldots, L$. Then, the pdfs of $\mathbf{y}$ under $\mathcal{H}_0$ and $\mathcal{H}_1$ are given by:

$$f_0(\mathbf{y}) = \prod_{j=1}^{L} \frac{e^{-(1/2)\mathbf{y}_j^T(\sigma_v^2\mathbf{A}_j\mathbf{A}_j^T + \sigma_w^2\mathbf{I})^{-1}\mathbf{y}_j}}{(2\pi)^{M/2}|\sigma_v^2\mathbf{A}_j\mathbf{A}_j^T + \sigma_w^2\mathbf{I}|^{1/2}} \quad \text{and}$$

$$f_1(\mathbf{y}) = \prod_{j=1}^{L} \frac{e^{-(1/2)(\mathbf{y}_j - \mathbf{A}_j\boldsymbol{\mu})^T(\mathbf{A}_j\mathbf{C}_s\mathbf{A}_j^T + \sigma_v^2\mathbf{A}_j\mathbf{A}_j^T + \sigma_w^2\mathbf{I})^{-1}(\mathbf{y}_j - \mathbf{A}_j\boldsymbol{\mu})}}{(2\pi)^{M/2}|\mathbf{A}_j\mathbf{C}_s\mathbf{A}_j^T + \sigma_v^2\mathbf{A}_j\mathbf{A}_j^T + \sigma_w^2\mathbf{I}|^{1/2}}$$

respectively. The LLR has the form:

$$\mathrm{LLR}(\mathbf{y}) = \sum_{j=1}^{L}\left[\frac{1}{2}\mathbf{y}_j^T(\mathbf{C}_{0j}^{-1} - (\mathbf{C}_{0j} + \mathbf{C}_{1j})^{-1})\mathbf{y}_j + \boldsymbol{\mu}^T\mathbf{A}_j^T(\mathbf{C}_{0j} + \mathbf{C}_{1j})^{-1}\mathbf{y}_j\right] + \bar{\kappa} \tag{2.38}$$

where $\mathbf{C}_{0j} = \sigma_v^2\mathbf{A}_j\mathbf{A}_j^T + \sigma_w^2\mathbf{I}$, $\mathbf{C}_{1j} = \mathbf{A}_j\mathbf{C}_s\mathbf{A}_j^T$, and $\bar{\kappa} = \frac{1}{2}\sum_{j=1}^{L}\big[\log(|\mathbf{C}_{0j}|/|\mathbf{C}_{1j} + \mathbf{C}_{0j}|) - \boldsymbol{\mu}^T\mathbf{A}_j^T(\mathbf{C}_{0j} + \mathbf{C}_{1j})^{-1}\mathbf{A}_j\boldsymbol{\mu}\big]$. It is noted that (2.38) contains a quadratic as well as a linear term of $\mathbf{y}$. Finding closed-form expressions for the detection performance with (2.38) is difficult in general. In the following, we consider a special case.

2.3.3.1 Zero mean Gaussian signals

When $\boldsymbol{\mu} = \mathbf{0}$, the LLR in (2.38) reduces to $\mathrm{LLR}(\mathbf{y}) = (1/2)\sum_{j=1}^{L}\mathbf{y}_j^T(\mathbf{C}_{0j}^{-1} - (\mathbf{C}_{0j} + \mathbf{C}_{1j})^{-1})\mathbf{y}_j + \bar{\kappa}_0$ where $\bar{\kappa}_0 = (1/2)\sum_{j=1}^{L}\big[\log(|\mathbf{C}_{0j}|/(|\mathbf{C}_{1j} + \mathbf{C}_{0j}|))\big]$. Defining the decision statistic $\Lambda(\mathbf{y}) \equiv \sum_{j=1}^{L}\mathbf{y}_j^T(\mathbf{C}_{0j}^{-1} - (\mathbf{C}_{0j} + \mathbf{C}_{1j})^{-1})\mathbf{y}_j$ and using the matrix inversion lemma, we can write:

$$\Lambda(\mathbf{y}) = \sum_{j=1}^{L}\mathbf{y}_j^T\mathbf{C}_{0j}^{-1}\mathbf{C}_{1j}(\mathbf{C}_{1j} + \mathbf{C}_{0j})^{-1}\mathbf{y}_j. \tag{2.39}$$

When the rows of the matrix $\mathbf{A}_j$ are selected to be orthonormal so that $\mathbf{A}_j\mathbf{A}_j = \mathbf{I}$, (2.39) reduces to:

$$\Lambda(\mathbf{y}) = \frac{1}{\sigma_{\tilde{w}}^2}\sum_{j=1}^{L}\mathbf{y}_j^T\mathbf{A}_j\underbrace{\mathbf{C}_s\mathbf{A}_j^T(\mathbf{A}_j\mathbf{C}_s\mathbf{A}_j^T + \sigma_{\tilde{w}}^2\mathbf{I})^{-1}\mathbf{y}_j}_{\hat{\mathbf{s}}_j} \tag{2.40}$$

where $\sigma_{\tilde{w}}^2 = \sigma_v^2 + \sigma_w^2$ and $\hat{\mathbf{s}}_j$ is the minimum mean squared error (MMSE) estimate of $\mathbf{s}_j$ based on $\mathbf{s}_j$ and $\mathbf{A}_j\hat{\mathbf{s}}_j$ is the corresponding estimate of the signal in the compressed domain. Thus, (2.40) has the form of estimator-correlator in the compressed domain similar to the well-known estimator-correlator as discussed in [52] with uncompressed data. Finding closed-form expressions for the probability of detection and probability of false alarm for the decision statistic (2.40) is difficult in general. In the special case where signal samples are also iid, so that $\mathbf{C}_s = \sigma_s^2\mathbf{I}$, (2.40) reduces to:

$$\Lambda(\mathbf{y}) = \frac{\sigma_s^2}{\sigma_{\tilde{w}}^2(\sigma_s^2 + \sigma_{\tilde{w}}^2)}\sum_{j=1}^{L}\mathbf{y}_j^T\mathbf{y}_j \tag{2.41}$$

where now $\mathbf{y}_j|\mathcal{H}_0 \sim \mathcal{N}(\mathbf{0}, \sigma_{\tilde{w}}^2\mathbf{I})$ and $\mathbf{y}_j|\mathcal{H}_1 \sim \mathcal{N}(\mathbf{0}, (\sigma_s^2 + \sigma_{\tilde{w}}^2)\mathbf{I})$. Letting $\tilde{\Lambda}(\mathbf{y}) = \sum_{j=1}^{L}\mathbf{y}_j^T\mathbf{y}_j$, the detection decision in this case is given by:

$$\tilde{\Lambda}(\mathbf{y}) \underset{\mathcal{H}_0}{\overset{\mathcal{H}_1}{\gtrless}} \tau_0. \tag{2.42}$$

Since the elements of $\mathbf{y}_j$ are Gaussian and uncorrelated, they are independent. Thus, we have $(1/\sigma_{\tilde{w}}^2)\tilde{\Lambda}(\mathbf{y})|\mathscr{H}_0 \sim \mathscr{X}_{ML}^2$ and $(1/(\sigma_s^2+\sigma_{\tilde{w}}^2))\tilde{\Lambda}(\mathbf{y})|\mathscr{H}_1 \sim \mathscr{X}_{ML}^2$ leading to the following expressions for probabilities of false alarm and probability detection:

$$P_f = \Pr(\tilde{\Lambda}(\mathbf{y}) \geq \tau_0|\mathscr{H}_0) = 1 - F\left(\frac{ML}{2}; \frac{\tau_0}{2\sigma_{\tilde{w}}^2}\right)$$

and

$$P_d = \Pr(\tilde{\Lambda}(\mathbf{y}) \geq \tau_0|\mathscr{H}_1) = 1 - F\left(\frac{ML}{2}; \frac{\tau_0}{2(\sigma_s^2+\sigma_{\tilde{w}}^2)}\right)$$

where $F(s;t)$ is as defined in (2.26). For NP detection with $P_f \leq \alpha$, the threshold τ_0 is selected as $\tau_0 = 2\sigma_{\tilde{w}}^2 F^{-1}((ML/2);(1-\alpha))$ where $F^{-1}(s;\cdot)$ denotes the inverse operation of $F(s;\cdot)$. With this selection of τ_0, P_d reduces to:

$$P_d = 1 - F\left(\frac{ML}{2}; \frac{\sigma_{\tilde{w}}^2}{(\sigma_s^2+\sigma_{\tilde{w}}^2)} F^{-1}\left(\frac{ML}{2};(1-\alpha)\right)\right). \tag{2.43}$$

2.3.3.2 Comparison with uncompressed data

In order to compare the performance of (2.42) with its uncompressed counterpart, we consider the Gaussian approximation of $\tilde{\Lambda}(\mathbf{y})$ which is a reasonable assumption when ML is large. Then, P_d (so that $P_f \leq \alpha$) associated with (2.42) is approximated by:

$$P_d \approx Q\left(\frac{\sigma_{\tilde{w}}^2}{\sigma_s^2+\sigma_{\tilde{w}}^2}\left(Q^{-1}(\alpha) - \sqrt{ML\frac{\sigma_s^4}{\sigma_{\tilde{w}}^4}}\right)\right). \tag{2.44}$$

It can be easily shown that if detection was performed based on uncompressed data, the probability of detection, P_d^u, of the NP detector can be approximated by:

$$P_d \approx Q\left(\frac{\sigma_{\tilde{w}}^2}{\sigma_s^2+\sigma_{\tilde{w}}^2}\left(Q^{-1}(\alpha) - \sqrt{\bar{\gamma}_{\text{av}}}\right)\right) \tag{2.45}$$

where $\bar{\gamma}_{\text{av}} = NL(\sigma_s^4/\sigma_{\tilde{w}}^4)$. In terms of $\bar{\gamma}_{\text{av}}$, (2.44) can be written as:

$$P_d \approx Q\left(\frac{\sigma_{\tilde{w}}^2}{\sigma_s^2+\sigma_{\tilde{w}}^2}\left(Q^{-1}(\alpha) - \sqrt{\frac{M}{N}\bar{\gamma}_{\text{av}}}\right)\right). \tag{2.46}$$

Comparing (2.45) and (2.46), it is observed that the impact of compression on the probability of detection with the NP detector lies in the quantity $\sqrt{(M/N)\bar{\gamma}_{\text{av}}}$. It is noted that $\bar{\gamma}_{\text{av}}$ is a squared function of the uncompressed (per node and per sample) SNR $(\sigma_s^2/\sigma_{\tilde{w}}^2)$. Thus, in contrast to deterministic signal detection with compressed data as discussed in Section 2.3.1 where the impact of compression on the probability of detection lies in a quantity of the form $(M/N)SNR$, here it is of the form $(M/N)f(SNR^2)$.

When the (uncompressed) signals or the noise vectors have non-diagonal covariance matrices, it is difficult to obtain probabilities of detection and false alarm in closed-form using compressed data. Thus, quantifying exactly the impact of the random projection scheme on the detection performance in such scenarios still remains an open problem.

2.3.4 CS-based detection with multimodal data with arbitrary pdfs

In the previous sections, we have considered the problem of detecting a signal in the presence of noise. In this section, we consider a general binary hypothesis testing problem where observations at a given node have arbitrary pdfs under the two hypotheses. Let us rewrite the detection problem with uncompressed data as follows:

$$\begin{aligned} \mathcal{H}_1 &: \mathbf{x}_j \sim f_1(\mathbf{x}_j) \\ \mathcal{H}_0 &: \mathbf{x}_j \sim f_0(\mathbf{x}_j) \end{aligned} \tag{2.47}$$

for $j = 1, \ldots, L$ where $f_i(\mathbf{x}_j)$ denotes the pdf of $\mathbf{x}_j$ under $\mathcal{H}_i$ for $i = 0, 1$. Suppose that $\mathbf{x}_j$s are not necessarily Gaussian and $\mathbf{x}_j$s for $j = 1, \ldots, L$ can be of different modalities with arbitrary dependence. In the case where the vectors $\mathbf{x}_1, \ldots, \mathbf{x}_L$ are independent under $\mathcal{H}_i$ for $i = 0, 1$, $f_i(\mathbf{x})$ can be written as $f_i(\mathbf{x}) = \prod_{j=1}^{L} f_i(\mathbf{x}_j)$ where $\mathbf{x} = [\mathbf{x}_1^T, \ldots, \mathbf{x}_L^T]^T$. Although widely used in practice, this independence assumption among different modalities (nodes) may not be realistic for certain applications. For example, consider the problem of detection of the presence of a common random phenomenon in a heterogeneous signal processing application where there are multiple sensors of different modalities. The data at different nodes may follow disparate marginal pdfs due to the differences in the physics that govern each modality. The presence of the common random phenomenon can change the statistics of the heterogeneous data and make the observations at different modalities dependent [69]. Thus, to detect the presence of the random phenomenon in the LR framework, computation of the joint pdf of data collected at the multiple nodes in the presence of intermodal dependencies is required. In this case, even if the marginal pdfs of $\mathbf{x}_j$s are available, computation of the joint pdf $\mathbf{x}$ is intractable (unless $\mathbf{x}_j$s are Gaussian). One of the commonly used suboptimal methods to solve (2.47) is to neglect the intermodal dependence and compute the likelihood functions based on the disparate marginal pdfs of each modality. This product approach leads to poor performance when the intermodal dependence is strong. Different techniques have been proposed to estimate the joint pdfs with dependent data such as histograms and kernel-based methods [70]. When the marginal pdfs of data of each modality are available (or can be estimated), copula theory [69,71] has been used to model intermodal complex dependencies. However, most of these approaches come at a high computational cost.

Now assume that we are interested in solving (2.47) with compressed measurements $\mathbf{y}_j = \mathbf{A}_j\mathbf{x}_j$ for $j = 1, \ldots, L$. We make the following assumptions to facilitate the discussion:

- $\mathbf{y}_j$s for $j = 1, \ldots, L$ are available at the FC without an error,
- $\mathbf{A}_j$ is an orthoprojector so that $\mathbf{A}_j\mathbf{A}_j^T = \mathbf{I}$ for $j = 1, \ldots, L$, and
- the elements of $\mathbf{x}_j$ are independent of each other for each j under both hypotheses while there is (spatial) dependence among $\mathbf{x}_1, \ldots, \mathbf{x}_L$ under $\mathcal{H}_1$ (i.e., there is spatial and temporal independence under $\mathcal{H}_0$ and temporal independence and spatial dependence under $\mathcal{H}_1$).

2.3.4.1 CS-based detection via Gaussian approximation

When the elements of $\mathbf{x}_j$ for given j are independent under each hypothesis, using the Lindeberg–Feller central limit theorem assuming that the required conditions are satisfied [72,73], each element in $\mathbf{y}_j$ can be approximated by a Gaussian random variable as $N \to \infty$. Once the mean vector and the covariance matrix of $\mathbf{x}$ are available, the joint pdf of $\mathbf{y} = [\mathbf{y}_1^T, \ldots, \mathbf{y}_L^T]^T$ can be computed as a multivariate Gaussian pdf. Let:

$$\boldsymbol{\beta}^i = [\boldsymbol{\beta}_1^{i\,T} \cdots \boldsymbol{\beta}_L^{i\,T}]^T \text{ and} \tag{2.48}$$

$$\mathbf{D}^i = \begin{bmatrix} \mathbf{D}_1^i & \mathbf{D}_{12}^i & \cdots & \mathbf{D}_{1L}^i \\ \mathbf{D}_{21}^i & \mathbf{D}_2^i & \cdots & \mathbf{D}_{2L}^i \\ \cdots & \cdots & \cdots & \cdots \\ \mathbf{D}_{L1}^i & \mathbf{D}_{L2}^i & \cdots & \mathbf{D}_L^i \end{bmatrix} \tag{2.49}$$

denote the $NL \times 1$ mean vector and the $NL \times NL$ covariance matrix of $\mathbf{x}$, respectively, under $\mathcal{H}_i$ for $i = 0, 1$ where $\boldsymbol{\beta}_j^i = \mathbb{E}\{\mathbf{x}_j|\mathcal{H}_i\}$, $\mathbf{D}_j^i = \mathbb{E}\{(\mathbf{x}_j - \boldsymbol{\beta}_j^i)(\mathbf{x}_j - \boldsymbol{\beta}_j^i)^T|\mathcal{H}_i\}$, $\mathbf{D}_{jk}^i = \mathbb{E}\{(\mathbf{x}_j - \boldsymbol{\beta}_j^i)(\mathbf{x}_k - \boldsymbol{\beta}_k^i)^T|\mathcal{H}_i\}$ for $j \neq k$ $k = 1, \ldots, L$ and $j = 1, \ldots, L$.

With Gaussian approximation, the joint pdf of $\mathbf{y}$ is given by $\mathbf{y}|\mathcal{H}_i \sim \mathcal{N}(\boldsymbol{\mu}^i, \mathbf{C}^i)$ where $\boldsymbol{\mu}^i$ and $\mathbf{C}^i$ are the notations used to define the mean vector and the covariance matrix of $\mathbf{y}$ which are analogous to the definitions in (2.48) and (2.49), respectively, with $\boldsymbol{\mu}_j^i = \mathbb{E}\{\mathbf{y}_j|\mathcal{H}_i\}$, $\mathbf{C}_j^i = \mathbb{E}\{(\mathbf{y}_j - \mathbb{E}\{\mathbf{y}_j\})(\mathbf{y}_j - \mathbb{E}\{\mathbf{y}_j\})^T|\mathcal{H}_i\}$, $\mathbf{C}_{jk}^i = \mathbb{E}\{(\mathbf{y}_j - \mathbb{E}\{\mathbf{y}_j\})(\mathbf{y}_k - \mathbb{E}\{\mathbf{y}_k\})^T|\mathcal{H}_i\}$ with $j \neq k$, $k = 1, \ldots, L$ and $j = 1, \ldots, L$ for $i = 0, 1$. We further denote by $\mathbf{D}_x(\mathbf{C}_y)$ the covariance matrix of $\mathbf{x}(\mathbf{y})$ where $\mathbf{D}_x = \mathbf{D}^1(\mathbf{C}_y = \mathbf{C}^1)$ under $\mathcal{H}_1$ and $\mathbf{D}_x = \mathbf{D}^0(\mathbf{C}_y = \mathbf{C}^0)$ under $\mathcal{H}_0$.

We have the following relationships among the first and second order statistics of compressed and uncompressed data:

$$\boldsymbol{\mu}_j^i = \mathbf{A}_j\boldsymbol{\beta}_j^i, \quad \mathbf{C}_j^i = \mathbf{A}_j\mathbf{D}_j^i\mathbf{A}_j^T, \quad \text{and} \quad \mathbf{C}_{jk}^i = \mathbf{A}_j\mathbf{D}_{jk}^i\mathbf{A}_k^T$$

for $j, k = 1, \ldots, L$ and $i = 0, 1$. Then, we can write, $\boldsymbol{\mu}^i = \mathbf{A}\boldsymbol{\beta}^i$ and $\mathbf{C}^i = \mathbf{A}\mathbf{D}^i\mathbf{A}^T$ where:

$$\mathbf{A} = \begin{pmatrix} \mathbf{A}_1 & \mathbf{0} & \cdot & \cdot & \mathbf{0} \\ \mathbf{0} & \mathbf{A}_2 & \cdot & \cdot & \mathbf{0} \\ \cdot & \cdot & \cdot & \cdot & \cdot \\ \mathbf{0} & \mathbf{0} & \cdot & \cdot & \mathbf{A}_L \end{pmatrix} \tag{2.50}$$

is a $ML \times NL$ matrix. With the assumption that $\mathbf{A}_j\mathbf{A}_j^T = \mathbf{I}_M$ for $j = 1, \ldots, L$, the decision statistic of the LLR-based detector is given by [65]:

$$\text{LLR}(\mathbf{y}) = \frac{1}{2}\mathbf{y}^T(\mathbf{C}^{0^{-1}} - \mathbf{C}^{1^{-1}})\mathbf{y} + (\boldsymbol{\mu}^{1^T}\mathbf{C}^{1^{-1}} - \boldsymbol{\mu}^{0^T}\mathbf{C}^{0^{-1}})\mathbf{y} + \tau_0 \tag{2.51}$$

where $\tau_0 = (1/2)\left(\log\left(|\mathbf{C}^0|/|\mathbf{C}^1|\right) + \boldsymbol{\mu}^{0^T}\mathbf{C}^{0^{-1}}\boldsymbol{\mu}^0 - \boldsymbol{\mu}^{1^T}\mathbf{C}^{1^{-1}}\boldsymbol{\mu}^1\right)$.

2.3.4.2 Comparison with uncompressed data

With the Gaussian approximation after compression, the intermodal dependence is captured only through the covariance matrix. Higher order dependencies of data are not taken into account in the compressed domain. In particular, $\mathbf{D}_x$ is compressed via $\mathbf{C}_y = \mathbf{A}\mathbf{D}_x\mathbf{A}^T$. To quantify the distortion of $\mathbf{D}_x$ due to compression, one measure is to consider the Frobenius norm of the covariance matrix. We have:

$$||\mathbf{C}_y||_F^2 = \text{tr}(\mathbf{A}\mathbf{D}_x^T\mathbf{A}^T\mathbf{A}\mathbf{D}_x\mathbf{A}^T) = \text{tr}(\mathbf{A}^T\mathbf{A}\mathbf{D}_x^T\mathbf{A}^T\mathbf{A}\mathbf{D}_x) \approx \frac{M^2}{N^2}||\mathbf{D}_x||_F^2$$

where the last approximation is due to $\mathbf{A}^T\mathbf{A} \approx (M/N)\mathbf{I}$. Thus, the Frobenius norm of the covariance matrix after compression is reduced approximately by a factor of $c_r = (M/N)$ compared to that with uncompressed data. In other words, the Gaussian approximation in the compressed domain can capture a compressed version of the covariance matrix of uncompressed data. In the following, an example is given to show the performance of CS-based detection with multimodal data compared to that with the product approach with uncompressed data.

Example 2.1. *Let $L = 3$ and a common random phenomenon causes a change in the statistics of heterogeneous data at the three sensors. The uncompressed data at the three nodes have the following marginal pdfs [69]:*

$$x_{n1}|\mathcal{H}_i \sim \mathcal{N}(0, \sigma_i^2),\ x_{n2}|\mathcal{H}_i \sim \text{Exp}(\lambda_i) \text{ and } x_{n3}|\mathcal{H}_i \sim \text{Beta}(a_i, b_i = 1)$$

for $i = 0, 1$. It is noted that $x \sim \text{Exp}(\lambda)$ denotes that x has an exponential distribution with $f(x) = \lambda e^{-\lambda x}$ for $x \geq 0$ and 0 otherwise, and $x \sim \text{Beta}(a, b)$ denotes that x has a beta distribution with pdf $f(x) = (1/\mathcal{B}(a, b))x^{a-1}(1-x)^{b-1}$ and $\mathcal{B}(a, b) = \Gamma(a)\Gamma(b)/\Gamma(a+b)$ is the beta function. The data under $\mathcal{H}_1$ is assumed to be dependent, and the following operations are used to generate dependent data. For the data at the second node, we use $x_{n2} = x_{n1}^2 + w^2$ for $n = 1, \ldots, N$ where $w \sim \mathcal{N}(0, \sigma_1^2)$. Then, we have $x_{n2} \sim \text{Exp}(\lambda_1)$ with $\lambda_1 = 1/2\sigma_1^2$. For the third node, the data under $\mathcal{H}_1$ is generated as

$$x_{n3} = \frac{u}{u + x_{n2}}$$

for $n = 1, \ldots, N$ where $u \sim \text{Gamma}(\alpha_1, \beta_1 = 1/\lambda_1)$. Then we have $x_{n3}|\mathcal{H}_1 \sim \text{Beta}(a_1, b_1 = 1)$ with $a_1 = \alpha_1$. It is noted that $x \sim \text{Gamma}(\alpha, \beta)$ denotes that x has a Gamma pdf with $f(x) = (1/(\beta^\alpha\Gamma(\alpha)))x^{\alpha-1}e^{-x/\beta}$ for $x \geq 0$ and $\alpha, \beta > 0$. Under $\mathcal{H}_0$, x_{n1}, x_{n2}, and x_{n3} are generated independently using the assumed marginal pdfs.

In Figure 2.4, the compressed detector with Gaussian approximation, "c:GA" is compared, in terms of the ROC curves, with the product approach (where dependence is ignored) with uncompressed data, "u:product." The parameter values are provided in the figure caption. With the considered parameter values under the two hypotheses, "u:product" with uncompressed data does not provide perfect detection, while "c:GA" (with compressed data) with fairly small value of c_r is capable of providing perfect detection. This implies that capturing some dependence among data (via the covariance matrix) can lead to better performance even after compression compared to completely neglecting dependence in the uncompressed domain.

Remark 2.1. *In the case where the optimal LR test cannot be performed with uncompressed data, the natural choice is to employ suboptimal detectors. When a widely*

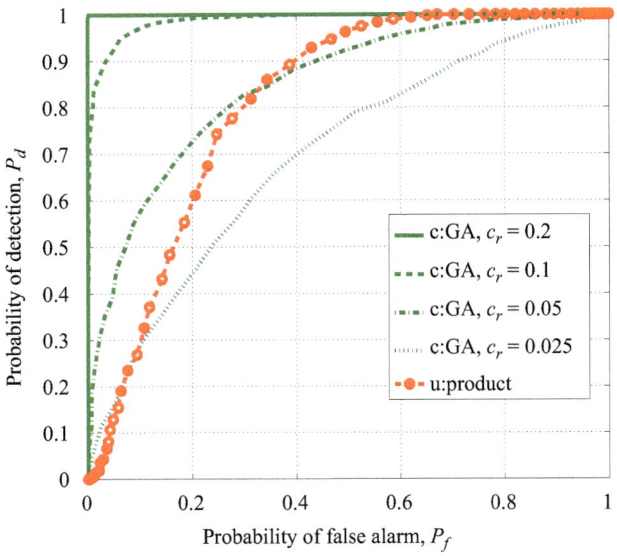

Figure 2.4 Performance of dependent data fusion for detection in Example 2.1: product approach with uncompressed data vs Gaussian approximation with compressed data: $N = 1{,}000$, $\sigma_0^2 = 5$, $\sigma_1^2 = 5.1$, $1/\lambda_0 = 10$, $1/\lambda_1 = 10.2$, $a_0 = 9.8$, $a_1 = 10$ *[73]*

considered suboptimal approach, the product approach, is employed for detection with uncompressed data, the intermodal dependence is completely ignored. On the other hand, with compressed data, the Gaussian approximation can be employed in the compressed domain at each node when the time samples are independent (even though different nodes have different marginal pdfs). With Gaussian approximation for compressed data of all the nodes, some dependence among different modalities is captured via a compressed covariance matrix. Thus, there is a possibility for CS-based detection to perform better than the product approach with uncompressed data.

To better understand the impact of the random projections scheme on dependent data fusion for detection (and also for other inference problems), further research is needed. For the case considered in this section, the temporal data at a given node is assumed to be independent so that the compressed data can be approximated by Gaussian random variables. With Gaussian approximation for multi-modal data, intermodal dependence is captured only via the covariance matrix. Thus, if the uncompressed data does not have a linear dependence among different modalities, no dependence is captured in the compressed domain when using CS. In such scenarios, degraded performance would be observed with compressed data compared to that with the product approach with uncompressed data. Further, when the temporal data is also dependent, the Gaussian approximation for compressed data at a given node may no longer be valid. Thus, further quantifying the performance of CS-based detection with different dependence models for uncompressed data beyond the model considered here needs further research.

2.4 Compressed parameter estimation

In Section 2.3, we have discussed the compressed detection (or binary hypothesis testing) problem considering different statistical models for the data under the two hypotheses. There, the goal was to understand how much information is retained in the compressed domain to make a reliable detection decision and analyze the detection performance with compressed data. In certain applications, we need to estimate some parameters associated with the PoI to better interpret what has been observed. Parameter estimation is another important inference problem having a variety of WSN applications. It is worth mentioning that the standard CS theory in fact deals with an estimation problem with specific constraints, i.e., to estimate a sparse signal based on compressed measurements. There, sparsity is an inherently discrete concept. In a more general setting, it is interesting to investigate how CS can be used to estimate a continuous valued parameter (or a set of parameters) that governs the signal being observed without reconstructing the signal. Sparsity of the signal is not a necessary requirement in such problems. Parameter estimation problems without any compression have been investigated quite extensively during last several decades [74]. With CS, we are interested in estimating the same set of parameters with a compressed version of the observed signal. Similar to the detection problem addressed in Section 2.3, the fundamental question here is the following: *how much information is retained in the compressed domain so that a reliable estimator can be obtained?*

Extension of the CS framework for parameter estimation has been studied by several authors in [46,75–79]. Early work such as [75] focuses on establishing properties and the dimension of the projection matrix to enable reliable estimation. In recent studies [78,79], the impact of the random projection scheme on estimation-theoretic bounds is studied considering specific models for the noise. In the following, we start with the classical parameter estimation problem in the presence of iid Gaussian noise and discuss the impact of the random projection scheme when the same parameters are estimated with compressed data.

Suppose that the signal of interest $\mathbf{s}_j \in \mathbb{R}^N$ at the jth node is parameterized by an unknown parameter vector, denoted by, $\boldsymbol{\omega} \in \Omega \in \mathbb{R}^K$. We assume, in general, $K < N$. Consider that we are interested in estimating $\boldsymbol{\omega}$ based on a noisy version of $\mathbf{s}_j$ for $j = 1, \ldots, L$. More specifically, the observation model (uncompressed) at the jth node is given by:

$$\mathbf{x}_j = \mathbf{s}_j(\boldsymbol{\omega}) + \mathbf{v}_j \tag{2.52}$$

for $j = 1, \ldots, L$ where $\mathbf{v}_j \in \mathbb{R}^N$ is the additive noise at the jth node. Specific approaches for estimating $\boldsymbol{\omega}$ from (2.52) depend on how the elements in $\boldsymbol{\omega}$ are modeled. If all the elements of $\boldsymbol{\omega}$ are assumed to be random elements, Bayesian estimation methods such as MMSE estimator, minimum mean absolute error estimator, and minimum a posteriori probability estimator are widely used. On the other hand, when $\boldsymbol{\omega}$

is modeled as nonrandom, widely used approaches are minimum variance unbiased estimate, ML estimator, and best linear unbiased estimator. These estimators along with their advantages and disadvantages are discussed in detail in [65,74].

Now, assume that we want to estimate $\boldsymbol{\omega}$ from a compressed version of $\mathbf{s}_j(\boldsymbol{\omega})$ for $j = 1, \ldots, L$. The compressed observation vector at the jth node can be expressed as:

$$\mathbf{y}_j = \mathbf{A}_j\mathbf{s}_j(\boldsymbol{\omega}) + \mathbf{v}_j \tag{2.53}$$

where $\mathbf{A}_j$ is the $M \times N$ random matrix with $M \leq N$ as defined before.

Remark 2.2. *The observation model at the jth node in* (2.53) *is equivalent to compressing a noiseless version of* (2.52). *If the noisy version as in* (2.52) *is compressed using* $\mathbf{A}_j$, *we would get* $\mathbf{y}_j = \mathbf{A}_j(\mathbf{s}_j(\boldsymbol{\omega}) + \mathbf{v}_j)$ *for* $j = 1, \ldots, L$. *For general* $\mathbf{A}_j$, *the two versions* ($\mathbf{y}_j = \mathbf{A}_j(\mathbf{s}_j(\boldsymbol{\omega}) + \mathbf{v}_j)$ *and* $\mathbf{y}_j = \mathbf{A}_j\mathbf{s}_j(\boldsymbol{\omega}) + \mathbf{v}_j$) *can lead to different performance. However, for specific selection of* $\mathbf{A}_j$ *(e.g., when* $\mathbf{A}_j$ *is chosen as a random orthoprojector so that* $\mathbf{A}_j\mathbf{A}_j^T = \mathbf{I}$), *the two versions can have the same performance. In the following, we restrict our discussion to the model in* (2.53) *while similar analysis with the other model is provided in [79] for* $L = 1$.

In order to understand how good the estimate of $\boldsymbol{\omega}$ based on (2.53) is instead of that based on (2.52), we consider fundamental bounds on the estimation error variance. For deterministic parameters, the Cramér–Rao Lower Bound (CRLB) sets a lower bound on the variance of any unbiased estimator (for vector-valued parameters, CRLB allows us to set a bound on the variance of each element) [74]. For random parameters, a Bayesian version of CRLB, known as BCRLB, is discussed in [80].

In order to establish the error bounds, it is important to understand certain properties of each $\mathbf{A}_j$ (which are different notions of the RIP property defined in (2.3)). For example, consider the ML estimate of $\boldsymbol{\omega}$ based on (2.53) when $\mathbf{v} \sim \mathcal{N}(\mathbf{0}, \sigma_v^2\mathbf{I})$ with $L = 1$ which is given by:

$$\hat{\boldsymbol{\omega}}_{ML} = \arg\min_{\tilde{\boldsymbol{\omega}}} ||\mathbf{y} - \mathbf{A}\mathbf{s}(\tilde{\boldsymbol{\omega}})|| = \arg\min_{\tilde{\boldsymbol{\omega}}} ||\mathbf{A}\mathbf{s}(\boldsymbol{\omega}) - \mathbf{A}\mathbf{s}(\tilde{\boldsymbol{\omega}}) + \mathbf{v}||$$

where we omit the use of subscripts on $\mathbf{A}$, $\mathbf{s}$, and $\mathbf{v}$ when $L = 1$. For general $\mathbf{A}$, it is possible that $||\mathbf{A}(\mathbf{s}(\boldsymbol{\omega}) - \mathbf{s}(\tilde{\boldsymbol{\omega}}))|| \approx 0$ if $\mathbf{A}$ has a large null space and M is very small. However, when $\mathbf{A}$ preserves pairwise distance, we have:

$$||\mathbf{A}(\mathbf{s}(\boldsymbol{\omega}) - \mathbf{s}(\tilde{\boldsymbol{\omega}}))|| \propto ||\mathbf{s}(\boldsymbol{\omega}) - \mathbf{s}(\tilde{\boldsymbol{\omega}})||, \ \forall \boldsymbol{\omega}, \tilde{\boldsymbol{\omega}} \in \Omega.$$

Pairwise δ-Isometry property of $\mathbf{A}$ is capable of capturing this pairwise distance preservation precisely.

Definition 2.3 (pairwise δ-isometry [78]). *A matrix* $\mathbf{A}$ *satisfies the pairwise* δ*-isometry property for* $\mathbf{x}(\boldsymbol{\omega})$ *if*

$$\sqrt{\frac{M}{N}}(1 - \delta) \leq \frac{||\mathbf{A}(\mathbf{x}(\omega_1) - \mathbf{x}(\omega_2))||}{||\mathbf{x}(\omega_1) - \mathbf{x}(\omega_2)||} \leq \sqrt{\frac{M}{N}}(1 + \delta) \tag{2.54}$$

for $\forall \omega_1, \omega_2 \in \Omega$ *where* $0 < \delta < 1$.

Another weaker notion of the pairwise δ-isometry property which is useful in establishing performance guarantees of parameter estimation is the tangent plane δ-isometry property as defined below.

Definition 2.4 (tangent plane δ-isometry [78]). *A matrix* $\mathbf{A}$ *satisfies the tangent plane δ-isometry property for* $\mathbf{x}(\boldsymbol{\omega})$ *if*

$$\sqrt{\frac{M}{N}}(1-\delta) \leq \frac{||\mathbf{A}(\sum a_m(\partial \mathbf{x}(\boldsymbol{\omega})|\partial \omega_m))||}{||\sum(\partial \mathbf{x}(\boldsymbol{\omega})|\partial \omega_m)||} \leq \sqrt{\frac{M}{N}}(1+\delta) \tag{2.55}$$

for $\forall\, [a_1, \ldots, a_K]^T \in \mathbb{R}^K \setminus \{\mathbf{0}\}$, $\forall\, \boldsymbol{\omega} \in \Omega$ *where* $0 < \delta < 1$.

2.4.1 Parameter estimation with compressed data with iid Gaussian noise

First we consider the case where $\mathbf{v}_j \sim \mathcal{N}(\mathbf{0}, \sigma_v^2\mathbf{I})$ for $j = 1, \ldots, L$. With this assumption, we have $\mathbf{y}_j|\boldsymbol{\omega} \sim \mathcal{N}(\mathbf{A}_j\mathbf{s}_j(\boldsymbol{\omega}), \sigma_v^2\mathbf{I})$, and only the mean vector of $\mathbf{y}_j$ contains unknown parameters.

2.4.1.1 Deterministic parameter estimation

Suppose that $\boldsymbol{\omega}$ is unknown but deterministic. For the measurement model (2.53), the conditional pdf of $\mathbf{y} = [\mathbf{y}_1^T, \ldots, \mathbf{y}_L^T]^T$ given $\boldsymbol{\omega}$ is given by:

$$f(\mathbf{y}|\boldsymbol{\omega}) = \frac{1}{(2\pi\sigma_v^2)^{ML/2}} e^{-(1/(2\sigma_v^2))\sum_{j=1}^{L} ||\mathbf{y}_j - \mathbf{A}_j\mathbf{s}_j(\boldsymbol{\omega})||_2^2}.$$

Denote $\hat{\boldsymbol{\omega}}(\mathbf{y})$ to be an estimate of $\boldsymbol{\omega}$. For any unbiased[1] estimator $\hat{\boldsymbol{\omega}}(\mathbf{y})$, the CRLB theorem for vector parameters states that the covariance matrix of $\hat{\boldsymbol{\omega}}(\mathbf{y})$ satisfies [74]

$$\mathrm{Cov}(\hat{\boldsymbol{\omega}}(\mathbf{y})) \succeq (\mathrm{FI}(\boldsymbol{\omega}))^{-1}$$

where $\mathbf{A} \succeq \mathbf{B}$ denotes that the matrix $\mathbf{A} - \mathbf{B}$ is positive semidefinite, and $\mathrm{FI}(\boldsymbol{\omega})$ is the *Fisher's information matrix* in which the (m, n)th element is given by:

$$\mathrm{FI}_{m,n}(\boldsymbol{\omega}) = \mathbb{E}_{\mathbf{y}|\boldsymbol{\omega}}\left\{\frac{\partial \ln f(\mathbf{y}|\boldsymbol{\omega})}{\partial \omega_m}\frac{\partial \ln f(\mathbf{y}|\boldsymbol{\omega})}{\partial \omega_n}\right\}. \tag{2.56}$$

For (2.53) with $\mathbf{v}_j \sim \mathcal{N}(\mathbf{0}, \sigma_v^2\mathbf{I})$ for $j = 1, \ldots, L$, (2.56) reduces to [74]:

$$\mathrm{FI}_{m,n}(\boldsymbol{\omega}) = \frac{1}{\sigma_v^2}\sum_{j=1}^{L}\left\{\left(\mathbf{A}_j\frac{\partial \mathbf{s}_j(\boldsymbol{\omega})}{\partial \omega_m}\right)^T \mathbf{A}_j\frac{\partial \mathbf{s}_j(\boldsymbol{\omega})}{\partial \omega_n}\right\}. \tag{2.57}$$

Further, let $\mathrm{FI}^u(\boldsymbol{\omega})$ denote the *Fisher's information matrix* with uncompressed data $\mathbf{x} = [\mathbf{x}_1^T, \ldots, \mathbf{x}_L^T]^T$ in which the (m, n)th element is given by:

$$\mathrm{FI}^u_{m,n}(\boldsymbol{\omega}) = \frac{1}{\sigma_v^2}\sum_{j=1}^{L}\left\{\left(\mathbf{I}_N\frac{\partial \mathbf{s}_j(\boldsymbol{\omega})}{\partial \omega_m}\right)^T \mathbf{I}_N\frac{\partial \mathbf{s}_j(\boldsymbol{\omega})}{\partial \omega_n}\right\} \tag{2.58}$$

where $\mathbf{I}_N$ is the $N \times N$ identity matrix.

[1] $\hat{\boldsymbol{\omega}}(\mathbf{y})$ is said to be unbiased if $\mathbb{E}_{\mathbf{y}|\boldsymbol{\omega}}\{\hat{\boldsymbol{\omega}}(\mathbf{y})\} = \boldsymbol{\omega}$.

In order to establish a relationship between $\mathrm{FI}(\boldsymbol{\omega})$ and its uncompressed counterpart $\mathrm{FI}^u(\boldsymbol{\omega})$, let us consider the term $\mathbf{a}\mathrm{FI}(\boldsymbol{\omega})\mathbf{a}^T$ for any $\mathbf{a} = [a_1, \ldots, a_K]^T \in \mathbb{R}^K$. $\mathbf{a}\mathrm{FI}(\boldsymbol{\omega})\mathbf{a}^T$ can be expressed as [78]:

$$\mathbf{a}\mathrm{FI}(\boldsymbol{\omega})\mathbf{a}^T = \frac{1}{\sigma_v^2}\sum_{j=1}^{L}\left\|\mathbf{A}_j\sum a_m\frac{\partial \mathbf{s}_j(\boldsymbol{\omega})}{\partial \omega_m}\right\|^2. \tag{2.59}$$

When each $\mathbf{A}_j$ for $j = 1, \ldots, L$ satisfies the tangent plane isometry property (2.55), (2.59) can be bounded by:

$$\frac{1}{\sigma_v^2}\frac{M}{N}(1-\delta)^2\sum_{j=1}^{L}\left\|\sum a_m\frac{\partial \mathbf{s}_j(\boldsymbol{\omega})}{\partial \omega_m}\right\|^2 \leq \mathbf{a}\mathrm{FI}(\boldsymbol{\omega})\mathbf{a}^T \leq \frac{1}{\sigma_v^2}\frac{M}{N}(1+\delta)^2\sum_{j=1}^{L}\left\|\sum a_m\frac{\partial \mathbf{s}_j(\boldsymbol{\omega})}{\partial \omega_m}\right\|^2 \tag{2.60}$$

resulting in

$$\frac{M}{N}(1-\delta)^2\mathbf{a}^T\mathrm{FI}^u(\boldsymbol{\omega})\mathbf{a} \leq \mathbf{a}\mathrm{FI}(\boldsymbol{\omega})\mathbf{a}^T \leq \frac{M}{N}(1+\delta)^2\mathbf{a}^T\mathrm{FI}^u(\boldsymbol{\omega})\mathbf{a} \tag{2.61}$$

for $\forall \mathbf{a} \in \mathbb{R}^K$ and $0 < \delta < 1$. Thus, $\mathrm{FI}(\boldsymbol{\omega})$ satisfies:

$$(1-\delta)^2\frac{M}{N}\mathrm{FI}^u(\boldsymbol{\omega}) \preceq \mathrm{FI}(\boldsymbol{\omega}) \preceq (1+\delta)^2\frac{M}{N}\mathrm{FI}^u(\boldsymbol{\omega}). \tag{2.62}$$

From (2.62), as $\delta \to 0$, we have $\mathrm{FI}(\boldsymbol{\omega}) \approx \frac{M}{N}\mathrm{FI}^u(\boldsymbol{\omega})$. Thus, each element in $\mathrm{FI}(\boldsymbol{\omega})$ is distorted approximately by a factor of M/N compared to the corresponding element in $\mathrm{FI}^u(\boldsymbol{\omega})$. From (2.58) and (2.57), it is observed that the impact of the random projections scheme on $\mathrm{FI}(\boldsymbol{\omega})$ lies in a scaled version of the noise variance σ_v^2. More specifically, $\mathrm{FI}(\boldsymbol{\omega})$ is approximately same as $\mathrm{FI}^u(\boldsymbol{\omega})$ when σ_v^2 in (2.52) is replaced by $(N/M)\sigma_v^2$. We have seen a similar observation in deterministic signal detection with compressed data in the presence of iid Gaussian noise as discussed in Section 2.3.1. There, the impact of the random projection scheme on the probability of detection is reflected in the term $(M/N)\gamma_{\mathrm{av}}$ where γ_{av} is the SNR of the uncompressed data. Thus, we can conclude that for both deterministic parameter estimation and the known deterministic signal detection problems in the presence of iid Gaussian noise, the impact of compression via random projections on the estimation/detection performance lies in a term expressed in terms of M/N times the uncompressed SNR.

2.4.1.2 Random parameter estimation with compressed data

Now suppose that $\boldsymbol{\omega}$ in (2.53) is random with a known prior pdf $f(\boldsymbol{\omega})$. The BCRLB satisfies:

$$\mathrm{Cov}(\hat{\boldsymbol{\omega}}(\mathbf{y})) \succeq (\mathrm{BI}(\boldsymbol{\omega}))^{-1}$$

where $\mathrm{BI}(\boldsymbol{\omega})$ denotes the *Bayesian information matrix*, in which the (m, n)th element is given by [80]:

$$\mathrm{BI}(\boldsymbol{\omega})_{m,n} = \mathbb{E}_{\omega}\{\mathrm{FI}_{m,n}(\boldsymbol{\omega})\} + \mathbb{E}_{\omega}\left\{\frac{\partial \ln f(\boldsymbol{\omega})}{\partial \omega_m}\frac{\partial \ln f(\boldsymbol{\omega})}{\partial \omega_n}\right\}. \tag{2.63}$$

The second term in the right hand side of (2.63) does not depend on $[\mathbf{A}_1, \ldots, \mathbf{A}_L]$. The impact of compression on $\mathrm{BI}(\boldsymbol{\omega})_{m,n}$ in (2.63) lies in the term $\mathrm{FI}_{m,n}(\boldsymbol{\omega})$. Thus, based on (2.62), it can be easily shown that:

$$(1-\delta)^2\frac{M}{N}\left(\mathrm{BI}^u(\boldsymbol{\omega}) - \bar{\mathbf{B}}\right) \preceq \mathrm{BI}(\boldsymbol{\omega}) - \bar{\mathbf{B}} \preceq (1+\delta)^2\frac{M}{N}\left(\mathrm{BI}^u(\boldsymbol{\omega}) - \bar{\mathbf{B}}\right)$$

for $0 < \delta < 1$ where the (m, n)th element of $\bar{\mathbf{B}}$ is given by $\mathbb{E}_{\omega}\{((\partial \ln f(\boldsymbol{\omega}))/(\partial \omega_m))((\partial \ln f(\boldsymbol{\omega}))/(\partial \omega_n))\}$, and $\mathrm{BI}^u(\boldsymbol{\omega})$ denotes the *Bayesian information matrix* with uncompressed data in (2.52) in which the (m, n)th element is given by:

$$\mathrm{BI}^u(\boldsymbol{\omega})_{m,n} = \mathbb{E}_{\omega}\{\mathrm{FI}^u_{m,n}(\boldsymbol{\omega})\} + \mathbb{E}_{\omega}\left\{\frac{\partial \ln f(\boldsymbol{\omega})}{\partial \omega_m}\frac{\partial \ln f(\boldsymbol{\omega})}{\partial \omega_n}\right\}. \tag{2.64}$$

Thus, as $\delta \to 0$, $\mathrm{BI}(\boldsymbol{\omega})$ can be approximated by $\mathrm{BI}(\boldsymbol{\omega}) \approx (M/N)\mathrm{BI}^u(\boldsymbol{\omega}) + ((N-M)/N)\bar{\mathbf{B}}$. In this case, each element of the compressed *Bayesian information matrix* is distorted by approximately a factor of M/N times the corresponding element of the uncompressed Bayesian information matrix plus some other quantity. Thus, in contrast to the deterministic parameter estimation problem, the impact of the compression on the BCRLB in this case is not directly related to a scaled version of σ_v^2.

2.4.2 Parameter estimation with compressed data with general Gaussian model

In Section 2.4.1, the noise vector $\mathbf{v}_j$ for $j = 1, \ldots, L$ in (2.53) is assumed to be iid Gaussian resulting a diagonal covariance matrix for $\mathbf{y}$. Now consider the general Gaussian case where the uncompressed observation vector at the jth node $\mathbf{x}_j \sim \mathcal{N}(\mathbf{m}_j(\boldsymbol{\omega}), \mathbf{C}_j(\boldsymbol{\omega}))$. Both the mean vector $\mathbf{m}_j(\boldsymbol{\omega}) \in \mathbb{R}^N$ and the covariance matrix $\mathbf{C}_j(\boldsymbol{\omega}) \in \mathbb{R}^{N\times N}$ for $j = 1, \ldots, L$ are parameterized by the unknown vector $\boldsymbol{\omega} \in \mathbb{R}^K$ with $K < N$. Further, assume that $\mathbf{x}_j$s for $j = 1, \ldots, L$ are independent vectors given $\boldsymbol{\omega}$. After compression with $\mathbf{y}_j = \mathbf{A}_j\mathbf{x}_j + \mathbf{v}_j$ where $\mathbf{v}_j \sim \mathcal{N}(\mathbf{0}, \sigma_v^2\mathbf{I})$, we have $\mathbf{y}_j \sim \mathcal{N}(\mathbf{A}_j\mathbf{m}_j(\boldsymbol{\omega}), \mathbf{A}_j\mathbf{C}_j(\boldsymbol{\omega})\mathbf{A}_j^T + \sigma_v^2\mathbf{I})$. In this case, *Fisher's information matrix* with compressed data can be expressed as [74]:

$$\begin{aligned}\mathrm{FI}(\boldsymbol{\omega}) &= \left(\frac{\partial \tilde{\mathbf{m}}(\boldsymbol{\omega})}{\partial \boldsymbol{\omega}}\right)^T \tilde{\mathbf{A}}^T(\tilde{\mathbf{A}}\tilde{\mathbf{C}}(\boldsymbol{\omega})\tilde{\mathbf{A}}^T + \sigma_v^2\mathbf{I})^{-1}\tilde{\mathbf{A}}\left(\frac{\partial \tilde{\mathbf{m}}(\boldsymbol{\omega})}{\partial \boldsymbol{\omega}}\right)\\ &\quad + \frac{1}{2}\mathrm{tr}\left[\left((\tilde{\mathbf{A}}\tilde{\mathbf{C}}(\boldsymbol{\omega})\tilde{\mathbf{A}}^T + \sigma_v^2\mathbf{I})^{-1}\frac{\partial \tilde{\mathbf{A}}\tilde{\mathbf{C}}(\boldsymbol{\omega})\tilde{\mathbf{A}}^T}{\partial \boldsymbol{\omega}}\right)^2\right]\end{aligned} \tag{2.65}$$

where $\tilde{\mathbf{m}}(\omega) = [\mathbf{m}_1(\omega)^T, \ldots, \mathbf{m}_L(\omega)^T]^T$, $\tilde{\mathbf{A}}$ is a $ML \times NL$ block diagonal matrix with jth block given by $\mathbf{A}_j$ and $\tilde{\mathbf{C}}(\omega)$ is a $NL \times NL$ block diagonal matrix with the jth block given by $\mathbf{C}_j(\omega)$ for $j = 1, \ldots, L$. The same with uncompressed data is given by:

$$\mathrm{FI}^u(\omega) = \left(\frac{\partial \tilde{\mathbf{m}}(\omega)}{\partial \omega}\right)^T (\tilde{\mathbf{C}}(\omega))^{-1} \left(\frac{\partial \tilde{\mathbf{m}}(\omega)}{\partial \omega}\right) + \frac{1}{2}\mathrm{tr}\left[\left((\tilde{\mathbf{C}}(\omega))^{-1}\frac{\partial \tilde{\mathbf{C}}(\omega)}{\partial \omega}\right)^2\right].$$

In order to quantify as to how good the CS-based parameter estimation with general Gaussian model is, further simplification of (2.65) is required. Thus, understanding the role of CS in parameter estimation beyond the simple model considered in Section 2.4.1 requires further research.

2.5 Summary

CS has emerged as a promising technique for high-dimensional data acquisition and processing. During the last decade, CS has gained considerable attention by researchers, engineers, and mathematicians in diverse fields. While CS is well motivated for resource constrained WSNs, to utilize it effectively in solving a variety of inference problems, further research is needed beyond the standard CS framework. In this chapter, our goal was to provide an overview and an up-to-date review on the use of CS in solving detection and parameter-estimation problems. As discussed in Sections 2.3 and 2.4, the impact of the random projection scheme on the detection and estimation performance largely depends on the particular signal and noise models of the uncompressed data. We have discussed the performance metrics for CS-based detection and parameter estimation and quantified the impact of the random projection scheme on these performance metrics in an easily interpretable manner in the presence of iid signal and noise models. We have emphasized the need for further research to quantify the impact of random projection scheme when there are intra- and inter-signal dependencies. We hope that this chapter will help readers get a thorough understanding on the potential of CS in solving detection and estimation problems in WSNs and motivate further research to promote CS in solving distributed inference problems in practical settings.

References

[1] Akyildiz IF, Su W, Sankarasubramaniam Y, *et al.* Wireless sensor networks: a survey. Comput Netw. 2002;38(4):393–422.

[2] Puccinelli D, and Haenggi M. Wireless sensor networks: applications and challenges of ubiquitous sensing. IEEE Circuits Syst Mag. 2005;5(3):19–31.

[3] Yick J, Mukherjee B, and Ghosal D. Wireless sensor network survey. Comput Netw. 2008;52(12):2292–2330.

[4] Stankovic JA, Wood AD, and He T. Realistic applications for wireless sensor networks. Theoretical Aspects of Distributed Computing in Sensor Networks. Springer. 2011; p. 853–863.

[5] Othman MF, and Shazali K. Wireless sensor network applications: a study in environment monitoring system. Eng Procedia. 2012;41:1204–1210.
[6] Chamberland JF, and Veeravalli VV. Wireless sensors in distributed detection applications. IEEE Signal Process Mag. 2007;24(3):16–25.
[7] Wimalajeewa T, and Jayaweera SK. Optimal power scheduling for correlated data fusion in wireless sensor networks via constrained PSO. IEEE Trans Wireless Commun. 2008;7(9):3608–3618.
[8] Duarte M, and Hu YH. Vehicle classification in distributed sensor networks. J Parallel Distributed Comput. 2004;64(7):826–838.
[9] Ozdemir O, Niu R, and Varshney PK. Tracking in wireless sensor networks using particle filtering: physical layer considerations. IEEE Trans Signal Process. 2009;57(5):1987–1999.
[10] Olfati-Saber R. Distributed Tracking for Mobile Sensor Networks with Information-Driven Mobility. In: Proc. of American Control Conf. New York City, NY; 2007. p. 4606–4612.
[11] Gungor VC, and Hancke GP. Industrial wireless sensor networks: challenges, design principles, and technical approaches. IEEE Trans Ind Electron. 2009;56(10):4258–4265.
[12] Alemdar H, and Ersoy C. Wireless sensor networks for healthcare: a survey. Comput Netw. 2010;54(15):2688–2710.
[13] Candès E, and Tao T. Near-optimal signal recovery from random projections: universal encoding strategies? IEEE Trans Inf Theory. 2006;52(12): 5406–5425.
[14] Candès E, and Wakin MB. An introduction to compressive sampling. IEEE Signal Process Mag. 2008;25(2):21–30.
[15] Donoho D. Compressed sensing. IEEE Trans Inf Theory. 2006;52(4): 1289–1306.
[16] Candès E, Romberg J, and Tao T. Robust uncertainty principles: exact signal reconstruction from highly incomplete frequency information. IEEE Trans Inf Theory. 2006;52(2):489–509.
[17] Duarte MF, Davenport MA, Takbar D, *et al.* Single-pixel imaging via compressive sampling. IEEE Signal Process Mag. 2008;25(2):83–91.
[18] Jiang H, Zhao S, Shen Z, *et al.* Surveillance video analysis using compressive sensing with low latency. Bell Labs Tech J. 2014;18(4):63–74.
[19] Berger C, Wang Z, Huang J, *et al.* Application of compressive sensing to sparse channel estimation. IEEE Signal Process Mag. 2010;48(11):164–174.
[20] Mohammadian R, Amini A, and Khalaj BH. Compressive sensing-based pilot design for sparse channel estimation in OFDM systems. IEEE Commun Lett. 2017;21(1):4–7.
[21] Haupt J, Bajwa WU, Rabbat M, *et al.* Compressed sensing for networked data. IEEE Signal Process Mag. 2008;25(2):92–101.
[22] Wang J, Tang S, Yin B, *et al.* Data Gathering in Wireless Sensor Networks Through Intelligent Compressive Sensing. In: IEEE Proc. INFOCOM. Orlando, FL; 2012. p. 603–611.
[23] Balouchestani M, Raahemifar K, and Krishnan S. Compressed sensing in wireless sensor networks: survey. Can J Multimedia Wireless Netw. 2011;2(1).

[24] Quer G, Masiero R, Pillonetto G, *et al.* Sensing, compression, and recovery for WSNs: sparse signal modeling and monitoring framework. IEEE Trans Wireless Commun. 2012;11(10):3447–3461.
[25] Wimalajeewa T, and Varshney PK. Compressive sensing for efficient data gathering and distributed inference in wireless sensor networks: a survey. IEEE Commun Surveys Tuts. 2018, under review, available online at https://arxivorg/pdf/170910401pdf.
[26] Sharma SK, Lagunas E, Chatzinotas S, *et al.* Application of compressive sensing in cognitive radio communications: a survey. IEEE Commun Surveys Tuts. 2016;18(3):1838–1860.
[27] Liang Q. Compressive Sensing for Radar Sensor Networks. In: IEEE Global Telecommunications Conference (GLOBECOM). Miami, FL; 2010. p. 1–5.
[28] Eldar YC, and Kutyniok G. Compressed sensing: theory and applications. New York: Cambridge University Press; 2012.
[29] Zhang Z, Xu Y, Yang J, *et al.* A survey of sparse representation: algorithms and applications. IEEE Access. 2015;3:490–530.
[30] Chen SS, Donoho DL, and Saunders MA. Atomic decomposition by basis pursuit. SIAM J Sci Comput. 1998;20:33–61.
[31] Tibshirani R. Regression shrinkage and selection via the LASSO. J Roy Stat Soc Ser B. 1996;58(1):267–288.
[32] Beck A, and Teboulle M. A fast iterative shrinkage-thresholding algorithm for linear inverse problems. SIAM J Imaging Sci. 2009;2:183–202.
[33] Bickel PJ, Ritov Y, and Tsybakov AB. Simultaneous analysis of LASSO and Dantzig selector. Ann Stat. 2009;37(4):1705–1732.
[34] Tropp J, and Gilbert A. Signal recovery from random measurements via orthogonal matching pursuit. IEEE Trans Inf Theory. 2007;53(12):4655–4666.
[35] Needell D, and Tropp J. CoSaMP: iterative signal recovery from incomplete and inaccurate samples. Appl Comput Harmon Anal. 2009;26:301–321.
[36] Dai W, and Milenkovic O. Subspace pursuit for compressive sensing signal reconstruction. IEEE Trans Inf Theory. 2009;55(5):2230–2249.
[37] Blumensath T, and Davies ME. Iterative hard thresholding for compressed sensing. Appl Comput Harmonic Anal. 2009;27(3):265–274.
[38] Ji S, Xue Y, and Carin L. Bayesian compressive sensing. IEEE Trans Signal Process. 2008;56(6):2346–2356.
[39] Wipf DP, and Rao BD. Sparse Bayesian learning for basis selection. IEEE Trans Signal Process. 2004;52(8):2153–2164.
[40] Babacan SD, Molina R, and Katsaggelos AK. Bayesian compressive sensing using Laplace priors. IEEE Trans Image Process. 2010;19(1):53–63.
[41] Tropp J, Gilbert A, and Strauss M. Algorithms for simultaneous sparse approximation. Part II: Convex relaxation. Signal Process. 2006;86(4):589–602.
[42] Duarte MF, Davenport MA, Wakin MB, *et al.* Sparse Signal Detection from Incoherent Projections. In: Proc. Acoust., Speech, Signal Processing (ICASSP). vol. III. Toulouse, France; 2006. p. 305–308.

[43] Haupt J, and Nowak R. Compressive Sampling for Signal Detection. In: Proc. Acoust., Speech, Signal Processing (ICASSP). vol. 3. Honolulu, Hawaii; 2007. p. III–1509–III–1512.

[44] Wang Z, Arce G, and Sadler B. Subspace Compressive Detection for Sparse Signals. In: Proc. Acoust., Speech, Signal Processing (ICASSP). Las Vegas, NV; 2008. p. 3873–3876.

[45] Meng J, Li H, and Han Z. Sparse Event Detection in Wireless Sensor Networks Using Compressive Sensing. In: 43rd Annual Conf. on Information Sciences and Systems (CISS). Baltimore, MD; 2009. p. 181–185.

[46] Davenport MA, Boufounos PT, Wakin MB, *et al.* Signal processing with compressive measurements. IEEE J Sel Topics Signal Process. 2010;4(2): 445–460.

[47] Rao BSMR, Chatterjee S, and Ottersten B. Detection of Sparse Random Signals using Compressive Measurements. In: Proc. Acoust., Speech, Signal Processing (ICASSP). Kyoto, Japan; 2012. p. 3257–3260.

[48] Kailkhura B, Wimalajeewa T, and Varshney PK. Collaborative compressive detection with physical layer secrecy constraints. IEEE Trans Signal Process. 2017;65(4):1013–1025.

[49] Wimalajeewa T, and Varshney PK. Sparse signal detection with compressive measurements via partial support set estimation. IEEE Trans Signal Inf Process Netw. 2017;3(1):46–60.

[50] Hariri A, and Babaie-Zadeh M. Compressive detection of sparse signals in additive white Gaussian noise without signal reconstruction. Signal Process. 2017;131:376–385.

[51] Razavi A, Valkama M, and Cabric D. Compressive detection of random subspace signals. IEEE Trans Signal Process. 2016;64(16):4166–4179.

[52] Kay SM. Fundamentals of Statistical Signal Processing Vol: II – Detection Theory. Upper Saddle River, NJ: Prentice Hall; 1998.

[53] Scharf LL, and Friedlander B. Matched subspace detectors. IEEE Trans Signal Process. 1994;42(8):2146–2157.

[54] McWhorter LT, and Scharf LL. Matched Subspace Detectors for Stochastic Signals. In: 11th Ann. Workshop on Adaptive Sensor Array Process. (ASAP). Lexington, MA; 2003.

[55] Jin Y, and Friedlander B. A CFAR adaptive subspace detector for second-order Gaussian signal. IEEE Trans Signal Process. 2005;53(3):871–884.

[56] Baraniuk RG. Compressive sensing. IEEE Signal Process Mag. 2007;24(4):118–121.

[57] Baron D, Wakin MB, Duarte MF, *et al.* Distributed compressive sensing. Rice Univ Dept Elect Comput Eng Houston, TX, Tech Rep TREE-0612. 2006 Nov.

[58] Duarte MF, Wakin MB, Baron D, *et al.* Measurement bounds for sparse signal ensembles via graphical models. IEEE Trans Inf Theory. 2013;59(7): 4280–4289.

[59] Rakotomamonjy A. Surveying and comparing simultaneous sparse approximation (or group-LASSO) algorithms. Signal Process. 2011;91(7):1505–1526.

[60] Cotter SF, Rao BD, Engan K, *et al.* Sparse solutions to linear inverse problems with multiple measurement vectors. IEEE Trans Signal Process. 2005;53(7):2477–2488.
[61] Tropp J, Gilbert A, and Strauss M. Algorithms for simultaneous sparse approximation. Part I: Greedy pursuit. Signal Process. 2006;86(4):572–588.
[62] Blanchard JD, Cermak M, Hanle D, *et al.* Greedy algorithms for joint sparse recovery. IEEE Trans Signal Process. 2014;62(7):1694–1704.
[63] Feng JM, and Lee CH. Generalized Subspace Pursuit for Signal Recovery from Multiple-Measurement Vectors. In: IEEE Wireless Communications and Networking Conference (WCNC). Shanghai, China; 2013. p. 2874–2878.
[64] Wipf D, and Rao B. An empirical Bayesian strategy for solving the simultaneous sparse approximation problem. IEEE Trans Signal Process. 2007;55(7): 3704–3716.
[65] Poor HV. An Introduction to Signal Detection and Estimation. New York: Springer-Verlag; 1994.
[66] Sankaran M. Approximations to the non-central chi-square distribution. Biometrika. 1963;50(1–2):199–204.
[67] Junger M, Liebling TM, Naddef D, *et al.* 50 Years of Integer Programming 1958–2008. Springer-Verlag: Berlin, Heidelberg; 2010.
[68] Leyffer S. Deterministic methods for mixed integer nonlinear programming. PhD dissertation University of Dundee Scotland, UK. 1993.
[69] Iyengar SG, Varshney PK, and Damarla T. A parametric copula-based framework for hypothesis testing using heterogeneous data. IEEE Trans Signal Process. 2011;59(5):2308–2318.
[70] Nandakumar K, Chen Y, Dass SC, *et al.* Likelihood ratio based biometric score fusion. IEEE Trans Pattern Anal Mach Intell. 2008;30:342–347.
[71] Nelsen RB. An Introduction to Copulas. 2nd ed. New York: Springer; 2006.
[72] Cramer H. Mathematical Methods of Statistics. Princeton, NJ: Princeton University Press; 1946.
[73] Wimalajeewa T, and Varshney PK. Compressive sensing based detection with multimodal-dependent data. IEEE Trans Signal Process. 2018;66(3):627–640.
[74] Kay SM. Fundamentals of Statistical Signal Processing Vol:I – Estimation Theory. Englewood Cliffs, NJ: Prentice Hall; 1993.
[75] Baraniuk R, and Wakin M. Random projections of smooth manifolds. Found Comput Math. 2009;9(21):51–77.
[76] Fannjiang A, and Liao W. Coherence pattern-guided compressive sensing with unresolved grids. SIAM J Imaging Sci. 2012;5(1):179–202.
[77] Duarte MF, and Baraniuk RG. Spectral compressive sensing. Appl Comput Harmonic Anal. 2013;35(1):111–129.
[78] Dinesh Ramasamy SV, and Madhow U. Compressive parameter estimation in AWGN. IEEE Trans Signal Process. 2014;62(8):2012–2027.
[79] Pakrooh P, Pezeshki A, Scharf LL, *et al.* Analysis of Fisher information and the Cramér–Rao bound for nonlinear parameter estimation after random compression. IEEE Trans Signal Process. 2005;63(23):6423–6428.
[80] Van-Trees HL, Bell KL, Tian Z Detection, Estimation, and Modulation Theory. Hoboken, NJ, USA: Wiley; 2013.

Chapter 3

Heterogeneous sensor data fusion by deep learning

Zuozhu Liu[1], Wenyu Zhang[2], Shaowei Lin[3], and Tony Q.S. Quek[1]

Heterogeneous sensor data fusion for decision-making is a challenging field that has gathered significant interest in recent years. In agriculture, for example, environmental conditions such as temperature, illuminance and humidity can be correlated with plant growth data, so that appropriate actions may be taken to maximize crop yield. In this chapter, we will provide an overview of heterogeneous sensor data fusion, including the background, basic deep learning techniques, and how these techniques can be used for sensor data fusion tasks. We will close this chapter with a detailed case study.

3.1 Introduction

Wireless sensor networks are widely deployed around many domains, and the sensor data collected can be used for many tasks, e.g., monitoring environmental changes, detecting infrastructural faults and improving physiological well-being [1,2]. Heterogeneous sensor data fusion refers to the statistical and machine-learning problems of combining data from different kinds of sensors so as to enable better inference, prediction and decision-making. For instance, smart cities can make use of a variety of signals from sensors, cameras and even social media to monitor the health of the urban infrastructures and to allocate resources more efficiently.

Many factors make sensor data fusion a difficult task. First, a sensor network is frequently composed of heterogeneous sensor nodes that capture multiple data modalities with potentially different statistical properties. Here **modality** measures different aspects of the same source of data. For example, illumination and temperature of the environment are two different modalities, and they are usually differently distributed. At the same time, wireless sensor data is often incomplete due to low battery, transmission loss or faulty sensors. These missing entries degrade the accuracy of data

[1]Information Systems Technology and Design (ISTD) Pillar, SUTD, Singapore
[2]Department of Statistical Science, Cornell University, USA
[3]Engineering Systems and Design (ESD) Pillar, SUTD, Singapore

fusion for decision-making. Moreover, as the number of sensor nodes and modalities increase in sensor networks, distributed sensor data processing would be required in many scenarios, and the need to compress sensor data for data transportation and storage becomes greater. The heterogeneity, incompleteness and high volume of sensor data raise new requirements for data fusion algorithms that are efficient, robust and scalable.

Many methods have been proposed to solve these problems. Computationally-simple algorithms like K-nearest neighbors (kNN) and sophisticated dimensionality reduction techniques like sparse-principle component analysis (PCA) and sparse coding techniques are widely used in many applications for missing data imputation and classification [3,4]. kNN adapts linear local computations for missing data and predicts the missing data with the mean value of the K nearest samples chosen with metrics like Euclidean distance. By extracting the sparse underlying structure of the data, sparse-PCA (S-PCA) reaches a more interpretable representation which explores different components for data reconstruction. However, most of these methods only consider intra-modal correlations, i.e., statistical correlations inside a single modality, but multimodal data provides the opportunity to also exploit intermodal correlations for enhancing the fusion performance.

In recent years, the effectiveness of deep learning has been successfully demonstrated in many real-world applications such as speech recognition, computer vision, natural language processing and so on [5]. Despite these incredible successes, the use of deep learning for heterogeneous sensor networks is still limited due to lots of challenges. The two main challenges are (1) learning from data with missing values and (2) finding shared representations for heterogeneous data to improve inference and prediction.

While the first challenge is still not fully investigated with deep learning methods, researchers already developed many frameworks to learn shared representations. Recent work showed that multimodal deep learning architecture is capable of learning a good joint representation for video and audio data that improves performance in discriminative tasks. The effectiveness of multimodal deep learning has subsequently been harnessed in other scenarios such as multimodal database query, content-based image retrieval tasks and robotics control [6,7]. However, these applications usually deal with complete datasets which may not be available in wireless sensor networks.

Inspired by these achievements, a new framework named deep multimodal encoder (DME), which will be introduced in this chapter later, was proposed for data fusion in multimodal large-scale sensor networks with missing data in a recent work [8]. The DME performs a two-stage training procedure to learn a shared representation that captures both the intra- and inter-modal correlations. It is designed to solve two problems: data compression and missing data imputation. To compress the sensor readings, unsupervised learning and finetuning are used to learn effective joint features of the data. A new learning objective function is incorporated to deal with incomplete data that embraces missing values and enables DME to capture latent features in the readings. Moreover, because the compressed encoding comes from a continuous transformation that preserves salient statistical properties of the raw data, new modalities may be predicted directly without decompressing the code. In a word,

DME takes both the intra- and inter-modal correlation into consideration to fill out the missing values, and such a complete dataset can be used for decision-making and inference tasks with robust performances.

3.2 Challenges in heterogeneous sensor data fusion

3.2.1 Compressive representation learning

Currently, the sensor networks become increasingly large, and so is the data collected. This raises requirements for fast data transmission and efficient data storage. Although we can always store and send all the collected information to a data center for analysis and decision-making, this would consume a large volume of disks and computational resources. Moreover, the distributed data processing in edge sensor nodes will be required in many scenarios, such as home automation, healthcare and Internet of things (IoT). Learning a compressive representation of the sensory data in the edge computing system can help a lot in these scenarios. Compressing the high-dimension original data into low-dimension representations could reduce the resource requirements for data transmission and storage. Meanwhile, decision-making over low-dimension data should be more accurate and faster than over high-dimension ones.

Many dimension reduction methods have been introduced in the past decades, from traditional PCA to kinds of neural network techniques. PCA converts n observations with k variables into a set of values of linearly uncorrelated variables called principal components. The number of distinct principle components is min $(n - 1, k)$. For compression purpose, we can select the first several principle components to represent the original data for decision-making or reconstruction tasks. For example, S-PCA, a technique extending PCA by adding a sparsity constraint on the input variables, reaches a sparse and interpretable representation which explores different components for data reconstruction. However, for sparse coding techniques, decoding the sparse code from the observation vector given the learned dictionary usually requires an iterative algorithm such as coordinate descent, which is computationally too intensive in real-time tasks [4,9].

Neural networks have been studied for a long time and received great success in the recent decades. Almost all of the neural network models such as multilayer perceptrons (MLP), restricted Boltzmann machines, autoencoders, recurrent neural networks (RNNs) and convolutional neural networks (CNNs) could be used for dimension reduction and representation learning [10]. For instance, autoencoders were proposed to represent data with nonlinear features which are more expressive than linear ones. When working with MLP, we can reduce the number of the hidden units in hidden layers sequentially to learn compressed representations. Currently, representation learning with neural networks has become an important and active research area. More details regarding neural networks would be listed in next section.

3.2.2 Missing data imputation

Wireless sensor data is often incomplete due to many reasons, such as low battery and transmission loss. Such low-quality raw data would degrade the performance

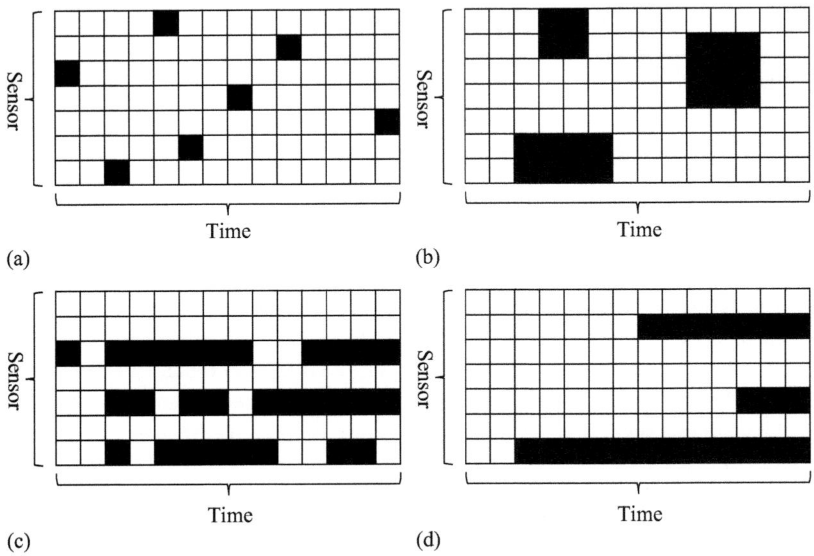

Figure 3.1 Data loss patterns: (a) random loss, (b) block random loss, (c) frequent loss in row and (d) successive loss © 2013. Reproduced, with permission, from Reference [12]

for decision-making and inference. Imputing the missing values to reconstruct the sensory data is the fundamental step to get in-depth understanding of the real-world physics and environments [11]. There are three steps to fill out the missing values in sensor data: (1) identifying data loss patterns, (2) selecting missing data imputation method and train an imputation model if necessary and (3) imputing the missing values for decision-making.

In the beginning, we need to identify which data loss pattern occurs in the original data. We consider four patterns of data loss [12], as shown in Figure 3.1:

- Random loss: Entries in the sensory data are dropped independently and randomly. Noise and collision might cause such a loss in wireless sensor networks.
- Block random loss: Entries from adjacent nodes in adjacent time slots are lost together. If lots of sensors are in congestion, such loss might occur.
- Element frequent loss in row: Entries from particular sensors are lost frequently. Unreliable links between the sensor and data receiver are the common reason for this pattern.
- Successive elements loss in row: Some nodes start losing from a particular time slot. If the sensor is out of energy or even damaged, it will suffer from this loss pattern.

In real-scenarios, data loss pattern might be a combination of the four types above.

Second, we need to select a proper missing value imputation method to fill out the missing entries. Different missing patterns might require different imputation methods. For example, in the random loss case, we can fill out with the mean over other non-missing samples or using kNN method. In this chapter, we will mainly introduce how to leverage on neural networks to impute the missing values in heterogeneous sensory data. Once we have imputed the missing values, we could finish decision-making tasks based on the complete dataset. A compressive representation might be learned from the data first. The final performance can be evaluated to test the effectiveness of imputation.

Let us formulate the missing data imputation problem mathematically since it will be one of the main tasks we will resolve in this chapter. In practice, many sensor nodes are distributed in certain areas, and they periodically sense the environmental data and send it to the server. If we have N sensors with each sampled for T time-slots, we can organize the collected data into an environmental matrix $\boldsymbol{X}_c \in \mathbb{R}^{N\times T}$ where the subscript "C" stands for the complete environmental matrix. Suppose we get the reconstruction $\hat{\boldsymbol{X}}_c$ from some compressed representation, the reconstruction problem of complete sensor data can be formulated as $\min\|\hat{\boldsymbol{X}}_c - \boldsymbol{X}_c\|_F^2$, where $\|\cdot\|_F$ is the Frobenius norm for matrices. This simple case is based on the assumption that the observed dataset $\boldsymbol{X}_c$ is complete; however, in real wireless sensor networks, this is usually not true. Hence, we need proposed new methods to deal with missing values.

Suppose we have an incomplete environmental matrix $\boldsymbol{X} = \boldsymbol{X}_c \cdot \boldsymbol{S}^x$ for a specific modality x, where $\boldsymbol{S}^x \in \mathbb{R}^{N\times T}$ is the indicator matrix and each entry $s_{n,t}^x$ is defined as

$$s_{n,t}^x = \begin{cases} 1, \text{if } x_{n,t} \text{ is observed,} \\ 0, \text{if } x_{n,t} \text{ is missing,} \end{cases}$$

for n from 1 to N and t from 1 to T, and $x_{n,t}$ is the tth reading in the nth sample. Now the task is to fill out the missing entries in $\boldsymbol{X}$ such that the imputation values are as close to the ground truth values as possible. Suppose we fill the missing values in $\boldsymbol{X}$ and get $\hat{\boldsymbol{X}}$, then the final goal is to minimize the cost function defined as:

$$\min\|(\hat{\boldsymbol{X}} - \boldsymbol{X}_c) \cdot (\boldsymbol{I}_{N\times T} - \boldsymbol{S}^x)\|_F^2, \tag{3.1}$$

where $\boldsymbol{I}_{N\times T}$ is a matrix of dimension $\mathbb{R}^{N\times T}$ with all ones, $\boldsymbol{A} \cdot \boldsymbol{B}$ denotes the element-wise product. However, since $\boldsymbol{X}_c$ is usually not available, the objective function is redefined as:

$$\min\|(\hat{\boldsymbol{X}} - \boldsymbol{X}) \cdot \boldsymbol{S}^x\|_F^2, \tag{3.2}$$

where **we only take the observed values into consideration.** The reason why we can use this objective function is that we assume by reconstructing the existing entries of the sensory data, the reconstruction model like neural networks can mine the underlying structures of the data. Subsequent imputation tasks can be finished based on this model, for example, feedforwarding the neural network.

The previous model only considers the intra-modal correlations, i.e., the correlations among data within the same modality. More advanced optimization of the objective function is provided in Section 3.3, where inter-modal correlations, the correlations among different data modalities, are taken into consideration.

3.2.3 Inter- and intra-modal correlations

When working with a dataset of multiple modalities, there are two possible statistical correlations we can take into consideration: intra-modal correlation and inter-modal correlation. Suppose we have two sensor data matrices $\boldsymbol{X}$ and $\boldsymbol{Y}$, each of them representing one modality, e.g., $\boldsymbol{X}$ for humidity on a farm and $\boldsymbol{Y}$ for temperature. Intra-modal correlation refers to the statistical correlation inside a single modality $\boldsymbol{X}$ or $\boldsymbol{Y}$. Inter-modal correlation refers to the correlation between different modalities, for example, temperature and humidity might be correlated. Multimodal data fusion techniques should consider both the intra-modal and inter-modal correlations. It can be used in many scenarios, such as robotics and database query, e.g., a robot needs to combine the information from visual inputs with voice signals to take interactive actions.

Now suppose $\boldsymbol{X}$ and $\boldsymbol{Y}$ are two incomplete matrices. In order to fill in the missing values, one straightforward method is to feedforward train models for these two modalities separately. This can be achieved by traditional methods like kNN, S-PCA or sparse autoencoder. However, given the multiple available modalities, could we try to mine the underlying correlations among modalities to impute the missing values? Actually, if we explore the relationship among modalities, we will find that some modalities are tightly correlated. For example, a higher temperature may lead to a lower humidity. Inspired by this, we should design new algorithms or models to take the inter-modal correlations into account as well.

Hierarchical framework is a common prototype to model both the intra-modal and inter-modal correlations in the fusion process. Assuming a hierarchical framework, usually the intra-model correlations will be learned in the bottom level, forming a compressive representation of the data on tops of which the inter-modal correlations could be captured. There are two advantages in this prototype. First, many techniques, such as PCA or neural networks, could be used for dimension reduction to learn a compressive representation. Second, the compressive representations will make the learning of inter-modal correlation much easier, while learning inter-modal correlations directly from concatenated raw data would be time-consuming and sometimes not accurate.

3.3 Deep learning techniques for heterogeneous sensor data fusion

In this section, we are going to introduce autoencoders, one of the building-blocks of deep neural networks. Stacked autoencoder is just a stacked version of autoencoders. After that, we will introduce the deep multimodal encoder (DME), the framework that can be used to learn both the intra- and inter-modal correlations for sensor data fusion. Other related architectures will be introduced at the end of this section.

3.3.1 Stacked autoencoder

Autoencoder is a neural network used for representation learning in an unsupervised style. It aims to learn an encoding for the input data, typically used for dimension

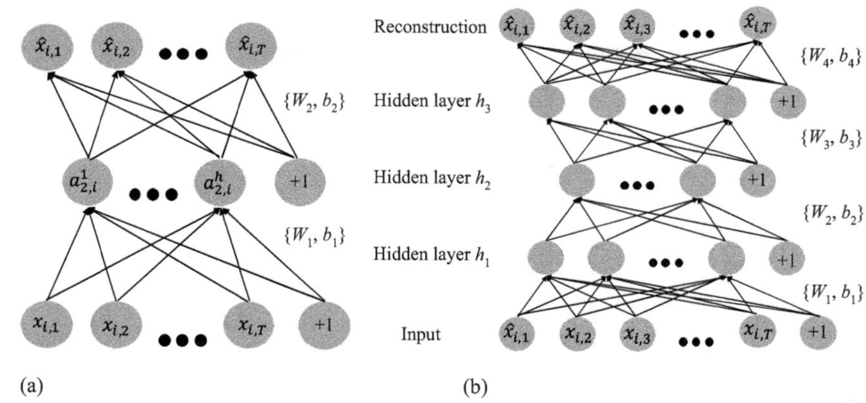

Figure 3.2 (a) Autoencoder and (b) Stacked autoencoder

reduction [13]. A single-layer autoencoder is a depth-two circuit with no directed loops. It consists of T input visible units, h hidden units and T output visible units. Given an N-samples training set $\boldsymbol{X} \in \mathbb{R}^{N\times T}$, where the ith row $\boldsymbol{x}_i^\top \in \mathbb{R}^T$ denotes the ith sample and $\boldsymbol{x}_{i,j}$ denotes the reading in time-slot j in the ith sample, an autoencoder learns parameters $\{\boldsymbol{W}_1, \boldsymbol{W}_2, \boldsymbol{b}_1, \boldsymbol{b}_2\} \in \{\mathbb{R}^{T\times h}, \mathbb{R}^{h\times T}, \mathbb{R}^T, \mathbb{R}^h\}$ such that:

$$\boldsymbol{h}_i^\top = \boldsymbol{a}_{2,i}^\top = f_1(\boldsymbol{x}_i^\top \boldsymbol{W}_1 + \boldsymbol{b}_1^\top), \tag{3.3}$$

$$\hat{\boldsymbol{x}}_i^\top = \boldsymbol{a}_{3,i}^\top = f_2(\boldsymbol{a}_{2,i}^\top \boldsymbol{W}_2 + \boldsymbol{b}_2^\top) \approx \boldsymbol{x}_i^\top. \tag{3.4}$$

Note that f_1 and f_2 are activation functions such as sigmoid, tanh and ReLU, and they are assumed to be executed element-wisely in the whole paper. $\boldsymbol{a}_{l,i}^\top$ is known as the activation in the lth layer for the ith sample, with $\boldsymbol{a}_{1,i}^\top = \boldsymbol{x}_i^\top$. The circuit simulates a neural network in the sense that a hidden unit with activation 1 (when using either sigmoid or tanh) corresponds to a firing neuron, and a hidden unit with activation 0 (with sigmoid) or -1 (with tanh) corresponds to a non-firing neuron. Using autoencoder, a representation is learned in the hidden layer. Hence, the autoencoder can be used for dimension reduction as long as $h < T$. A typical autoencoder can be seen in Figure 3.2(a).

The core part of autoencoder is to learn good parameters of the model. To do this, the autoencoder is trained using backpropagation to minimize the squared reconstruction error. That is, we are trying to learn the parameters $\{\boldsymbol{W}, \boldsymbol{b}\} = \{\boldsymbol{W}_1, \boldsymbol{W}_2, \boldsymbol{b}_1, \boldsymbol{b}_2\}$ that minimize the cost function defined as:

$$J(\boldsymbol{W}, \boldsymbol{b}) = \frac{1}{2N} \sum_{i=1}^{N} \|\hat{\boldsymbol{x}}_i^\top - \boldsymbol{x}_i^\top\|^2. \tag{3.5}$$

A regularization term of the weight is added to prevent overfitting. When learning sparse features, the cost function is modified to include sparsity constraints. The new cost function becomes:

$$J_{\text{sparse}}(\boldsymbol{W},\boldsymbol{b}) = J(\boldsymbol{W},\boldsymbol{b}) + \frac{\lambda}{2}\sum_{l=1}^{2}\sum_{p=1}^{m_l}\sum_{q=1}^{m_{l+1}}(\boldsymbol{W}_l^{p,q})^2 + \beta\sum_{j=1}^{m_2} KL(\rho\|\hat{\rho}_j), \tag{3.6}$$

where m_l is the number of units in the lth level and $\boldsymbol{W}_l^{p,q}$ is the weight between the pth node in the lth layer and the qth node in the $(l+1)$th layer. The decay weight λ and sparsity weight β control the relative contributions of the terms to the cost function. For the definition of $\rho^j(\rho^j)$, we compute it as the average activation over all the input samples at the jth hidden unit. Mathematically, it is defined as:

$$\hat{\rho}_j = \frac{1}{N}\sum_{i=1}^{N}\boldsymbol{a}_{2,i}^j, \tag{3.7}$$

where $\boldsymbol{a}_{2,i}$ is the activation of the ith sample in the hidden layer, and $\boldsymbol{a}_{2,i}^j$ is the jth entry of it. $KL(\rho\|\hat{\rho}_j)$ is the Kullback–Leibler divergence between the desired sparsity ρ and $\hat{\rho}_j$ and is defined as:

$$KL(\rho\|\hat{\rho}_j) = \rho\log\frac{\rho}{\hat{\rho}_j} + (1-\rho)\log\frac{1-\rho}{1-\hat{\rho}_j}. \tag{3.8}$$

In the following, we will refer to λ, β, ρ as hyper-parameters.

A stacked autoencoder is a deeper circuit of multiple layers of autoencoders in which the hidden units of each autoencoder are the inputs to the successive autoencoder. A typical stacked autoencoder with three hidden layers is illustrated in Figure 3.2(b). The activation output of the lth layer is:

$$\boldsymbol{a}_{l,i}^{\top} = f_l(\boldsymbol{a}_{l-1,i}^{\top}\boldsymbol{W}_{l-1} + \boldsymbol{b}_{l-1}^{\top}), \tag{3.9}$$

while $\boldsymbol{a}_{1,i} = \boldsymbol{x}_i$. By defining $r_{\boldsymbol{W},\boldsymbol{b}}(\boldsymbol{x}_i^{\top}) = \boldsymbol{a}_{n_l,i}$, the cost function can be defined similarly as:

$$J(\boldsymbol{W},\boldsymbol{b}) = \frac{1}{2N}\sum_{i=1}^{N}\|r_{\boldsymbol{W},\boldsymbol{b}}(\boldsymbol{x}_i^{\top}) - \boldsymbol{x}_i\|^2, \tag{3.10}$$

and

$$J_{\text{sparse}}(\boldsymbol{W},\boldsymbol{b}) = J(\boldsymbol{W},\boldsymbol{b}) + \frac{\lambda}{2}\sum_{l=1}^{n_l-1}\sum_{p=1}^{m_l}\sum_{q=1}^{m_{l+1}}(\boldsymbol{W}_l^{p,q})^2 + \sum_{l=2}^{n_l-1}\sum_{j=1}^{m_l}\beta^l KL(\rho^l\|\hat{\rho}_j^l), \tag{3.11}$$

where n_l is the total number of layers in the network, and β^l and ρ^l are respective hyper-parameters in the lth layer.

Learning in a stacked autoencoder consists of two steps. The first step is the greedy layer-wise training of the parameters for each individual autoencoder to initialize the parameters near a local minimum of the cost function. For example, for a

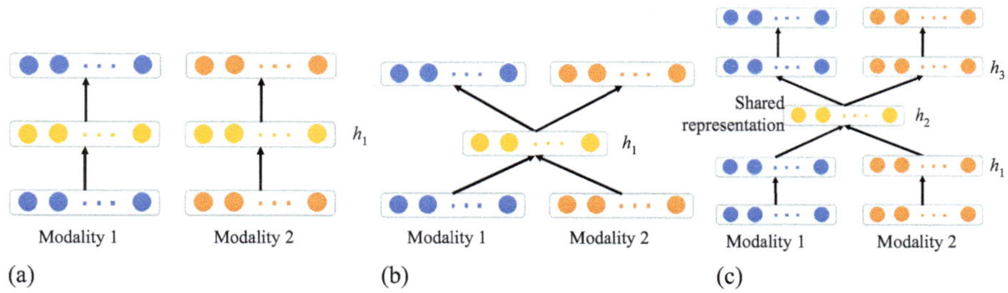

Figure 3.3 Different neural network models for missing data imputation: (a) UAE, (b) CAE and (c) DME

stacked autoencoder with three hidden layers, parameters $\{\boldsymbol{W}_1, \boldsymbol{W}_4, \boldsymbol{b}_1, \boldsymbol{b}_4\}$ are first learned greedily using an autoencoder with the raw data $\boldsymbol{X}$ as input. Using the hidden unit activations as input for the next autoencoder, $\{\boldsymbol{W}_2, \boldsymbol{W}_3, \boldsymbol{b}_2, \boldsymbol{b}_3\}$ are learned. The second step is the finetuning using back propagation. During finetuning, all model parameters are changed simultaneously to improve the overall results of the algorithm.

3.3.2 Deep multimodal encoder

As we stated above, missing data imputation is an important task for heterogeneous sensor data fusion. In this section, we will introduce the DME model which is proposed to resolve this task effectively [8]. Suppose we have two incomplete environmental matrices $\boldsymbol{X}$ and $\boldsymbol{Y}$, the goal of this task is to recover the missing values. Usual methods like unimodal autoencoder (UAE) learn separate models for different modalities as shown in Figure 3.3(a) by optimizing over the objective function as defined in (3.2) independently for each modality. Then we can fill the missing values in $\boldsymbol{X}$ through the feedforwarding outputs of the neural network. The idea behind is that as long as the autoencoder can reconstruct the observed values, it captures the important features of the sensor data. Hence, when the model is trained, we can compute the activations in each hidden layer to get the reconstruction $\hat{\boldsymbol{X}}$ of the incomplete input $\boldsymbol{X}$ by following the feedforward process we described above, and any missing entry $\boldsymbol{x}_{n,t}$ in the input can be imputed with the reconstruction output $\hat{\boldsymbol{x}}_{n,t}$.

The UAE model works well, but it only considers the intra-modal correlations and ignores the inter-modal correlations. The concatenated autoencoder (CAE) model in Figure 3.3(b) shows a possible design to incorporate inter-modal correlations. However, this concatenated model simultaneously learns the intra- and inter-modal correlations in one layer, which obfuscates the signals with different statistical properties and prevents either of them from being learned accurately.

In contrast to UAE and CAE, DME aims at capturing both the intra- and inter-modal correlations to resolve the issues above. It is a specific implementation of the deep learning architecture with autoencoders along with a new objective function to be introduced in the next subsection. The basic framework is illustrated in Figure 3.3(c). The concept behind DME is to learn the intra-modal correlations of each modality individually in the first hidden layer and then learn the inter-modal correlations in the second hidden layer. Consistency in filling out the missing values among the different modalities is then achieved by exploiting the inter-modal correlations.

DME is also different from other multimodal frameworks: it adapts to missing entries with a new objective function. Many proposed multimodal deep learning frameworks only learn from complete datasets, the DME, on the other hand, is designed to fill in the missing values in a dataset. To adapt to the inherent incompleteness of wireless sensor datasets, DME modifies the original loss function of the autoencoder in both the intra- and inter-modal learning. Finally, the missing values can be filled with the values predicted by feedforwarding the DME model.

3.3.2.1 Intra-modal learning

Given two incomplete input datasets $\boldsymbol{X} \in \mathbb{R}^{N\times T^x}$ and $\boldsymbol{Y} \in \mathbb{R}^{N\times T^y}$, where N is the number of samples and T^x, T^y are the dimension of each sample in $\boldsymbol{X}$ and $\boldsymbol{Y}$, respectively. The intra-modal learning objective function is devised with a new normalized factor which is related to the missing nature of the input data. In (3.6), the least square error and KL-divergence are normalized by the number of samples N. In DME, these two terms are redefined with a new vector normalizer. More formally, the square error loss for input $\boldsymbol{X}$ becomes:

$$\tilde{J}(\boldsymbol{W},\boldsymbol{b}) = \frac{1}{2}\left\| \frac{1}{\mathbb{1}^N \otimes \boldsymbol{\theta}^{x\top}} \cdot (\hat{\boldsymbol{X}} - \boldsymbol{X}) \cdot \boldsymbol{S}^x \right\|_F^2, \tag{3.12}$$

where $\mathbb{1}^N$ is the N-dimension column vector $[1, 1, \ldots, 1]^\top$, $\boldsymbol{S}^x$ is the missing indicator matrix as defined before, $\hat{\boldsymbol{X}}$ is the output of the autoencoder, $\otimes$ is the Kronecker product and all calculations are done element-wisely. The central part here is the term $1/\boldsymbol{\theta}^x$ which means that we are not simply normalizing the square loss by a single scalar but with respect to a vector. More formally, $\boldsymbol{\theta}^{x\top} \in \mathbb{R}^{T^x}$ is defined as:

$$\boldsymbol{\theta}_t^{x\top} = \sum_{n=1}^{N} s_{n,t}^x\,, \text{ for } t \in 1, 2, \ldots, T^x, \tag{3.13}$$

where we can regard each entry $\boldsymbol{\theta}_t^{x\top}$ as the number of observed samples in modality x of a certain dimension t.

The sparsity penalty term is also redefined with the same idea. First, the mean activation term is modified as $\hat{\boldsymbol{\rho}}^x \in \mathbb{R}^{T^x\times m_x}$, where m_x is the number of units in the first hidden layer h_1 for modality x. With the hidden layer activation function f, the output of the hidden layer is $\boldsymbol{A}^x = f(\boldsymbol{X}\boldsymbol{W}^x + \mathbb{1}^N \otimes \boldsymbol{b}^{x\top})$, where $\boldsymbol{W}^x$ and $\boldsymbol{b}^x$ are the weight and bias in this intra-modal autoencoder for modality x. Then the definition of $\hat{\boldsymbol{\rho}}^x$ becomes:

$$\hat{\boldsymbol{\rho}}^x = \left[(\boldsymbol{A}^x)^\top \boldsymbol{S}^x \cdot \frac{1}{\mathbb{1}^{m_x} \otimes \boldsymbol{\theta}^{x\top}} \right]^\top, \tag{3.14}$$

and the new sparsity penalty term is redefined as:

$$\frac{\beta^x}{T^x} \| KL(\hat{\boldsymbol{\rho}}^x \| \rho^x) \|_1 = \frac{\beta^x}{T^x} \sum_{m=1}^{m_x} \sum_{t=1}^{T^x} KL(\hat{\rho}_{m,t}^x \| \rho^x), \tag{3.15}$$

where β^x and ρ^x are the sparse weight and predefined sparsity for the autoencoder of modality x, respectively, and $\|\cdot\|_1$ is the l^1-norm.

The new normalizer can better take the missing data property into account. The intuition is that dimensions with different numbers of missing entries should be weighted differently in the objective function. Together with the unchanged decay weight regularization term, the new objective function $L^x(\boldsymbol{X})$ for modality x is given by:

$$\min L^x(\boldsymbol{X}) = \frac{1}{2}\left\| \frac{1}{\mathbb{1}^N \otimes \boldsymbol{\theta}^{x\top}} \cdot (\hat{\boldsymbol{X}} - \boldsymbol{X}) \cdot \boldsymbol{S}^x \right\|_F^2 + \lambda^x \|\boldsymbol{W}^x\|_F^2 + \frac{\beta^x}{T^x} \sum_{m=1}^{m_x} \sum_{t=1}^{T^x} KL(\hat{\rho}^x_{m,t} \| \rho^x). \tag{3.16}$$

If there is no missing data, this reduces to the usual case in Section 3.3.1. In the multimodal case, DME learns autoencoders for each modality individually to explore features in the missing data. The final objective function $L^{xy}(\boldsymbol{X}, \boldsymbol{Y})$ in the intra-modal learning procedure is defined as:

$$\min L^{xy}(\boldsymbol{X}, \boldsymbol{Y}) = L^x(\boldsymbol{X}) + L^y(\boldsymbol{Y}), \tag{3.17}$$

where $L^y(\boldsymbol{Y})$ for modality y is defined with the same rules as $L^x(\boldsymbol{X})$. The hyper-parameters, i.e., $\rho^x, \rho^y, \beta^x, \beta^y, \lambda^x, \lambda^y$, can vary among different modalities for better performance. Like other deep neural networks, hyper-parameter search is needed to achieve a good performance with DME.

3.3.2.2 Inter-modal learning

One main advantage of the DME model is the ability to learn the inter-modal correlations among different modalities. After learned the above autoencoders, the hidden layer outputs $\boldsymbol{H}_1^x, \boldsymbol{H}_1^y$ can be extracted for another layer-wise learning procedure to capture the inter-modal correlations. The framework is illustrated as Figure 3.3(c).

A different objective function is employed to learn the inter-modal correlations. The input for inter-modal correlations learning is the concatenation of $\boldsymbol{H}_1^x, \boldsymbol{H}_1^y$, i.e., $\boldsymbol{H} = [\boldsymbol{H}_1^x, \boldsymbol{H}_1^y]$. This new input is used to train the second autoencoder with the original loss function defined in (3.6). More specifically, the objective function $L_2(\boldsymbol{H})$ is:

$$L_2(\boldsymbol{H}) = \frac{1}{2N}\|\hat{\boldsymbol{H}} - \boldsymbol{H}\|_F^2 + \lambda^{xy}\|\boldsymbol{W}^{xy}\|_F^2 + \beta^{xy}\sum_{j=1}^{m_2} KL(\rho^{xy}\|\hat{\rho}_j), \tag{3.18}$$

where $\hat{\boldsymbol{H}}$ is the reconstruction of $\boldsymbol{H}$, $\boldsymbol{W}^{xy}$ is the matrix of weight parameters, $\hat{\rho}_j$ is the average activation of the jth neuron, m_2 is the number of hidden units, and λ^{xy}, β^{xy} and ρ^{xy} are the respective hyper-parameters.

This inter-modal learning procedure benefits from the intra-model correlations which were learned previously, allowing higher order structures in the data to be captured more accurately.

3.3.3 More neural network architectures

Besides autoencoder, there are lots of other popular architectures for deep learning, such as CNNs and RNNs. CNN is extremely suitable for image and video processing. It takes the idea from biology that individual neurons only responds to stimuli in a small visual field known as receptive field to learn multiple filters for feature extraction. Convolution neural network differs from traditional multilayer perceptrons in two aspects. First, it is not a fully connected network. The input is convoluted with every filter independently with predefined padding and strides. Second, it takes a convolutional step and a pooling step besides simple feedforward activation. The design pattern makes CNN good at extracting local feature from image, video and other spatial-correlated data. In contrast to convolution neural network, recurrent network takes the temporal features into consideration. In a traditional RNN, all of the information from the past will be accounted when computing the activation of current state. Advanced recurrent models like long-short-term memory are proposed with different types of gates to control the flow of past information for efficient and robust learning. More about these models is out of the scope of this chapter and discussed in modern deep learning books and research papers.

3.4 A case study

In this section, we present a case study with real-world sensor data on two tasks: sensor data compression and reconstruction, missing data imputation. We first introduce the formulation of each task and subsequently list the experimental details and results.

3.4.1 Dataset

The dataset used for the experiments consists of three modalities: temperature (in degrees Celsius), humidity (relative humidity in %) and illuminance (light integral lux), collected through an agriculture sensor network of 40 sensors deployed in different locations over 4 months. Each sensor samples every 5 min and has sensing components to measure all three environmental factors. The readings are binned into 10-min intervals to reduce the percentage of missing readings. Since these environmental factors do not vary significantly over a short time period, lowering the resolution does not compromise the integrity of the data by much. Daily time series are selected as samples in which there are no missing data across all three modalities. There are a total of 3,306 samples, 306 of which are randomly selected as validation set, 600 for test set and the remaining 2,400 samples for the training set. That is, for each modality, the training set is a $2,400 \times 144$ matrix, where each row is a sample consisting of 144 readings, with timestamps ranging from 00.00 to 23.50 at 10-min intervals. The test set for each modality has dimensions 600×144. The validation data is used for hyper-parameter selection to prevent overfitting, while the test data is used to measure the accuracy of the model predictions after the hyper-parameters have been selected. Thus, the dataset is partitioned based on samples rather than on sensor nodes, for example, some readings from a specific node can be selected for training while some other reading from this same node can be used for test or validation.

Table 3.1 Dataset statistics

	Temp.	Hum.	Illum.
Min	21.16	9.58	0
Max	60.95	100.00	98,295.30
Lower quartile	25.90	72.63	0
Median	27.60	84.42	29.68
Upper quartile	31.28	90.87	2,411.29
Standard deviation	5.03	16.16	6,635.97

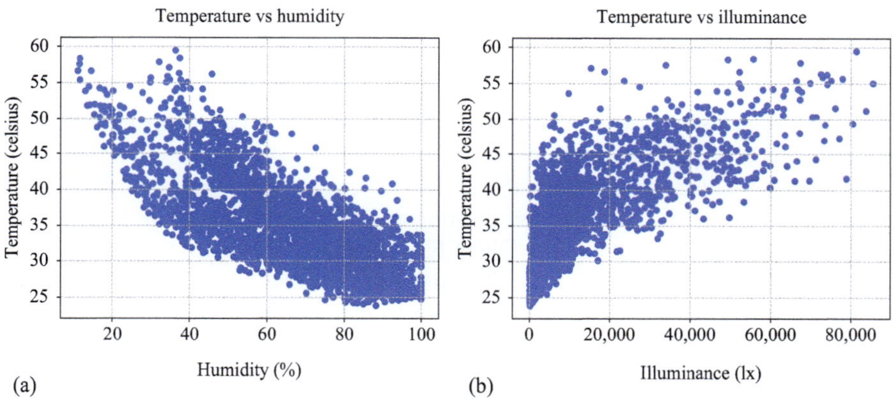

Figure 3.4 Modality correlation: (a) temp vs hum and (b) temp vs illum

Table 3.1 shows the basic statistics of the training sets. Humidity is known to be bounded at $[0, 100]$, and temperature can be assumed to be bounded at $[15, 65]$ where the sensors are deployed. Illuminance is a heavy-tailed distribution, and most readings taken during night conditions are near zero. Figure 3.4 displays scatter plots of temperature against humidity and temperature (HT) against illuminance at noon when the environmental factors demonstrate great variability. The plots suggest that it is difficult to predict temperature using only readings from a single modality.

3.4.2 Data preprocessing

Consider the raw dataset of a single modality with sample size N, each sample of length t, as an $N \times t$ matrix. Before training the network, raw data should be corrected for associated biases. We do this by standardizing the datasets with zero-mean columns or features. This preserves the magnitude of the pairwise covariance and linear correlations of the columns, which is important because the network cannot learn from a sample of uncorrelated readings.

The resulting dataset is scaled to have a standard deviation s. This enables datasets of all modalities to have the same spread, such that the network is not biased toward

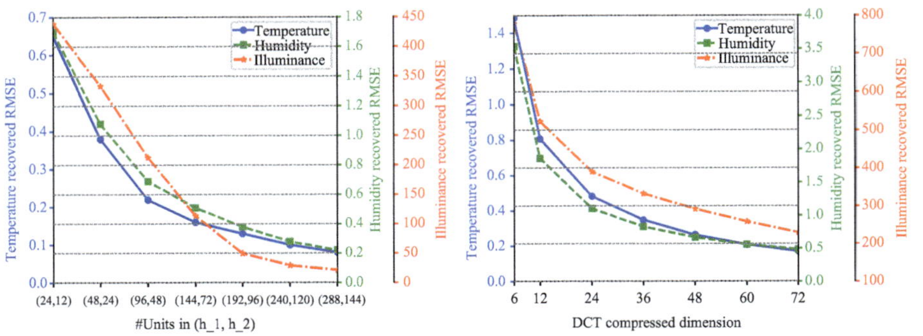

(a) (b)

Figure 3.5 Recovered RMSE of DME (a) and discrete cosine transform (b)

learning parameters to reduce the reconstruction error of a particular dataset with more spread. $s = 0.17$ for the experiments.

For heavy-tailed distributions, the DME is trained for the truncated dataset. In particular, we first truncate the raw illuminance datasets to the range $[LQ - 1.5 \cdot IQR, UQ + 1.5 \cdot IQR]$ where LQ, UQ and IQR are the lower quartile, upper quartile and interquartile range, respectively. This method is popular for removing outliers in neural network approaches. The aforementioned data standardization steps are then applied to the truncated dataset.

3.4.3 Task 1: sensor data compression and reconstruction

We first show that DME can definitely reconstruct the **complete** multimodal data $\boldsymbol{X}_c$ with low root-mean-square error (RMSE). This can be done by training the DME model but with $s^x_{n,t} = 1$ everywhere. In this step, we vary the number of hidden units in the two autoencoders, i.e., the number of units in layer h_1 and h_2. Remember that m_i is the number of hidden unit in the ith hidden layer. (m_1, m_2) take values in {(24,12), (48,24), (96,48), (144,72), (192,96), (240,120), (288,144)}. We conducted two data-reconstruction experiments: one for HT and another for humidity and illuminance (HI). To demonstrate the efficiency and accuracy of DME, we compared the performance of DME with a traditional data compression technique, discrete cosine transformation (DCT), implemented with Python's `scipy` package. Since the performance of DCT is tested for each modality individually, in order to ensure the same compressed dimensions are tested, we use half the number of units in layer h_2, i.e., [6, 12, 24, 36, 48, 60, 72], in DCT as compressed dimensions. The results of the data recovering RMSE are illustrated in Figure 3.5. Note there is no missing data in the datasets here.

As shown in the results, the RMSE in both DME and DCT varies a lot among modalities, i.e., large error for illuminance and low error for HT. This is mainly due to different statistical distributions of original data, e.g., illuminance is of long-tail distribution which has a larger variance and is much harder to reconstruct. We can also observe that as the compressed dimension increases, the error will decrease. This phenomenon is obvious in illuminance modality in DME, whose long-tail distribution incurs large reconstruction error with only little hidden units. For HT, the overall error difference in DME is always small despite decrease, i.e., less than 2 for humidity and 0.7 for temperature.

Compared with DCT, DME can achieve better performance, although the performance gap would decrease when the compressed dimension increases. As we can see, the RMSE is less than 1 for temperature in DME, while for DCT it is a bit larger. The same results also hold for humidity. For illuminance, the performance is much worse when using DCT. This may result from its long-tail distribution. For both DCT and DME, as the compressed dimension increases, the performance improves significantly initially but tapers off at larger dimensions. Motivated by the great performance of DME for data reconstruction, we will show how DME deals with incomplete datasets.

3.4.4 Task 2: missing data imputation

We use the DME model proposed in the previous section for missing data imputation task and evaluate the performance in this section. In the experiments, we use the RMSE as the metric for performance. More specifically, we define the RMSE e for $\boldsymbol{X}$ as:

$$e_x := \sqrt{\frac{\|(\hat{\boldsymbol{X}} - \boldsymbol{X}_c) \cdot (\boldsymbol{I}_{N\times T} - \boldsymbol{S}^x)\|_F^2}{\|\boldsymbol{I}_{N\times T} - \boldsymbol{S}^x\|_F^2}}. \tag{3.19}$$

This is the error metric with one modality. In the multimodal case, the same measure should be used for other modalities as well. For example, in this case study, we consider two modalities, then the same error measure e should be computed for both modality x and y, as shown in the results.

Five algorithms, namely, UAE, CAE, kNN, S-PCA and DME, are compared in this section for missing data imputation. Experiments with kNN and S-PCA are conducted by using Python's `scikit-learn` package. The parameter k in kNN is selected from $[20, 30, 49, 60, 70, 80]$, and it is finally chosen as $\lceil\sqrt{n}\rceil = 49$ where n is the number of training samples. For S-PCA, we extract 20 sparse atoms and set the sparsity controlling parameter as 0.2. The sparse atom is chosen among $[10, 20, 30]$ and the sparsity is chosen among $[0.1, 0.2, 0.3, 0.4, 0.5, 0.6]$, and in total, 18 combinations are tested and the best one is chosen. The neural network architectures of UAE and CAE are shown in Figure 3.3(a) and (b), respectively. The revised objective function for missing data is also used in UAE and CAE.

Two different multimodal experiments are conducted: one for HT and another one for HI. We vary the missing rate from 10% to 90% in the experiment with HT, and 10% to 80% for HI. The 90% missing rate is not used for illuminance because illuminance data has a heavy tail with zeros; a high missing rate may lead to too many entries in the dataset being 0, thus making the experiment meaningless. To get datasets with missing values from the original complete environmental matrix, the indicator matrix $\boldsymbol{S}$ for each modality is randomly generated with respect to the missing rate. For example, under 10% missing rate, each entry in $\boldsymbol{S}$ will be set to 0 with probability 0.1. More formally, we are exploring the element random loss pattern in the environmental matrix.

Hyper-parameters play a pivotal role in the performance of deep learning models, and there are many studies on how to find appropriate hyper-parameters. In this experiment, the learning rate is set to 0.06, and the other hyper-parameters such as the decay weights and sparse weights are selected with a simplified version of grid search [14]. We first fix the decay weight at a reasonable value such as 0.001 and find the best sparse weight, i.e., the sparse weight with the least RMSE, from the set $\{10^{-1}, 10^{-2}, \ldots, 10^{-15}\}$. The best sparse weight found is then fixed while we search for the best decay weight. This alternating process is continued until there are no more changes to either weight. With this resulting pair of weights, the simplified grid search is repeated by zooming into a smaller neighborhood of the weights. Despite not exhaustively searching the space of hyper-parameters, we still achieved reasonably good performance for the DME model and saved a fair amount of computational time.

The RMSE for different algorithms are shown in Tables 3.2 and 3.3. We let UAE-6 and CAE-6 denote the corresponding neural networks with 6 hidden units. Similarly, DME-6 and DME-2 are respectively the DME models with 6 and 2 hidden units in the fused representation hidden layer. Note that for DME-6, the compression ratio is $6/288$ where 288 is the total input dimension of the two modalities. Fair comparison is achieved by using the same compression ratio across UAE, CAE and DME.

(1) Humidity–temperature: First, we observe that DME-6 outperforms all other models in all the experiments. In fact, the RMSE for kNN and S-PCA increases dramatically as the missing rate increases. The humidity RMSE of kNN is two to seven times larger than DME-6, and the corresponding error of S-PCA is also three to six times larger than DME-6. For temperature, DME-6 outperforms kNN and S-PCA by an order of magnitude when the missing rate is greater than 50%.

If we compare the DME-6 against the linear methods with the metric:

$$\text{Relative RMSE} = \text{RMSE}_{\text{Algorithm}} - \text{RMSE}_{\text{DME-6}}, \tag{3.20}$$

we will find that the relative RMSE increases significantly with the missing rate. This demonstrates that DME-6 is more robust than the linear methods, especially for large missing rates. The main reason is that linear methods like kNN and S-PCA are not able to capture the underlying data distribution accurately when a lot of data is missing. However, DME employs nonlinear transformations, producing models which are much more expressive. This observation is also strongly supported by the good performance of UAE and CAE.

Among the three deep-learning-based models, the best performer turns out to be DME-6. The performance gain can be credited to the higher order inter-modal features captured by the deeper DME model. Learning such features seems to be challenging for the shallower UAE and CAE models. Another interesting observation is that when the missing rate becomes larger, the improvement also increases. This phenomenon is even more obvious when comparing DME-6 to kNN and S-PCA.

Table 3.2 RMSE humidity and temperature

Miss rate	**Humidity**									**Temperature**								
	0.1	**0.2**	**0.3**	**0.4**	**0.5**	**0.6**	**0.7**	**0.8**	**0.9**	**0.1**	**0.2**	**0.3**	**0.4**	**0.5**	**0.6**	**0.7**	**0.8**	**0.9**
KNN	8.31	14.56	20.98	27.90	35.41	43.56	52.36	61.85	71.96	2.45	4.66	7.12	9.81	12.71	15.83	19.13	22.70	26.48
S-PCA	15.45	17.88	19.09	21.79	26.62	33.66	42.97	54.40	66.85	6.09	7.28	7.88	8.67	10.08	12.49	15.90	19.96	24.68
UAE-6	5.09	5.98	6.79	7.73	8.13	9.00	10.07	11.06	13.87	2.03	2.32	2.48	2.71	2.95	3.10	3.27	3.69	4.52
CAE-6	**4.64**	5.98	6.40	7.36	8.05	8.87	9.75	11.07	13.83	**1.73**	2.07	2.32	2.52	2.78	**2.93**	3.28	3.81	5.09
DME-2	5.15	5.87	6.58	7.41	7.97	8.71	9.70	10.95	13.94	2.00	2.19	2.33	2.56	2.79	3.02	3.21	3.66	3.92
DME-6	**4.64**	**5.79**	**6.37**	**7.15**	**7.85**	**8.62**	**9.50**	**10.69**	**11.96**	**1.73**	**2.04**	**2.29**	**2.45**	**2.68**	**2.93**	**3.14**	**3.65**	**3.34**

Bold values indicate the best performance among the 6 compared methods.

Table 3.3 RMSE humidity and illuminance

Miss rate	**Humidity**								**Illuminance**							
	0.1	**0.2**	**0.3**	**0.4**	**0.5**	**0.6**	**0.7**	**0.8**	**0.1**	**0.2**	**0.3**	**0.4**	**0.5**	**0.6**	**0.7**	**0.8**
KNN	8.31	14.56	20.98	27.90	35.41	43.56	52.36	61.85	**290.63**	**515.26**	762.50	1,043.95	1,315.64	1,604.82	1,887.74	2,169.86
S-PCA	15.45	17.88	19.09	21.79	26.62	33.66	42.97	54.40	389.38	614.82	870.16	1,081.57	1,312.71	1,540.43	1,797.90	2,069.87
UAE-6	5.29	5.98	6.78	7.62	8.01	8.96	10.04	11.13	683.01	715.33	753.04	793.97	833.18	890.74	959.53	**1,022.86**
CAE-6	5.24	5.88	6.49	**7.11**	7.88	8.96	9.53	10.81	624.06	679.22	727.41	768.66	**806.65**	891.63	972.08	1,095.98
DME-2	5.27	6.15	6.65	7.67	8.22	8.86	9.69	10.86	701.36	722.66	758.22	789.42	829.23	889.15	**943.58**	1,027.02
DME-6	**4.83**	**5.73**	**6.16**	7.15	**7.45**	**8.30**	**9.31**	**10.61**	615.17	651.43	**696.10**	**739.20**	811.31	**863.50**	955.17	1,027.21

Bold values indicate the best performance among the 6 compared methods.

The result further demonstrates that DME is more robust than the other models at learning underlying structures in the data, even when many values are missing.

To better understand DME, we also investigated the performance of DME-2 in Table 3.2. When the number of fusion units is increased from 2 to 6, the performance improves. In fact, DME-2 cannot always dominate the performance, since it is a little difficult for these two hidden units to learn the inter-correlations very well. The performance of DME-2 is not necessarily at least as good as UAE-6 and CAE-6 since the second hidden layer is trying to capture some correlations and learn compressed representations from the first hidden layer rather than simply learning an identity map from the first hidden layer. This is consistent with many empirical results where deeper neural networks cannot always beat shallow ones if extra tricks like residual network are not used. Although the performance of DME-2 did not dominate, it falls behind UAE-6 and CAE-6 only in a few cases. Indeed, DME-2 shows comparable performance with UAE and CAE in most of the situations.

(2) Humidity–illuminance: In these experiments, DME-6 mostly outperforms other models in reconstructing humidity; in the few cases where it is not the state-of-the-art, the performance is only slightly worse. However, for illuminance which has a long-tailed distribution, DME-6 is much worse than kNN and S-PCA when the missing rate is 10% or 20%. This is because the high variability in long-tailed distributions is not easily captured by neural networks with few hidden units. Meanwhile, when the missing rate is low, kNN is able to estimate the missing values just by querying the many neighboring sensors which do have the same readings. When the missing rate becomes larger, the performance of kNN degrades due to the lack of neighbors, while DME-6 is better able to capture the diminishing information using its 6 fusion units.

In summary, our DME model does better than traditional linear algorithms such as kNN and S-PCA as well as shallower neural network models such as UAE and CAE. In particular, when the missing rate is high, DME proves itself to be more robust at discovering underlying features in the data.

(3) Generalization performance evaluation

Since DME requires a separate network for each missing rate, a new experiment for cross generalization test is conducted to test its ability to generalize to different missing rates. According to the previous subsections, for each missing rate i from 0.1 to 0.9, we have trained an associated DME model M_i. In the experiment, we test the missing data imputation performance for each model M_i at three different missing rates $[i - 10\%, i, i + 10\%]$. The experiment results are shown in Figure 3.6, where we denote the performance of model M_i, $M_{i-10\%}$ and $M_{i+10\%}$ at missing rate i as the baseline, downward test and upward test, respectively. For missing rate 10%, the downward test is meaningless and hence ignored. So is the upward test for missing rate 90% in HT and 80% in HI.

As we can see in the results, when the missing rate is low, i.e., less than 50%, the cross-generalization results, although a little worse than the baseline, still lie in reasonable ranges. We can also notice that the upward test is closer to the baseline than the downward test, which can be credited to the regularization effect. In fact, the RMSE difference between the upward test and baseline is about 0.6–3.2 for HT

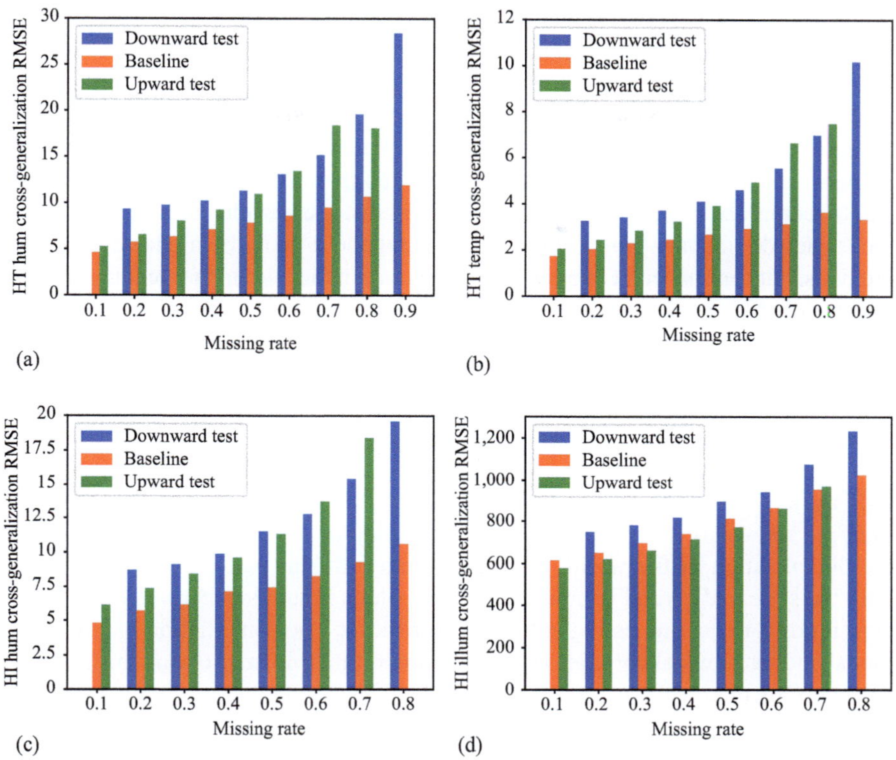

Figure 3.6 ***Cross-generalization test.*** *The baseline is the performance of our trained model for missing rate i. The upward test involves predicting data with missing rate i using our trained model for missing rate i + 10%. The downward test involves predicting data with missing rate i using our trained model for missing rate i − 10%. (a) HT humidity, (b) HT temperature, (c) HI humidity and (d) HI illuminance*

humidity and 0.3–1 for HT temperature. For HI, the performance degrades a little due to the long-tail illuminance distribution. When the missing rate is getting greater than 50%, the cross generalization performance is much worse than the baseline. This phenomenon is consistent with our previous results – when the missing rate i is getting larger, it is much harder for DME to impute the missing values with the concrete trained model M_i, and even more so for the models $M_{i-10\%}$ and $M_{i+10\%}$ which are not specifically trained for missing rate i.

3.5 Summary

Currently, many challenges exist in heterogeneous sensor data fusion. With the development of IoT techniques and widely distributed sensors, we believe more problems would be raised from data collection, data processing, representation learning and decision-making. In this chapter, we illustrate several challenges and present the DME framework for missing data imputation and multimodal data fusion in wireless sensor networks. The framework computes a shared representation based on deep neural networks for the multimodal data that may be used directly in other predictive tasks. There are more interesting tasks we can explore with different types of data or models. Regarding the sensor data, it is believed that successive loss or block loss would require more complex models and algorithms. Meanwhile, the agricultural data modalities do not vary drastically over time and could easily be modeled by simple methods, but dealing with complex sensor data such as accelerometer readings used in human activity recognition and biomedical sensor readings is much more challenging. Regarding the models, the DME framework is proposed years ago and composed of only autoencoders, while many more powerful models such as CNN, RNN, LSTM and GRU are available nowadays. These advanced neural networks should be employed for more complex environments such as healthcare and robotics.

Multimodal learning is one of the hardest research problems we want to highlight again. Taking human beings as an example, we make daily decisions based on multiple sources of information. With the development of sensor technologies and the digitalization of the world, more data from different modalities will flow into both the edge and central computing system, which requires faster and more robust algorithms for data processing and decision-making. The combination of both intra-modal and inter-modal correlations plays an important role in resolving this problem. Advanced methods such as CNN and RNN should be used in future systems to extract such correlations for a good shared representations of the data. Afterwards, powerful reasoning and interactive algorithms should be employed for decision-making.

Finally, designing an energy-efficient pattern for distributed processing in heterogeneous sensor networks would be a new trend. As more and more sensors are located in the edge system, it would be even harder to send all the sensory data to a central system for processing. Currently, mobile computing and edge computing have received significant success, as well as fog computing and cloud computing. All of these paradigms could be integrated into a unified efficient distributed computing system. Meanwhile, proper data processing framework makes a difference in distributed processing as well. For example, this multilayer hierarchical DME framework is well suited for distributed processing in sensor networks with tree topologies. We leave this as a main direction for future research.

Acknowledgments

We thank Dr. Domenico Ciuonzo and Dr. Pierluigi Salvo Rossi for inviting us to write this chapter. We also thank Sky Greens Pte. Ltd. for providing the data and thank NVIDIA for the computational resources.

References

[1] Khaleghi B, Khamis A, Karray FO, and Razavi SN. Multisensor data fusion: a review of the state-of-the-art. Information Fusion. 2013;14(1):28–44.

[2] Alsheikh MA, Lin S, Niyato D, *et al.* Machine learning in wireless sensor networks: algorithms, strategies, and applications. IEEE Communications Surveys & Tutorials. 2014;16(4):1996–2018.

[3] Cover TM, and Hart PE. Nearest neighbor pattern classification. IEEE Transactions on Information Theory. 1967;13(1):21–27.

[4] Jenatton R, Obozinski G, and Bach F. Structured sparse principal component analysis. In: *Proc. 13th Int. Conf. on Artificial Intelligence and Statistics*; 2010. p. 366–373.

[5] LeCun Y, Bengio Y, and Hinton G. Deep learning. Nature. 2015;521(7553): 436–444.

[6] Ngiam J, Khosla A, Kim M, *et al.* Multimodal deep learning. In: *Proc. 28th Int. Conf. on Machine Learning*; 2011. p. 689–696.

[7] Wang W, Yang X, Ooi BC, *et al.* Effective deep learning-based multi-modal retrieval. The VLDB Journal. 2016;25(1):79–101.

[8] Liu Z, Zhang W, Lin S, *et al.* Heterogeneous sensor data fusion by deep multimodal encoding. IEEE Journal of Selected Topics in Signal Processing. 2017;11(3):479–491.

[9] Donoho DL. Compressed sensing. IEEE Transactions on Information Theory. 2006;52(4):1289–1306.

[10] Hinton GE, and Salakhutdinov RR. Reducing the dimensionality of data with neural networks. Science. 2006;313(5786):504–507.

[11] Wong LZ, Chen H, Lin S, *et al.* Imputing missing values in sensor networks using sparse data representations. In: *Proc. 17th ACM Int. Conf. on Modeling, Analysis and Simulation of Wireless and Mobile Systems*; 2014. p. 227–230.

[12] Kong L, Xia M, Liu XY, *et al.* Data loss and reconstruction in sensor networks. In: *INFOCOM, 2013 Proc. IEEE*; 2013. p. 1654–1662.

[13] Ng A. UFLDL Tutorial; 2013. Available from: http://ufldl.stanford.edu/wiki/index.php/UFLDL_Tutorial.

[14] Bergstra J, and Bengio Y. Random search for hyper-parameter optimization. The Journal of Machine Learning Research. 2012;13(1):281–305.

Part II

Reporting channel uncertainty

Chapter 4

Energy-efficient clustering and collision-aware distributed detection/estimation in random-access-based WSNs

Seksan Laitrakun[1], Deepa Phanish[2], and Edward J. Coyle[2]

Sporting events, such as football, baseball, and soccer games, attract large numbers of fans. Each team participating in the event will have or bring many personnel, including coaches, trainers, and players. The stadium hosting the game will have a large staff to provide security, safety, food, media, and other event services. Thus, there may be 50,000–100,000 people attending a game and many thousands of team and event staff working in the stadium during the game.

Almost every fan and staff member at a game now has a smartphone that has cellular, WiFi, and Bluetooth connectivity. Each of those devices also has accelerometers, a microphone, a camera, and possibly other sensors. The media people in the stadium will have high definition video cameras, two-way radios for coordination, and wireless links with flying cameras. Each team will often have its own wireless system to enable coaches on the field to talk with other coaches in booths in the upper decks of the stadium.

The stadium itself will nowadays have sophisticated infrastructure to handle the short-term but very large demands for communication and sensing from all of the people and systems at the game. A modern stadium will thus have a distributed-antenna cellular system, hundreds of WiFi access points, and many Bluetooth systems for sensing, tracking, and short-range communications. It will also have sensors for structural monitoring, monitoring and localization of radio frequency (RF) sources in the stadium, and audio monitoring of the venue for safety and other purposes.

Thus, large sporting events in modern stadiums, in just 1 square kilometer, have wireless communication and sensing requirements of much greater complexity than found almost anywhere else. They are thus an ideal environment to study these systems and develop the ideas and technologies for the future Internet of People and Things (IoPT). We have thus developed, over many years, the Stadium IoPT test bed at Bobby Dodd stadium at Georgia Tech. It includes several systems in the stadium that we have designed and studied that gather and process data from the game, from the venue and

[1]School of ICT, SIIT, Thammasat University, Thailand
[2]School of ECE, Georgia Institute of Technology, USA

Figure 4.1 Picture of the Bobby Dodd Stadium at Georgia Institute of Technology showing the large-scale extreme emitter density environment during a football game. The scale and spatial-density of the sensing devices pose a challenge to the effective collection of sensor data for the purposes of event operations, infotainment, safety, and security

from fans, in order to deliver unique services or develop new capabilities that are available during the game.

In this chapter, we focus exclusively on our work on gathering data from a large number of sources in the stadium. In the theoretical work reported here, we are particularly interested in minimizing the time and energy required to gather data from many different sensor nodes. There is a parallel effort, reported elsewhere, to take the results of our theoretical research and turn them into real systems that we deploy in the stadium and study when they are used during the sporting events paces [1–3]. Our current efforts include

- Gathering video and data from the game to share with fans in the stands via web applications that enable on-demand access to multimedia content, including video-clips of plays, visualization of game events, and game/player stats [4].
- Developing wireless sensor networks to monitor structural vibrations of the stadium and audio of the crowd [5].
- An extreme emitter density test bed for RF spectrum sensing and cross-layer localization of wireless devices [6].
- Analytical models and associated algorithms for collecting, processing, and communicating very large amounts of data for detection, estimation, and other tasks.

To return to the theoretical issues covered in this paper, consider trying to gather sensed data from the phones and/or other devices that the fans and stadium personnel have with them in the stands. Each fan, as can be seen in the photo in Figure 4.1 of a game in our football stadium, has a seat and occupies about 1 square meter of space. The fans are thus arranged in a dense grid pattern on the "bowl" that supports the stadium seats. Assume that each fan has and will use their phone for a particular sensing application, such as detecting how exciting a play was by monitoring the acceleration of the fan's phone or FitBit, with some probability p. We further assume that each fan's decision to participate is independent of the decisions of other fans. It is then clear that the distribution of active sensors for any given application will form a spatial Poisson process in the stands of the stadium. When all of these sensors use the wireless connectivity of the phones that host them, the result is clearly a very large-scale wireless sensor network, one with many thousands of nodes covering a square kilometer. The question is then: How do we best organize the collection of data from all of these devices that make up this very large sensor network?

The results in this chapter provide answers to this question in two steps. We first ask, assuming that we want to collect data from every node in the spatial Poisson process, if clustering of the sensor nodes is effective and how many levels of clustering are needed. We summarize recent work we have completed that provides a clustering algorithm to minimize the aggregate energy used for nodes' transmissions and receptions to collect all of this data [7,8]. In fact, the optimal clustering strategy is now known for any specified level of aggregation at each level of the clustered architecture.

At the lowest level of this hierarchically clustered network are multi-hop clusters whose clusterheads (CHs) gather data from the nodes in their clusters. Because this is where the number of transmissions and receptions will be the greatest, it is natural to ask the following question: Given we are trying to make a decision or produce an estimate in a timely fashion with the data that we gather, how can we order the collection of data so that the highest quality data is collected first and the poorest quality data is not collected at all? Rephrased, this question is about how to censure poor quality data when the data is collected over a wireless channel that is using a random-access protocol [9–11], such as the variants of the CSMA/CA protocol that are found in WiFi, Bluetooth, and ZigBee networks.

Here we provide new results in the context of a single-hop network performing distributed detection/estimation, with the goal of generalizing to a multi-hop cluster in future work. In this single-hop context, we show how a population splitting algorithm can be used within a random-access network to achieve the desired result. Perhaps, the most interesting result contained here is that when we collect data this way, the collisions that occur on the random-access channel are found to contain a significant amount of information [12–15]. Exploiting this information leads to very efficient data gathering algorithms and the very interesting result that channel loadings that produce many collisions are optimal.

This chapter is organized as follows. We first review our prior work on fast, energy efficient clustering algorithms. In this work, we showed, using results from stochastic geometry, how to rapidly organize the nodes into a hierarchy of clusters that minimizes the energy used to gather data from every participating node. Of particular interest is that this work accounts for different levels of data aggregation at each level of the clustered architecture. We then focus on a single cluster in the architecture to study population-splitting algorithms to reduce the amount of data that must be collected to make a decision or estimate of a specified quality. This allows faster decisions and estimates to be produced and further reduces the energy the system spends in gathering data. The form that the population splitting algorithm takes in a random-access network is very interesting when collisions can be detected. We show that the channel loading that is optimal is much higher than the loading that yields maximum throughput in the traditional view of these networks. We conclude by summarizing our results and pointing out the future work that is needed, such as the extension of the population splitting-based algorithm to the case of multi-hop clusters.

4.1 Clustering in wireless sensor networks

Clustering of nodes within a wireless sensor network keeps communication local when appropriate and provides energy-efficient strategies for collecting/distributing data from/to a large numbers of nodes. Instead of using direct long-range communications to the processing/fusion center, the sensed data from cluster members can first be collected at the central CH, aggregated and reduced in size before transmission to the processing/fusion center, thereby reducing the total energy cost.

Nodes can be efficiently organized into clusters with a distributed algorithm [16], where the nodes elect themselves as CHs with a certain probability. Once elected, the CHs broadcast a Route-REQuest (RREQ) packet to initiate the clustering process. On receiving a RREQ packet, the node associates with the corresponding CH. The RREQ is then forwarded if the hop count is within the specified upper bound. To cluster the nodes with the nearest CH, a node can be reassigned whenever it receives a RREQ with hop count fewer than its preceding assignment. This process of clustering is repeated at CHs to achieve multiple levels of hierarchy. The resulting architecture of the hierarchically clustered network is shown in Figure 4.2. It is a multi-hop, multilevel, data-gathering network. The illustration shows two levels before data is sent to the processing/fusion center.

The aggregated data size at the CH depends on the application and signal processing schemes in place, as illustrated in Figure 4.3. In an event-detection application, forwarding a single decision bit along the levels of the hierarchy could be sufficient. However, multiple bits are typically necessary in order to include performance statistics or location information [17]. In applications monitoring spatial–temporal variation of state variables in a multi-sensor environment, the data may accumulate along the hierarchy, the size of which is determined by the aggregation schemes and sensing systems in place [18]. On the other hand, in data-mining applications, the forwarded data size reduces at each level of the hierarchy [19,20].

An important question that is critical to achieving energy-efficient clustering is as follows: Given the network size and other characteristics, what is the optimal probability of choosing CHs in order to minimize the total communication cost? [8].

4.1.1 Communication cost in multi-hop multilevel clusters

Consider a homogeneous ad hoc network where all the nodes, including CHs, have the same fixed range of communication, as shown in Figure 4.2. Consequently, the routes between CHs on consecutive levels are multi-hop via the end (level-0) nodes.

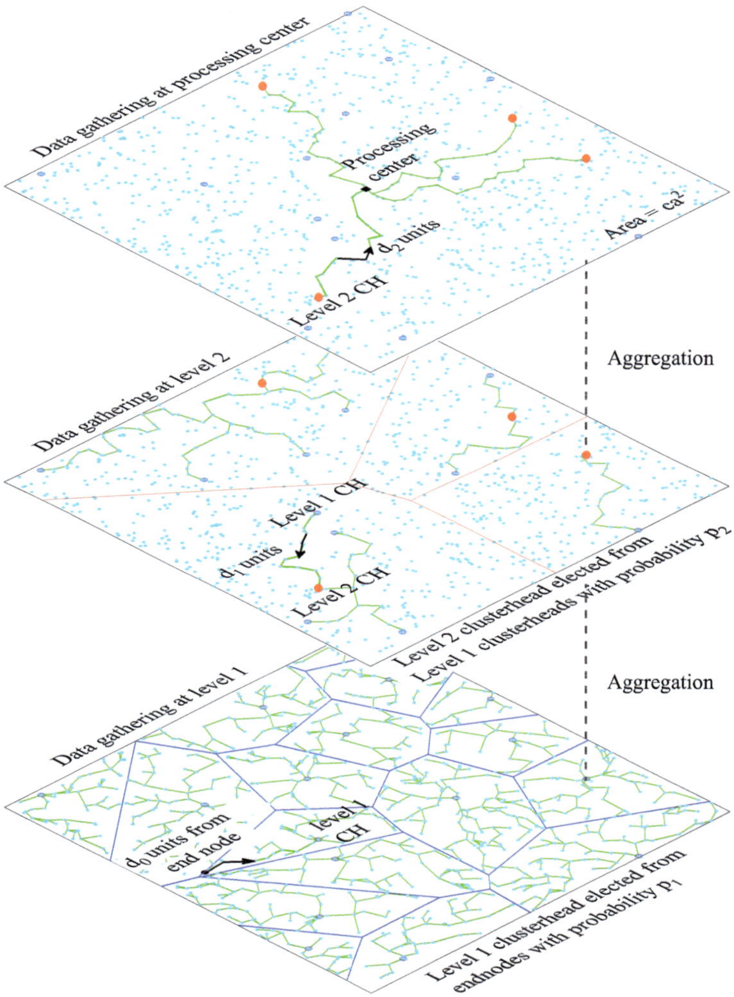

Figure 4.2 Architecture of a multi-hop, two-level data gathering clustered network. The clusterheads (CHs) at level-1 are chosen independently from the end nodes. Similarly, the CHs at level-2 are chosen independently from the level-1 CHs. Association of the respective nodes to their nearest CH results in the Voronoi tessellations shown in both levels

A level-1 cluster consists of end nodes and their associated level-1 CH. A level-n cluster consists of the level-$(n-1)$ CHs and their associated level-n CH $\forall n > 1$. Further, the amount of data generated by the nodes at any level are variable and independent of that at the other levels, i.e., each end-node sends d_0 units to its respective level-1 CH, each level-1 CH sends d_1 units to its respective level-2 CH, which, in turn, sends

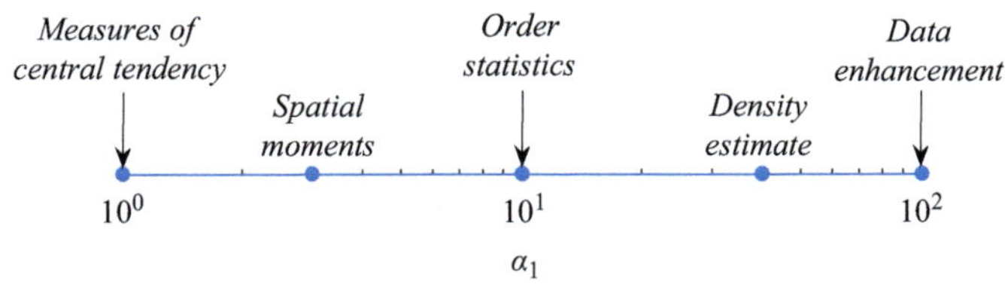

Figure 4.3 An example of the variation of the data ratio with the statistical processing schemes: Consider a unit of data gathered from air-quality monitoring CO_2 sensors in a cluster. The aggregated data can be either the mean/median/mode, first- and second-order geometric moments, the ten highest values, or a histogram with 40 bins. Even supplementary temporal data sensed between successive rounds can be appended at the clusterhead. Further, the data size at the next level clusterhead could be smaller, equal, or larger depending on whether there is a decision, a pooled estimate, or enhancement, respectively

d_2 units to the processing center. To model the total communication cost, we make the following assumptions:

1. The nodes are distributed according to a homogeneous spatial Poisson process with intensity λ in a 2D space of area $A = ca^2$.
2. The communication cost increases linearly with the data size.

The architecture we have considered has hop-by-hop routes, both within a cluster and between the CHs of any two levels, as demonstrated in Figure 4.2. In this case (i) the range of communication r is a fixed constant; (ii) the energy required to transmit (ε_{TX}) and receive (ε_{RX}) a unit of data over one hop—although determined by the electronics—are also fixed constants; and (iii) the true length of the multi-hop shortest route (L_r) can be approximated to the scaled Euclidean distance between the source and destination (L_e). In fact, the scaling factor between their means has been shown to be $4/\pi$ [21]. Approximating the hop length to the range of communication r, the total cost of communicating a unit of data over a route of length L_r is (# hops)×(Energy per hop) = $(((4/\pi)L_e)/r)\,(\varepsilon_{TX} + \varepsilon_{RX})$. By eliminating the scaling factors, minimization of the total communication cost reduces to the minimization of the sum of the Euclidean distances.

When the level-i CHs are chosen independently from level-$(i-1)$ CHs with probability p_i (the "clusterheads" on "level 0" are indeed all the nodes in the network), their locations are governed by the thinned homogeneous spatial Poisson process $\mathbf{PP1}_i$ of intensity $\lambda_{1i} = \lambda \prod_{j=1}^{i} p_j$. The non-CHs of level-$i$ are also governed by a thinned Poisson process $\mathbf{PP0}_i$ of intensity $\lambda_{0i} = \lambda_{1(i-1)} - \lambda_{1i}$. Suppose L_i is the total length of all the segments connecting the points of $\mathbf{PP0}_i$ process to the nearest point of a $\mathbf{PP1}_i$ process, then [22]:

$$\mathbb{E}[L_i] = \frac{\lambda_{0i}}{2\lambda_{1i}^{3/2}} = \frac{(1-p_i)}{2p_i\sqrt{\lambda \prod_{j=1}^{i} p_j}} \tag{4.1}$$

Table 4.1 Scaling factor ρ for a specified location of the processing center

Deployment region $A = ca^2$	**Processing center location**	ρ'	$\rho = 2\rho'/\sqrt{c}$
Square $2a \times 2a$	Center	0.765	0.765
Square $2a \times 2a$	Corner	1.530	1.530
Rectangle $2a \times 4a$	Center	1.186	0.838
Rectangle $a \times 2a$	Corner	1.186	1.677
Circle radius a	Center	0.666	0.751

Since the expected total number of level-i CHs $\mathbb{E}[TCH_i] = \lambda_{1i}A$, the expected communication cost to gather a unit of data from all the nodes of level $i < h$ at their respective CHs in a network with h levels of hierarchy is given by [16]:

$$\begin{aligned} \mathbb{E}[C_i] &\approx \mathbb{E}[L_i]\,\mathbb{E}[TCH_i] \\ &= \lambda A(1 - p_i)\left[\prod_{j=1}^{i-1} p_j\right]\left[\frac{1}{2\sqrt{\lambda \prod_{j=1}^{i} p_j}}\right] \end{aligned} \tag{4.2}$$

The above approximation due to the bounded area over which the nodes are distributed is better as the number of nodes increases. The network simulations verify that it accurately models the cost when $\mathbb{E}[TCH_i] \geq 1$.

The expected distance between a node in a region of area $A = ca^2$ and the processing center at the origin is given by:

$$\mathbb{E}[D] = \frac{1}{A}\iint\limits_{A} \sqrt{u^2 + v^2}\, du\, dv = \rho' a \tag{4.3}$$

where ρ' for various reference cases are computed in Table 4.1. Therefore, the additional cost of communicating a unit of data from the CHs at level h to the processing center is:

$$\mathbb{E}[C_h] = \mathbb{E}[TCH_h]\,\mathbb{E}[D] = \lambda A\left[\rho' a \prod_{i=1}^{h} p_i\right] \tag{4.4}$$

For an absolute aggregation at the CHs, the number of data units transmitted at any level-i, d_i, is same as the number of data units generated by the end nodes, d_0. But, for any partial aggregation scheme, $d_i > d_0 \quad \forall\, i = 1, 2, \ldots, h$. The aggregated data size is predetermined and independent of cluster size within a broad class of data fusion algorithms. Assuming a linear scaling of the communication cost with the data size, $\mathbb{E}[C_i]$ is pre-multiplied by d_i. Therefore, the total communication cost of collecting

Table 4.2 Notations and definitions used in Section 4.1

Notation	Definition
n	Number of nodes
$A = ca^2$	Area of deployment
λ	Density of nodes
h	Number of levels of hierarchy
$\rho' a$	Expected distance from a node to the processing center
C	Total communication cost for gathering data
f	Cost objective function parameterized by n
$\hat{p}_i$	ith coordinate of the critical point of f
$p_i^\star$	ith coordinate of the unconstrained minimum of f
p_i^{min}	ith coordinate of the constrained minimum of f in $(0, 1)$
d_0	Number of data units generated by the end node
d_i	Number of data units forwarded by the level-i CH
$\alpha_i = \frac{d_i}{d_0}$	Data ratio variable at level-i
α, β, γ	Deterministic values taken by the data ratios α_is

data from all the nodes and forwarding the aggregated data to the processing center follows from (4.2) and (4.4), given by:

$$\begin{aligned} \mathbb{E}[C] &= \sum_{i=1}^{h} \mathbb{E}[C_i] \\ &= \lambda A \left[\sum_{i=1}^{h} \frac{d_{i-1}(1-p_i)\prod_{j=1}^{i-1} p_j}{2\sqrt{\lambda \prod_{j=1}^{i} p_j}} + \rho' a d_h \prod_{i=1}^{h} p_i \right] \end{aligned} \tag{4.5}$$

The symbols used in Section 4.1 and their definitions are summarized in Table 4.2. The optimal number of levels and the optimal probabilities of choosing the CHs that minimize $\mathbb{E}[C]$ are studied in relation to the number of nodes n and the data ratios $\alpha_i = d_i/d_0$ within the hierarchy. Note that we are minimizing only the communication cost for successful transmissions in the network. In addition to $\mathbb{E}[C]$, idle energy at the nodes and data retransmissions constitute the total communication cost. However, both of these factors largely depend on the protocols in place for medium access (MAC) control and hop-by-hop routing. For instance, a synchronized spatially sampled data gathering scheme [23] can effectively minimize this overhead. However, a system-level reduction in the number of transmissions favorably reduces interference, idle energy, and retransmissions as well.

4.1.2 Optimal probabilities of electing clusterheads

The goal is to choose the probabilities p_i of electing CHs at levels $i = 1, 2, \ldots, h$ that minimize $\mathbb{E}[C]$. This is formulated as a constrained optimization problem:

$$\left(p_1^{\min}, p_2^{\min}, \ldots, p_h^{\min}\right) = \underset{p_1, p_2, \ldots, p_h}{\arg\min}\ \mathbb{E}[C] \quad s.t. \quad 0 \le p_1, p_2, \ldots, p_h \le 1 \tag{4.6}$$

For a given number of nodes n, we have $\lambda = n/A = n/ca^2$. In this new perspective, the optimal probabilities $p_i^{\min}$ are viewed as functions of n, making further analysis tractable:

$$\min_{p_1,p_2,\ldots,p_h} \mathbb{E}[C] = \min_{p_1,p_2,\ldots,p_h} f(p_1,p_2,\ldots,p_h) \tag{4.7}$$

where the cost objective f parameterized by n is given by:

$$f(p_1,p_2,\ldots,p_h) = \sum_{i=1}^{h} \frac{\alpha_{i-1}(1-p_i)\prod_{j=1}^{i-1} p_j}{\sqrt{\prod_{j=1}^{i} p_j}} + \rho\alpha_h\sqrt{n}\prod_{i=1}^{h} p_i \tag{4.8}$$

Here $\rho = 2\rho'/\sqrt{c}$ (Table 4.1) and the "data ratio" $\alpha_i = d_i/d_0$ is the ratio of the number of data units forwarded by each CH on level i to the number of data units generated at each end node ($\alpha_0 = 1$). Suppose that each end node sends 1 unit of data to its level-1 CH. Then, each level-1 CH sends α_1 units of data to its level-2 CH, and each level-2 CH sends α_2 units of data to its level-3 CH, and so on. The level-h CHs at the top all send α_h units of data to the processing center. It is assumed that $\alpha_i \geq 1$, i.e., the size of the aggregated data at the CH at any level d_i is at least the size of data d_0 generated by an end node. However, it is still possible to have $\alpha_{i+1} < \alpha_i$, $\alpha_{i+1} = \alpha_i$ or $\alpha_{i+1} > \alpha_i$. Figure 4.3 demonstrates how α_is vary with the application requirements.

Note that in the rest of this section, $p_i^{\min}$ denotes the value of the constrained minimum, whereas $p_i^\star$ denotes the value of the unconstrained minimum:

$$\left(p_1^\star,p_2^\star,\ldots,p_h^\star\right) = \arg\min_{p_1,p_2,\ldots,p_h} \mathbb{E}[C] = \arg\min_{p_1,p_2,\ldots,p_h} f \tag{4.9}$$

In a general h-level hierarchical network, the optimal probabilities of CH election at each level is obtained by minimizing the communication cost $\mathbb{E}[C]$ in (4.5). Specifically, for a given number of nodes n, we need to minimize the cost objective f in (4.8). By equating the partial derivatives of f w.r.t $p_1,p_2,\ldots,p_h$ to zero, we obtain the following system of h simultaneous equations indexed by $k = 1,2,\ldots,h$:

$$\alpha_{k-1}\left[\prod_{j=0}^{k-1} x_j\right]\left(1+x_k^2\right) = \sum_{i=(k+1)\leq h}^{h} \alpha_{i-1}\frac{1-x_i^2}{x_i}x_k\prod_{j=0}^{i-1} x_j + 2\rho\sqrt{n}\alpha_h x_k\prod_{j=0}^{h} x_j^2$$

where $\alpha_0 = 1$, $p_0 = 1$, and $x_i = \sqrt{p_i}$ $\forall i = 1,2,\ldots,h$. The solution ($p_1 = p_1^\star, p_2 = p_2^\star,\ldots,p_h = p_h^\star$) satisfying (4.10) provides the unconstrained minimum. It warrants a case-by-case analysis to determine the conditions that bound $p_i^\star$ to (0,1).

In the following theorem, we perform an unconstrained optimization of $\mathbb{E}[C]$ for a single level of hierarchy, $h = 1$, and derive the conditions on α_i for which the critical solution satisfying the derivative test is unique and is a minimum within constraints [8].

Theorem 4.1. *In a single level hierarchical cluster, there exists a unique* $p_1^\star$ *that minimizes* $\mathbb{E}[C]$*. Further,* $0 < p_1^\star < 1$ *when* $\rho\alpha_1\sqrt{n} > 1$.

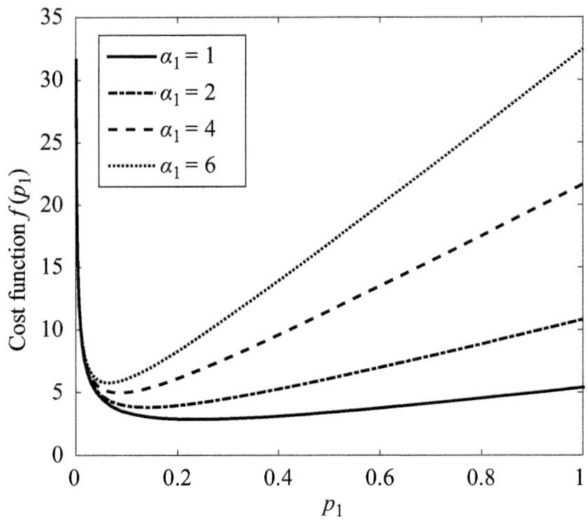

Figure 4.4 Variation of energy objective with the probability of CH election. There are $n = 50$ nodes and $\rho = 0.765$ (processing center located at the center of a square). Note the relatively flat minimum for the data ratio $\alpha_1 = 1$ and the increasing sharpness of it as α_1 increases

Corollary 4.1. *When $\rho\alpha_1\sqrt{n} > 1$, the optimal probability of CH election in a single level cluster is given by the square of the real-valued solution of $2\alpha_1 dx^3 - x^2 - 1 = 0$ where $d = \rho\sqrt{n}$. Note that $p_1^{\min}$ can be expressed in closed form.*

The variation of the objective function f with p_1 (in (4.8)) for different values of α_1 is shown in Figure 4.4. The results indicate that as the aggregated data size decreases, the optimal number of CHs in a network will increase. The reason is that the energy savings from each aggregation is larger. Also, the smaller α_1 is, the less sensitive the energy usage in the network is to variations in the probability of choosing a CH. In other words, because the minimum is more acute for larger aggregated data sizes, the probability needs to be more accurately implemented within the clustering algorithm. The variation in the cost is more rapid with a variation in the aggregated data size compared to variations due to the number of nodes. Overall, as $\alpha_1\sqrt{n}$ increases, the minimum communication cost $\mathbb{E}[C]$ increases, the optimal probability of choosing a CH decreases (Figure 4.5), and the network's sensitivity to error in the probability of CH selection increases. Therefore, the reduction in energy with a reduction in the aggregated data size is more pronounced when the size of the network is relatively large.

Similar results can be obtained for a two-level hierarchy with $h = 2$ [8].

Theorem 4.2. *In a two-level hierarchical cluster, when $\rho^2\alpha_2^2 n > \alpha_1^2(2\alpha_1 - 1)$, there exist unique $p_1^\star, p_2^\star \in (0, 1)$ that minimize the communication cost $\mathbb{E}[C]$.*

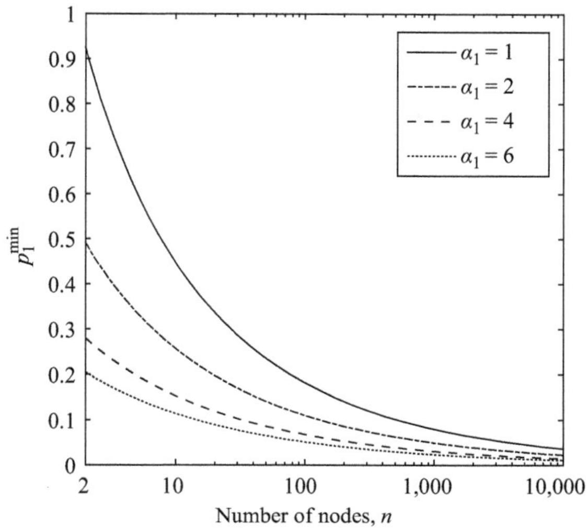

Figure 4.5 The optimal probability of CH election across the different size-scales of a single-level network. Here $\rho = 0.765$ (processing center located at the center of a square)

Corollary 4.2. *When $\rho^2\alpha_2^2 n > \alpha_1^2(2\alpha_1 - 1)$ in a two-level hierarchical cluster, the optimal probability of level-2 CH election is given by the square of the real-valued solution x^* of $4\alpha_2^2 d^2 x^7 + \alpha_1^2 x^5 - 2\alpha_1^3 x^4 + 2\alpha_1^2 x^3 - 4\alpha_1^3 x^2 + \alpha_1^2 x - 2\alpha_1^3 = 0$ and the optimal probability of level-1 CH election is given by $\alpha_1^2\left(x^{*2} + 1\right)^2/4\rho^2\alpha_2^2 n x^{*6}$.*

So far, we have studied how to optimally cluster nodes with a distributed algorithm. In the rest of the chapter, we focus on how to gather data efficiently within a single-hop cluster of nodes for detection/estimation applications.

4.2 Histogram-frame-based random access

In addition to energy-efficient clustering, we will study, design, and evaluate *collision-aware* distributed detection/estimation when collisions are useful and exploited in making final decisions/estimates. A key innovation is that—with a properly revised transmission strategy and a data collection strategy in which packet collisions will always involve packets carrying the same data—a packet collision indicates that two or more sensor nodes have the same data. As a result, the number of packet collisions can be used in making a final decision/estimation. Generally, the design of collision-aware distributed detection/estimation consists of two steps:

- First, we design a transmission protocol such that collisions are informative. Specifically, we design a transmission protocol based on a population splitting algorithm called *histogram-frame-based random access* (HFRA), which is

a combination of a histogram-based transmission strategy and a frame-based random-access protocol. We assume throughout the chapter that the sensor nodes use only the HFRA to send their data to the FC. This idea can be applied in a similar way to design other collision-aware transmission protocols.
- Then, we derive corresponding *fusion rules* for distributed detection and estimation tasks where packet collisions are exploited in making a final decision or estimate. Fusion rules in the following scenarios are derived: distributed detection in a simple hypothesis testing problem, distributed detection in a composite hypothesis testing problem, and distributed estimation.

In traditional data networks [24] and conventional WSNs using random access [9,10,25], a network traffic load that maximizes the average data throughput will optimize the performances of these systems. In the collision-aware distributed detection/estimation, on the other hand, we will show that a network traffic load causing frequent packet collisions will optimize the performance of collision-aware schemes.

We study distributed detection and estimation tasks being performed by a large number, N, of sensor nodes that are organized into a one-hop cluster. A collection period consisting of $T < N$ time slots is provided for the CH to gather data, over a single channel, from the nodes in the cluster. If two or more sensor nodes send data packets in the same time slot, a collision occurs and the FC cannot decode anything transmitted in this time slot. Since T is less than N, the FC can collect some local data but not all of it. For example, in distributed detection/estimation using a transmission strategy based on time division multiple access (TDMA), the FC will be able to collect at most T packets of local data. Therefore, the collection period of T slots will limit the performance of the distributed detection/estimation algorithm.

Given the above situation, we focus on jointly designing a transmission strategy, a multiple access protocol, and fusion rules at the FC to improve the performance of distributed detection and estimation algorithms.

As an example, we design a transmission protocol called HFRA, which is a combination of the *histogram* of the observations and a *frame*-based random access protocol. We divide sensor nodes' observations into M levels (i.e., the number of sensor nodes in each level represents the histogram of the observations) and the collection period of T slots into M frames. The sensor nodes whose observations are in the mth level will send their data in the mth frame with a *transmission probability* ρ_m (keeping silent with the probability $1 - \rho_m$). Since the frame itself indicates the observation level (i.e., a packet found in the mth frame corresponds to an observation in the mth level), we assume that every sensor node transmitting in this frame sends a data bit 1.

In addition, because we do not know which sensor nodes will send their data bits in the mth frame, a frame-slotted ALOHA protocol is used. The sensor nodes that are to send their data bits in the mth frame will randomly choose a time slot in that frame to send their bits. By using a frame-slotted ALOHA protocol, collision time slots, when two or more sensor nodes send their data bits at the same time slot, will occur. However, unlike [9–11,25], in the subsequent sections, we design *collision-aware* distributed detection/estimation whose FC is aware of collisions and utilize them in making a final decision or estimate.

This section is organized as follows. First, we introduce our system model in Section 4.2.1. The HFRA protocol is described in Section 4.2.2. Its corresponding mathematical model is shown subsequently in Section 4.2.3.

4.2.1 System model

We consider WSNs with the following assumptions.

- *Centralized fusion system:* There are N sensor nodes deployed in an area to monitor an event of interest. The FC will broadcast an inquiry about the existence of this event in the monitored area to start the local-decision collection process. Each sensor node will make an observation, compute a data bit, and send it to the FC via a single-hop wireless channel according to the transmission protocol proposed in Section 4.2.2.
- *Transmission channel:* We assume that the sensor nodes share a transmission channel when sending their data bits to the FC. The channel is divided into time slots, with the FC and sensor nodes knowing when a time slot begins and ends (i.e., synchronous time). A collision-channel model is assumed, i.e., a data bit will be successfully sent to the FC in a time slot if it is the only one transmitted in that slot; otherwise, the slot is idle or a collision occurs. We assume that the collisions are solely from the transmissions of the nodes in the considered network. The length of each time slot is equal to the packet containing a data bit.
- *Observation model:* We assume that the noisy observation at a sensor node, x, is governed by the following binary hypothesis model:

  $$H_0 : x \sim f_X(x|H_0) \quad \text{and} \quad H_1 : x \sim f_X(x|H_1)$$

  where $f_X(x|H_i)$ is the conditional probability density function (PDF) of x. The observations are assumed to be *independent and identically distributed* (IID) given H_i, for $i = 0, 1$, and among the sensor nodes and time slots. The prior probability that H_0 happens, $\Pr(H_0)$, is equal to P_0, and, the prior probability that H_1 happens, $\Pr(H_1)$, is equal to $P_1 = 1 - P_0$.

4.2.2 Protocol description

The HFRA protocol is described below.

Histogram-frame-based random access

The collection time T is divided into M frames, where each frame consists of K time slots, as shown in Figure 4.6. Given a set of thresholds $\boldsymbol{\tau} = (\tau_0, \tau_1, \ldots, \tau_M)$, where $\tau_0 = -\infty$ and $\tau_M = \infty$, the sensor nodes perform the following steps:

1. At the beginning of the mth frame, each sensor draws a new observation x.

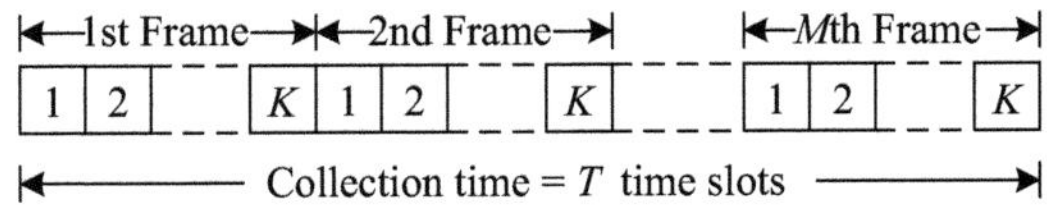

Figure 4.6 A collection period of T slots is divided into M frames. Each frame consists of K slots

2. Each sensor node will make a data bit b, which is obtained from $b = \mathbb{1}_{\{x \in (\tau_{m-1}, \tau_m]\}}$, where $\mathbb{1}_{\{\cdot\}}$ is the indicator function. Specifically, the data bit b will be equal to 1 if $x \in (\tau_{m-1}, \tau_m]$ and 0 otherwise.
3. The sensor nodes with the data bit $b = 1$ will decide to send their data bits in this frame (specifically, the mth frame) with a transmission probability ρ_m (i.e., $0 \leq \rho_m \leq 1$).
4. The sensor nodes deciding to send their data bits will choose a time slot in the mth frame to send their data bits with a probability $1/K$. The other sensor nodes keep silent.

The steps above are repeated from $m = 1$ until $m = M$.

We have the following remarks on the design of the HFRA:

- Each sensor node first quantizes its observation x into one of M levels by using a given set of thresholds $\boldsymbol{\tau}$. Since each frame will be used to indicate a quantization level, a data bit sent in the mth frame is enough to represent which quantization level the observation is in.
- Let n_m be the number of sensor nodes whose observation $x \in (\tau_{m-1}, \tau_m]$. Since each sensor node will draw a new observation at the beginning of the mth frame, the random variables $n_1, n_2, \ldots, n_M$ are independent.
- The number and identity of sensor nodes that have observations $x \in (\tau_{m-1}, \tau_m]$ are unknown. A *frame-slotted ALOHA* protocol is used to handle the resulting multiple access problem. On average, a sensor whose observation $x \in (\tau_{m-1}, \tau_m]$ will send its data bit at a time slot in the mth frame with the probability ρ_m/K.
- Consequently, each time slot can be classified as having one of the following *states*: *idle*, no data bit is sent in the slot; *successful*, exactly one data bit is sent in the slot; or, *collision*, when two or more data bits are sent in the time slot).
- Recall that we design the HFRA protocol such that only the sensor nodes with observations $x \in (\tau_{m-1}, \tau_m]$ send their data bits (i.e., $b = 1$) in the mth frame. Similar to [12–14], a collision time slot is meaningful and recognized. Specifically, a collision time slot indicates that two or more sensor nodes had observations $x \in (\tau_{m-1}, \tau_m]$. Therefore, we will devise a distributed detection algorithm that is *aware* of these collision time slots as well as the successful and idle time slots. The FC will exploit the number of each type of time-slot state in making a final decision.

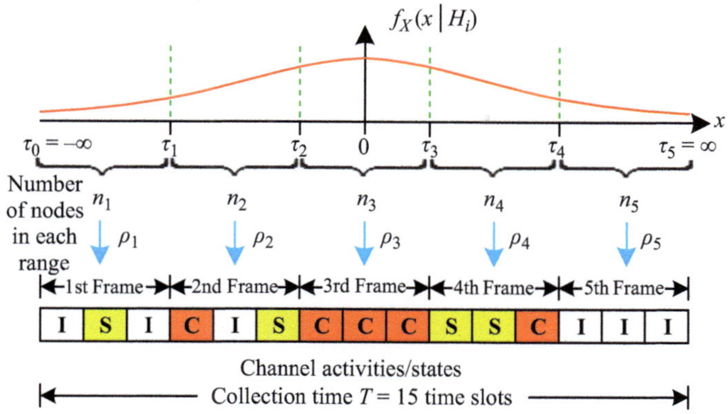

Figure 4.7 An example illustrates the idea of the HFRA protocol when $T = 15$, $M = 5$, $K = 3$, and $\boldsymbol{\tau} = (\tau_0, \tau_1, \ldots, \tau_5)$. Assume that there are n_m sensor nodes with $x \in (\tau_{m-1}, \tau_m]$, for $m = 1, 2, \ldots, 5$. They will send their data bits in the mth frame. If these nodes decide to send their data bits (with the probability ρ_m), they each randomly choose a time slot in which to send their bits. As a result, idle time slots (I), successful time slots (S), and collision time slots (C) may all occur.

- The transmission probability ρ_m is a key parameter since it controls the number of transmitting sensor nodes and the network traffic. As a result, the number of idle, successful, and collision time slots depend on ρ_m. A large value of ρ_m induces a large number of transmitting nodes and a large number of collision time slots. There will be a transmission probability ρ_m^* maximizing the performance of the corresponding distributed detection/estimation.
- Note that we are not yet interested in finding a set of optimal thresholds $\boldsymbol{\tau}^* = (\tau_0^*, \tau_1^*, \ldots, \tau_M^*)$, which is known to be a difficult problem [11,26–28].

An example showing the main idea of the HFRA is provided in Figure 4.7. Assume that the event H_i is happening. The distribution of observations is $f_X(x|H_i)$. Given a set of thresholds $\boldsymbol{\tau}$, there are n_m sensor nodes whose observations $x \in (\tau_{m-1}, \tau_m]$. According to the HFRA protocol, these sensor nodes will decide to send their data bits in the *m*th frame with the probability ρ_m by using frame-slotted ALOHA. As a result, the FC observes the *channel states* $\mathbf{z}_m = (z_{0,m}, z_{s,m}, z_{c,m})$ in the *m*th frame, where $z_{0,m}$, $z_{s,m}$, and $z_{c,m}$ are the number of idle, successful, and collision time slots, respectively. The FC will exploit these channel states $\mathbf{z}_m$, for $m = 1, 2, \ldots, M$, in making a final decision. Note that if a parallel access channel (PACs)[1] is assumed, the FC would clearly see the number of nodes n_m, for $m = 1, 2, \ldots, M$, and would directly use them in making a final decision instead.

[1]An example of a PACs is frequency division multiple access. Each sensor node exploits its own frequency channel to send the decision. However, using the PAC is not a bandwidth-efficient approach in this scenario.

4.2.3 Mathematical model

The mathematical model of distributed detection using the HFRA protocol can be explained as follows. Consider the mth frame. The number of sensor nodes whose observations $x \in (\tau_{m-1}, \tau_m]$, denoted by n_m, can be modeled as a binomial random variable. The probability mass function (PMF) of n_m given H_i can be expressed as:

$$\Pr(n_m|H_i) = \binom{N}{n_m} q_{m|i}^{n_m}(1 - q_{m|i})^{N-n_m} \tag{4.10}$$

where $0 \leq n_m \leq N$ and $q_{m|i} = \int_{\tau_{m-1}}^{\tau_m} f_X(x|H_i)\, dx$. On average, these n_m sensor nodes will send their data bits in a time slot with the probability ρ_m/K.

At the end of each time slot, the FC observes the state of that time slot. Let $d_{k,m}$ be the *time-slot state* of the kth time slot in the mth frame. We have $d_{k,m} \in \{0, s, c\}$, where $d_{k,m} = 0$, $d_{k,m} = s$, and $d_{k,m} = c$ indicate the idle time slot, the successful time slot, and the collision time slot, respectively. Therefore, a time slot will be an idle, successful, or collision time slot with the following conditional probabilities:

$$\Pr(d_{k,m} = 0|n_m) = p_{0,m} = \left(1 - \frac{\rho_m}{K}\right)^{n_m} \tag{4.11}$$

$$\Pr(d_{k,m} = s|n_m) = p_{s,m} = n_m\left(\frac{\rho_m}{K}\right)\left(1 - \frac{\rho_m}{K}\right)^{n_m-1} \tag{4.12}$$

$$\Pr(d_{k,m} = c|n_m) = p_{c,m} = 1 - p_{0,m} - p_{s,m} \tag{4.13}$$

The probabilities $p_{0,m}$, $p_{s,m}$, and $p_{c,m}$ are called the probability of no transmission, the probability of successful transmission, and the probability of a collision, respectively.

At the end of the mth frame, the FC observes the following time-slot states $d_{1,m}, d_{2,m}, \ldots, d_{K,m}$. Since $d_{k,m} \in \{0, s, c\}$, the joint conditional PMF that the FC will obtain $z_{0,m}$ time slots whose $d_{k,m} = 0$, $z_{s,m}$ time slots whose $d_{k,m} = s$, and $z_{c,m}$ time slots whose $d_{k,m} = c$ given n_m is[2]:

$$\Pr(\mathbf{z}_m|n_m) = \frac{K!}{z_{0,m}!z_{s,m}!z_{c,m}!} p_{0,m}^{z_{0,m}} p_{s,m}^{z_{s,m}} p_{c,m}^{z_{c,m}} \tag{4.14}$$

where $\mathbf{z}_m = (z_{0,m}, z_{s,m}, z_{c,m})$. Note that $z_{0,m} + z_{s,m} + z_{c,m} = K$. In addition, we can write the joint conditional PMF of $z_{0,m}$, $z_{s,m}$, and $z_{c,m}$ given H_i as:

$$\Pr(\mathbf{z}_m|H_i) = \sum_{n_m=0}^{N} \Pr(\mathbf{z}_m|n_m)\Pr(n_m|H_i) \tag{4.15}$$

[2]Recall that $z_{0,m}$, $z_{s,m}$, and $z_{c,m}$ are the numbers of idle time slots, successful time slots, and collision time slots in the mth frame, respectively.

At the end of the collection time, the FC has observed a set of the channel states $\mathbf{z}_1, \mathbf{z}_2, \ldots, \mathbf{z}_M$. The joint conditional PMF of $\mathbf{z} = (\mathbf{z}_1, \mathbf{z}_2, \ldots, \mathbf{z}_M)$ given H_i can be expressed as:

$$\Pr(\mathbf{z}|H_i) = \prod_{m=1}^{M} \Pr(\mathbf{z}_m|H_i) = \prod_{m=1}^{M} \sum_{n_m=0}^{N} \Pr(\mathbf{z}_m|n_m)\Pr(n_m|H_i) \tag{4.16}$$

Thereafter, the FC exploits the channel states $\mathbf{z}$ to decide whether H_0 or H_1 is happening.

4.3 Collision-aware fusion rule for distributed detection in a simple binary hypothesis testing problem

The optimal fusion rule for distributed detection using the HFRA can be derived directly from the Bayesian approach [29]:

$$\sum_{m=1}^{M} \log \left[\frac{\sum_{n_m=0}^{N} \Pr(\mathbf{z}_m|n_m)\Pr(n_m|H_1)}{\sum_{n_m=0}^{N} \Pr(\mathbf{z}_m|n_m)\Pr(n_m|H_0)} \right] \underset{H_0}{\overset{H_1}{\gtrless}} \log \left(\frac{P_0}{P_1} \right)$$

Unfortunately, an analytical derivation of the corresponding optimal fusion rule appears intractable, so no insight can be discovered or proved via that approach. We thus aim to devise a suboptimal distributed detection algorithm that does allow us to derive meaningful properties.

A problem when using a random access protocol is that the number of sensor nodes in each level, n_m, is hidden beneath the channel states. A simple fusion rule can be derived if we know the numbers of sensor nodes $n = (n_1, n_2, \ldots, n_M)$. Therefore, we propose a *collision-aware* distributed detection[3] when using the HFRA as shown in Figure 4.8. The proposed distributed detection algorithm is structured as a

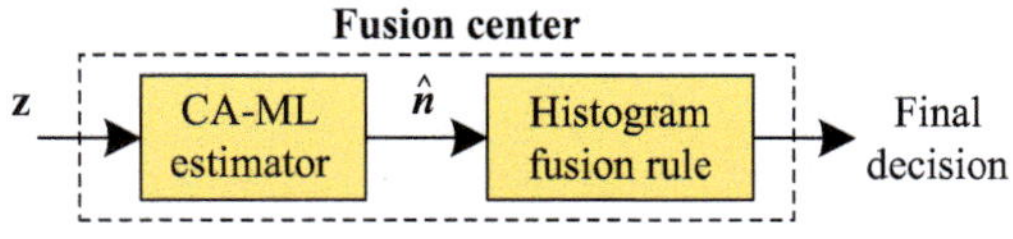

Figure 4.8 The structure of the proposed collision-aware distributed detection scheme. Stage 1: A collision-aware maximum likelihood estimator (CA-MLE). Stage 2: A histogram-based fusion rule, where $\mathbf{z}$ *is a set of channel states and* $\hat{\mathbf{n}}$ *is the set of estimated numbers of transmitting nodes*

[3]We call the scheme as a *collision-aware* distributed detection algorithm since, unlike many other random-access-based schemes, the FC will also exploit the collision time slots in making final decisions.

joint estimation/detection scheme[4] that consists of two stages: a collision-aware maximum-likelihood (CA-ML) estimator (CA-MLE) of the node numbers followed by a histogram-based fusion rule. First, the FC will use the CA-MLE to compute estimates of $n_1, n_2, \ldots, n_M$, which are denoted as $\hat{n}_1, \hat{n}_2, \ldots, \hat{n}_M$, respectively, from the observed channel states $\mathbf{z} = (\mathbf{z}_1, \mathbf{z}_2, \ldots, \mathbf{z}_M)$. Thereafter, the FC exploits these estimates $\hat{\mathbf{n}} = (\hat{n}_1, \hat{n}_2, \ldots, \hat{n}_M)$ in the proposed histogram-based fusion rule. The CA-ML estimator and the histogram-based fusion rule are described subsequently.

We first estimate the number of sensor nodes n_m in each frame by exploiting the CA-MLE. The details are shown in the proposition below. From now on we omit the fact that n_m is an integer and thus consider n_m to be real valued. We can derive and prove some useful properties.

Proposition 4.1 (CA-MLE of n_m). *The CA-ML estimate of the number of sensor nodes whose $x \in (\tau_m, \tau_{m+1}]$ based on the observed channel states $\mathbf{z}_m$ is:*

$$\hat{n}_m = \underset{n_m \in [0, N]}{\arg\max} \left[z_{0,m} \log (p_{0,m}) + z_{s,m} \log (p_{s,m}) + z_{c,m} \log (p_{c,m}) \right] \tag{4.17}$$

The asymptotic PDF of $\hat{n}_m$ can be expressed as a Gaussian PDF: $\hat{n}_m \overset{a}{\sim} \mathcal{N}(n_m, \upsilon_m^2)$, where $\overset{a}{\sim}$ denotes "asymptotically distributed according to" and, for a large n_m and $n_m \gg K$,

$$\upsilon_m^2 \approx \left\{ K \left(\frac{1 - p_{c,m}}{p_{c,m}} \right) \left[\log \left(1 - \frac{\rho_m}{K} \right) \right]^2 \right\}^{-1} \tag{4.18}$$

Proof. The MLE of the variable n_m is obtained from $\hat{n}_m = \underset{n_m \in [0, N]}{\arg\max} \log \Pr(\mathbf{z}_m | n_m)$, where $\Pr(\mathbf{z}_m | n_m)$ is shown in (4.14). Since $\log (K!/z_{0,m}! z_{1,m}! z_{c,m}!)$ is not a function of n_m, there is no need to include this term in (4.17). The variance υ_m^2 is equal to the Cramer–Rao lower bound (CRLB) of the estimate $\hat{n}_m$. To compute υ_m^2, we omit the fact that n_m is an integer and, then, consider n_m as a real number. The variance υ_m^2 is obtained from $\upsilon_m^2 = (1/(-\mathbb{E}\left\{(\partial^2/\partial n_m^2) \log \Pr(\mathbf{z}_m | n_m)\right\}))$. For a large n_m and $n_m \gg K$, the variance υ^2 can be approximated as shown in (4.18). □

Based on the fact that $\log \Pr(\mathbf{z}_m | n_m)$ is a concave function of n_m, we can alternatively find the estimate $\hat{n}_m$ as shown in Corollary 4.3 below.

Corollary 4.3 (Equivalent CA-ML estimate of n_m). *The CA-ML estimate $\hat{n}_m$ obtained from* (4.17) *is equal to the value n_m satisfying:*

$$\frac{1 - p_{c,m}}{p_{c,m}} = \frac{K - z_{c,m}}{z_{c,m}} \tag{4.19}$$

Note that n_m is inside $p_{c,m}$.

[4]Joint estimation/detection has been used as suboptimal distributed detection in many papers [30,31].

Proof. We assume that n_m is real-valued. We can show that $(\partial^2/\partial n_m^2)\log\Pr(\mathbf{z}_m|n_m) \leq 0$, which means the term $\log\Pr(\mathbf{z}_m|n_m)$ is a concave function of n_m. Therefore, the CA-ML estimate $\hat{n}_m$ is the value n_m satisfying $(\partial/\partial n_m)\log\Pr(\mathbf{z}_m|n_m) = 0$, which is equivalent to (4.19). □

Interestingly, Corollary 4.3 shows that only the number of collision time slots $z_{c,m}$ is needed to estimate n_m. As a result, the input to the diagram of the proposed scheme, shown in Figure 4.8, can be the number of collision time slots in each frame $\mathbf{z}_c = (z_{c,1}, z_{c,2}, \ldots, z_{c,M})$ instead of $\mathbf{z}$.

After obtaining the estimates $\hat{\boldsymbol{n}}$ from the CA-MLE, the FC will use the histogram-based fusion rule to compute a final decision. The histogram fusion rule is described in the proposition below.

Proposition 4.2 (Histogram-based fusion rule given $\hat{\boldsymbol{n}}$). *Given a set of estimated numbers of sensor nodes, $\hat{\mathbf{n}} = (\hat{n}_1, \hat{n}_2, \ldots, \hat{n}_M)$. The histogram-based fusion rule is expressed as:*

$$T_H = \sum_{m=1}^{M} \hat{n}_m \left[\log\left(\frac{1-q_{m|0}}{q_{m|0}}\right) - \log\left(\frac{1-q_{m|1}}{q_{m|1}}\right)\right] \underset{H_0}{\overset{H_1}{\gtrless}} \gamma_H \tag{4.20}$$

Proof. The test statistic T_H is directly obtained from the following log-likelihood ratio: $T_H = \Pr(\hat{n}_1, \hat{n}_2, \ldots, \hat{n}_M|H_1)/\Pr(\hat{n}_1, \hat{n}_2, \ldots, \hat{n}_M|H_0)$. Since the random variables $\hat{n}_1$, $\hat{n}_2$, ..., $\hat{n}_M$ are independent, we have $\Pr(\hat{n}_1, \hat{n}_2, \ldots, \hat{n}_M|H_i) = \Pr(\hat{n}_1|H_i)\Pr(\hat{n}_2|H_i)\cdots\Pr(\hat{n}_M|H_i)$. From (4.10) and after some mathematical arrangement, we can write T_H as shown in (4.20). □

Now, we will study the characteristics of and evaluate the performance of the proposed collision-aware distributed detection algorithm. Throughout this study, we assume that the observation x is governed by the following model:

$$H_0 : x \sim \mathcal{N}(0, 5) \quad \text{and} \quad H_1 : x \sim \mathcal{N}(1, 5). \tag{4.21}$$

Note that the observation signal-to-noise ratio is equal to −7 dB.

First, we show the effect of the transmission probabilities ρ_m on the probability of detection (P_D) of the proposed scheme and the network traffic in Figure 4.9. The other parameters are set as shown in the caption of the figure. In addition, we plot the corresponding network traffic measured as probabilities $p_{0,m}$, $p_{s,m}$, and $p_{c,m}$, which are the probabilities of no transmission, successful transmission, and collisions, respectively. We vary ρ_m in a way that $\rho_1 = \rho_2 = \cdots = \rho_M$ (i.e., they are the same). Since a set of thresholds $\boldsymbol{\tau}$ is chosen such that $q_{m|1} = 1/M$, for all m, we have $p_{0,1} = p_{0,2} = \cdots = p_{0,M}$ and $p_{s,1} = p_{s,2} = \cdots = p_{s,M}$ and $p_{c,1} = p_{c,2} = \cdots = p_{c,M}$. We observe that there is a transmission probability $\rho_m^* = 0.45$ maximizing P_D. It is interesting that the optimal transmission probability ρ_m^* for the collision-aware distributed detection does not maximize the network throughput (equivalently, maximize the probability of successful transmission $p_{s,m}$).

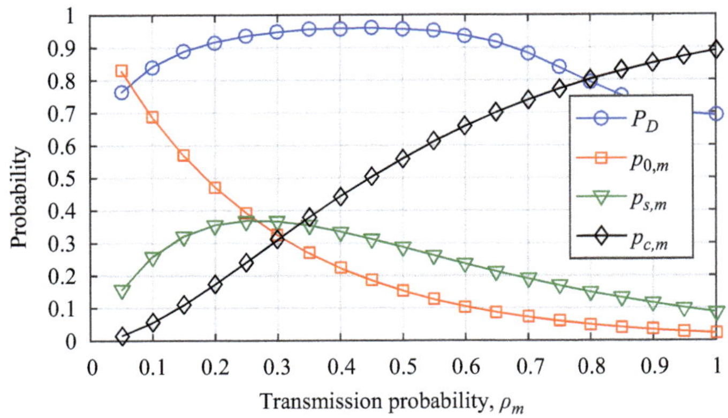

Figure 4.9 Collision-aware distributed detection. A plot of the probabilities P_D, $p_{0,m}$, $p_{s,m}$, and $p_{c,m}$ versus the transmission probability ρ_m when $N = 300$, $T = 80$, $M = 8$, and $K = 10$. A set of thresholds $\boldsymbol{\tau} = (\tau_0, \tau_1, \ldots, \tau_M)$ is chosen such that $q_{m|1} = 1/M$, for all m

Second, in Figure 4.10, we show the optimal transmission probability ρ_m^* (maximizing P_D) and its corresponding probabilities $p_{0,m}^*$, $p_{s,m}^*$, and $p_{c,m}^*$ versus N. We see that, in the proposed distributed detection, the probability ρ_m^* decreases as N increases to *preserve* $p_{0,m}^* \approx 0.21$, $p_{s,m}^* \approx 0.33$, and $p_{c,m}^* \approx 0.46$. The results have confirmed that the proposed distributed detection favors a transmission probability ρ_m that causes *higher collision rates* compared to that in a data network [24] or conventional distributed detection using slotted ALOHA [9–11,24,25]. The reason is that, unlike these networks where collisions are non-informative and thus neglected, in the proposed distributed detection, the collision time slots can be exploited in making a final decision.

Lastly, we compare the probability of error P_E of the proposed distributed detection to those of other schemes in Figure 4.11. The schemes that we compare with are listed as follows.

- *TDMA*: A conventional distributed detection using TDMA is a distributed detection whose sensor nodes will send their binary decisions to the FC by using a TDMA protocol.[5] Since the collection time is fixed to T time slots, there will be T sensor nodes that send their binary decisions to the FC.
- *HFRA-S*: The proposed scheme with an assumption that we use the number of successful time slots (in the mth frame) instead of $\hat{n}_m$ in (4.20). We show this one to see how much the performance has improved in the proposed scheme which also exploits the collision time slots.

[5]In a TDMA protocol, each node will be assigned an exclusive time slot to send its decision.

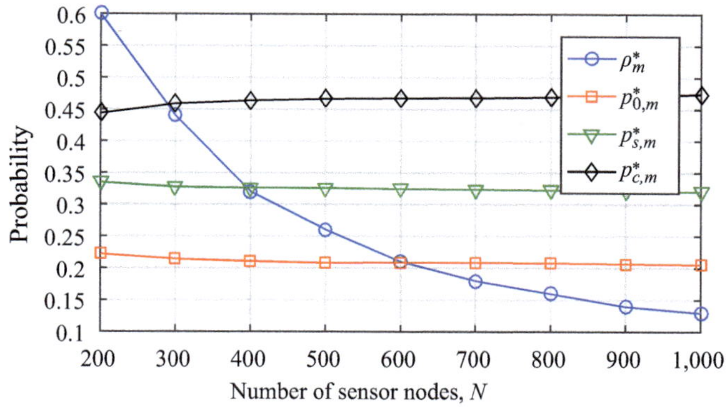

Figure 4.10 *Collision-aware distributed detection. A plot of the optimal transmission probability ρ^*_m (maximizing P_D) and its corresponding probabilities $p^*_{0,m}$, $p^*_{s,m}$, and $p^*_{c,m}$ versus N when $T = 80$, $M = 8$, and $K = 10$. A set of thresholds $\boldsymbol{\tau} = (\tau_0, \tau_1, \ldots, \tau_M)$ is chosen such that $q_{m|1} = 1/M$, for all m*

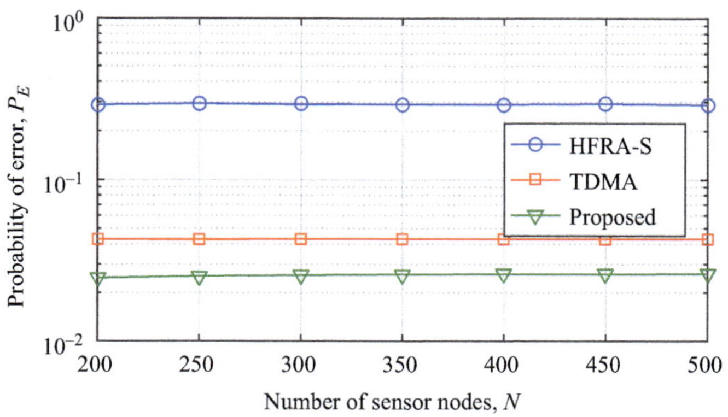

Figure 4.11 *Collision-aware distributed detection. We compare the probability of error P_E of the collision-aware distributed detection to those of other schemes: TDMA and HFRA-S. The parameters are $T = 80$, $M = 8$, $K = 10$, and $P_0 = P_1 = 1/2$. The thresholds $\boldsymbol{\tau} = (\tau_0, \tau_1, \ldots, \tau_M)$ are chosen such that $(q_{m|0}P_0 + q_{m|1}P_1) = 1/M$, for all m*

The probabilities of error P_E are plotted versus N. The other parameters are set as shown in the caption of the figure. As shown in Figure 4.11, we see that the probability P_E of the proposed scheme is better than those of TDMA and HFRA-S around two times and ten times, respectively.

4.4 Collision-aware fusion rule for distributed detection in a composite hypothesis testing problem

In the previous section, the distributions of the observations taken by the sensor nodes under both hypotheses (H_0 and H_1) are assumed to be known. The distribution of the observations under H_0 (i.e., the event does not happen) can be acquired by applying a learning method when WSNs are initially deployed in the area. Unfortunately, finding the distribution of the observations under H_1 (i.e., the event is happening) is problematic since this distribution would typically depend on, at least, the location and the strength of the event. Therefore, distributed detection in composite hypothesis testing problems [29] is also of interest [28,32–35].

According to the details in Section 4.2, we can cast our detection problem as a composite hypothesis testing problem with an unknown parameter as follows:

$$H_0 : n_m \sim \binom{N}{n_m} q_{m|0}^{n_m}(1-q_{m|0})^{N-n_m} \quad \text{and} \quad H_1 : n_m \sim \binom{N}{n_m} q_{m|1}^{n_m}(1-q_{m|1})^{N-n_m}$$

where $q_{m|0} \neq q_{m|1}$ for $m = 1, 2, \ldots, M$. Recall that $q_{m|i} = \int_{\tau_{m-1}}^{\tau_m} f_X(x|H_i)\, dx$ is the probability that a sensor node will have the observation $x \in (\tau_{m-1}, \tau_m]$ given H_i. Let $\mathbf{q}_i$ is a set of the conditional PMFs $q_{m|i}$ given H_i, i.e., $\mathbf{q}_i = (q_{1|i}, q_{2|i}, \ldots, q_{M|i})$ and $\mathbf{q}_0 \neq \mathbf{q}_1$. Since $f_X(x|H_0)$ is known, the PMF vector $\mathbf{q}_0$ is known while the PMF vector $\mathbf{q}_1$ is unknown. Note that the model here is similar to those in [36,37] except that, in [36,37], the PMF vector $\mathbf{q}_1$ is known.

Many tests have been proposed and studied for composite hypothesis testing problems [28,29,32–35]. Here, we are interested in designing a Rao test [29] for distributed detection using the HFRA protocol. A reason is that, unlike a general likelihood ratio test and a Wald test, a Rao test does not require maximum likelihood estimates of the unknown parameters (i.e., $\mathbf{q}_1$). Since this test is aware of collisions, it is called the *collision-aware* Rao test (CA-Rao test), which is shown below.

Proposition 4.3 (CA-Rao test). *At the end of the collection time, the FC has observed the channel states* $\mathbf{z} = (\mathbf{z}_1, \mathbf{z}_2, \ldots, \mathbf{z}_M)$*, where* $\mathbf{z}_m = (z_{0,m}, z_{s,m}, z_{c,m})$*. The CA-Rao test* T_R *used by the FC to decide whether* H_0 *or* H_1 *is happening can be expressed as:*

$$T_R = \sum_{m=1}^{M} \left[\frac{\left(\mathbb{E}\{n_m|\mathbf{z}_m, H_0\} - Nq_{m|0}\right)^2}{\mathbb{V}ar\left\{\mathbb{E}\{n_m|\mathbf{z}_m, H_0\}\middle|H_0\right\}} \right] \underset{H_0}{\overset{H_1}{\gtrless}} \gamma_R \tag{4.22}$$

where $\mathbb{E}\{n_m|\mathbf{z}_m, H_0\}$ *is a conditional expectation of* n_m*,* $\mathbb{V}ar\{\mathbb{E}\{n_m|\mathbf{z}_m, H_0\}\}$ *is the variance of* $\mathbb{E}\{n_m|\mathbf{z}_m, H_0\}$*, and* γ_R *is a decision threshold.*

Proof. The Rao test is derived from [29]:

$$T_R = \frac{\partial}{\partial \mathbf{q}_i} \log \Pr(\mathbf{z}|H_i)\Big|_{\mathbf{q}_i=\mathbf{q}_0}^{T} \mathbf{I}^{-1}(\mathbf{q}_0) \frac{\partial}{\partial \mathbf{q}_i} \log \Pr(\mathbf{z}|H_i)\Big|_{\mathbf{q}_i=\mathbf{q}_0} \tag{4.23}$$

where $(\cdot)^T$ is the transpose operator, $\mathbf{I}(\mathbf{q}_0)$ is the $M \times M$ Fisher information matrix (FIM), and $\mathbf{I}^{-1}(\mathbf{q}_0)$ is the inverse of the FIM. By following similar steps shown in [14], we will get (4.22). □

According to [29], the CA-Rao test (4.22) has the following asymptotic distribution:

$$T_R \overset{a}{\sim} \begin{cases} \mathscr{X}_M^2 & \text{under } H_0 \\ \mathscr{X}_M'^2(\lambda) & \text{under } H_1 \end{cases} \tag{4.24}$$

where "a" denotes an asymptotic distribution, $\mathscr{X}_M^2$ denotes a chi-square PDF with M degree of freedom, and $\mathscr{X}_M'^2(\lambda)$ denotes a noncentral chi-square PDF with M degrees of freedom and noncentrality parameter λ defined as [29]:

$$\begin{aligned} \lambda &= (\mathbf{q}_1 - \mathbf{q}_0)^T \mathbf{I}(\mathbf{q}_0)(\mathbf{q}_1 - \mathbf{q}_0) \\ &= \sum_{m=1}^{M} \left[\frac{(q_{m|1} - q_{m|0})^2 \mathbb{V}ar\{\mathbb{E}\{n_m|\mathbf{z}_m, H_0\}|H_0\}}{q_{m|0}^2(1 - q_{m|0})^2} \right] \end{aligned} \tag{4.25}$$

Note that the value of λ depends on the unknown $\mathbf{q}_1$. Asymptotic performance of the CA-Rao test can be measured as a probability of detection $P_D = \Pr(T_R > \gamma_R|H_1)$ and a probability of false alarm $P_F = \Pr(T_R > \gamma_R|H_0)$.

Now, we will study characteristics and evaluate the performance of the CA-Rao test. The observation model in (4.21) is used. A set of thresholds $\boldsymbol{\tau} = (\tau_0, \tau_1, \ldots, \tau_M)$ is chosen such that $q_{m|0} = 1/M$, for all m. First, we study the effect of the transmission probabilities ρ_m on the probability of detection P_D when the probability of false alarm P_F is fixed to 0.01 in Figure 4.12, where $\bar{p}_0$, $\bar{p}_s$, and $\bar{p}_c$ are the average probabilities of no transmission, successful transmission, and collisions, respectively. Note that we vary ρ_m in a way that $\rho_1 = \rho_2 = \cdots = \rho_M$ (i.e., they are the same). In Figure 4.12, there is a transmission probability $\rho_m^* = 0.65$ maximizing $P_D^* = 0.9949$. In addition, this ρ_m^* induces a high packet collision rate ($\bar{p}_c^* = 0.64$). Second, we plot ρ_m^*, $\bar{p}_0^*$, $\bar{p}_s^*$, and $\bar{p}_c^*$ versus N in Figure 4.13. The transmission probability ρ_m^* is smaller as N increases to *preserve* $p_{0,m}^* \approx 0.13$, $p_{s,m}^* \approx 0.21$ and $p_{c,m}^* \approx 0.66$. Therefore, similar to the collision-aware distributed detection in Section 4.3, the CA-Rao test prefers a transmission probability ρ_m causing a high number of packet collisions.

4.5 Collision-aware fusion rule for distributed estimation

In this section, we study distributed estimation of an unknown parameter when the WSNs is using the HFRA protocol. The observation model has been revised as follows. Each sensor node makes a noisy observation x of an unknown parameter θ, where $x = \theta + y$ and y is a random variable with the PDF $f_Y(y)$. The observation x is assumed to be IID (given θ). The sensor nodes exploit the HFRA protocol described in Section 4.2 to send their data to the FC. The FC observes the channel states $\mathbf{z}$ and applies these

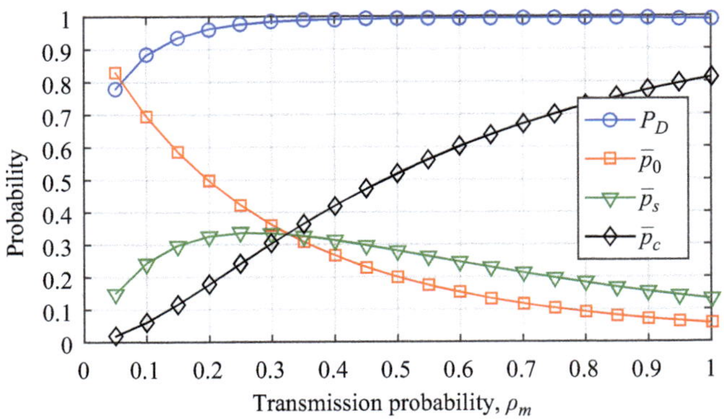

Figure 4.12 Collision-aware Rao test. A plot of the probabilities P_D, $\bar{p}_0$, $\bar{p}_s$, and $\bar{p}_c$ versus the transmission probability ρ_m when $P_F = 0.01$, $N = 300$, $T = 80$, $M = 8$, and $K = 10$. A set of thresholds $\boldsymbol{\tau} = (\tau_0, \tau_1, \ldots, \tau_M)$ are chosen such that $q_{m|0} = 1/M$, for all m. Note that the PMF $\boldsymbol{q}_1$ is unknown

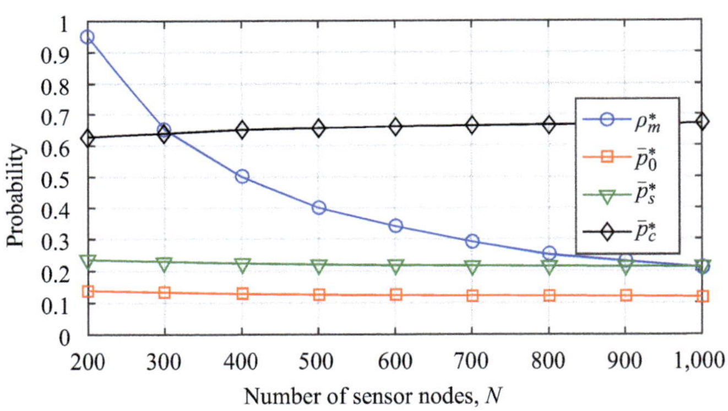

Figure 4.13 Collision-aware Rao test. A plot of the optimal transmission probability ρ_m^* (maximizing P_D) and its corresponding probabilities $\bar{p}_0^*$, $\bar{p}_s^*$, and $\bar{p}_c^*$ versus N when $P_F = 0.01$, $T = 80$, $M = 8$, and $K = 10$. A set of thresholds $\boldsymbol{\tau} = (\tau_0, \tau_1, \ldots, \tau_M)$ are chosen such that $q_{m|0} = 1/M$, for all m. Note that the PMF $\boldsymbol{q}_1$ is unknown

channel states $\mathbf{z}$ to a fusion rule to estimate the parameter θ. The fusion rule to estimate θ is derived from the maximum likelihood estimation approach [38].

In general, an MLE entails high computation complexity except in some special cases [38–40]. However, deriving an MLE is a traditional first step to devising suboptimal estimators and it provides a benchmark of the variance of the estimation error [26,39,41]. Similarly, we derive an MLE whose input is the set of the channel states $\mathbf{z}$ during the collection time. As a result, we call this estimator a *collision-aware* maximum likelihood estimator (CA-MLE). The corresponding CRLB is derived and can be used as a benchmark for comparison with the performance of suboptimal estimators.

Proposition 4.4 (CA-MLE of θ). *The MLE of the parameter θ based on the observed channel states $\mathbf{z} = (\mathbf{z}_1, \mathbf{z}_2, \ldots, \mathbf{z}_M)$, where $\mathbf{z}_m = (z_{0,m}, z_{s,m}, z_{c,m})$, is:*

$$\hat{\theta} = \arg\max_{\theta} \log \Pr(\mathbf{z}) = \arg\max_{\theta} \sum_{m=1}^{M} \log \Pr(\mathbf{z}_m) \tag{4.26}$$

The CRLB for any unbiased estimator $\hat{\theta}$ based on the observed channel states $\mathbf{z}$ is given by:

$$CRLB = \left[\sum_{m=1}^{M} \left[\frac{f_Y(\tau_m - \theta) - f_Y(\tau_{m+1} - \theta)}{q_m(1 - q_m)} \right]^2 \mathbb{V}ar\{\mathbb{E}\{n_m|\mathbf{z}_m\}\} \right]^{-1} \tag{4.27}$$

where $q_m = \int_{\tau_{m-1}}^{\tau_m} f_X(x)\, dx$ and $\mathbb{V}ar\{\mathbb{E}\{n_m|\mathbf{z}_m\}\}$ is the variance of $\mathbb{E}\{n_m|\mathbf{z}_m\}$, which is computed from $\mathbb{E}\{(\mathbb{E}\{n_m|z_m\})^2\} - (Nq_m)^2$.

Proof. The MLE of the parameter θ is directly obtained from (4.16). The CRLB is derived from $\text{CRLB} = 1/(-\sum_{m=1}^{M} \mathbb{E}\left\{\frac{\partial^2}{\partial\theta^2} \log \Pr(\mathbf{z}_m)\right\})$. The derivation of this CRLB is similar to that shown in Appendix A in [15] (but note that the variable x in [15] is the observation noise). □

When considering the CA-MLE shown in (4.26), for each candidate of θ, we need to compute (4.16). This is the reason that the CA-MLE has high computational complexity.

Now, we study the characteristics of and evaluate the performance of the CA-MLE of θ. The unknown parameter θ is assumed to be 1 and the noise y is modeled as $y \sim \mathcal{N}(0, 5)$. A set of thresholds $\boldsymbol{\tau} = (\tau_0, \tau_1, \ldots, \tau_M)$ is chosen such that $\int_{\tau_{m-1}}^{\tau_m} f_Y(y)\, dy = 1/M$, for all m. Note that the PDF of x is unknown since it depends on the parameter θ. First, we study the effect of the transmission probabilities ρ_m on the CRLB in Figure 4.14, where $\bar{p}_0, \bar{p}_s$, and $\bar{p}_c$ are the average probabilities of no transmission, successful transmission, and collisions, respectively. Note that we vary ρ_m in a way that $\rho_1 = \rho_2 = \cdots = \rho_M$ (i.e., they are the same). We see that there is a transmission probability $\rho_m^* = 0.62$ minimizing the CRLB. In addition, this ρ_m^* induces a high number of packet collisions ($\bar{p}_c^* = 0.77$). Second, we plot ρ_m^*, $\bar{p}_0^*$, $\bar{p}_s^*$, and $\bar{p}_c^*$ versus N in Figure 4.15. The transmission probability ρ_m^* is smaller as N increases to *preserve* $p_{0,m}^* \approx 0.1$, $p_{s,m}^* \approx 0.2$, and $p_{c,m}^* \approx 0.7$. Therefore, similar to the collision-aware distributed detection in Section 4.3 and the CA-Rao test in Section 4.4, the CA-MLE prefers a transmission probability ρ_m causing a high number of packet collisions.

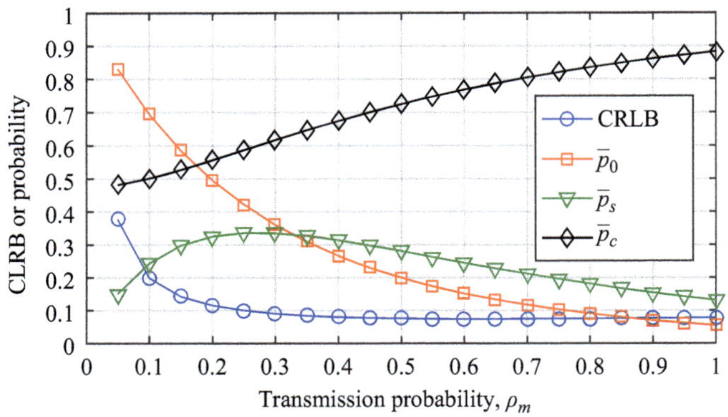

Figure 4.14 *Collision-aware MLE. A plot of CRLB and the probabilities $\bar{p}_0$, $\bar{p}_s$, $\bar{p}_c$ versus the transmission probability ρ_m when N = 300, T = 80, M = 8, and K = 10. A set of thresholds $\boldsymbol{\tau} = (\tau_0, \tau_1, \ldots, \tau_M)$ are chosen such that $\int_{\tau_{m-1}}^{\tau_m} f_Y(y)\, dy = 1/M$, for all m*

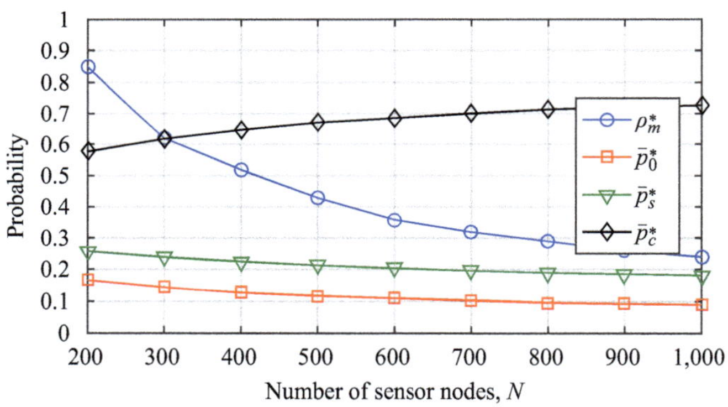

Figure 4.15 *Collision-aware MLE. A plot of the optimal transmission probability ρ_m^* and its corresponding probabilities $\bar{p}_0^*$, $\bar{p}_s^*$, $\bar{p}_c^*$ versus N when T = 80, M = 8, and K = 10. A set of thresholds $\boldsymbol{\tau} = (\tau_0, \tau_1, \ldots, \tau_M)$ are chosen such that $\int_{\tau_{m-1}}^{\tau_m} f_Y(y)\, dy = 1/M$, for all m*

4.6 Conclusions and extensions

We have studied energy-efficient data gathering in high density large-scale wireless networks. We have used a stochastic geometry based ad hoc network communication model to derive the optimality conditions for the design of hierarchical networks. This study shows how the application- and aggregation-dependent data sizes across the levels of hierarchy influence the optimal number of clusters at each level and the number of levels required. The study can be extended to design algorithms for adapting the hierarchical structure depending on the application. Further, we have studied fusion rules for efficient data aggregation within a cluster in detection/estimation applications.

We have studied distributed detection/estimation in a collision-aware framework that is proposed for bandwidth-constrained distributed detection/estimation, where the sensors must share a single collision channel to send their data to the FC. Under this circumstance, properly designing a transmission protocol applied by the sensors to send their data to the FC could make data-packet collisions informative. As an example, a transmission protocol called HFRA has been proposed such that packet collisions are useful to the FC in decision/estimate making. Three fusion rules for detection/estimation are derived and evaluated. The numerical results show that, unlike data networks, the maximum-throughput transmission rate does not optimize the performance of the collision-aware distributed detection/estimation. On the other hand, the optimum performance is achieved at the transmission rate that induces a high packet-collision rate. There are still many problems for further study in the collision-aware framework:

- *Combinations of sensors and FC's detection/estimation:* Many combinations of sensors' tests/estimations and the FC's test/estimation are not yet investigated in the collision-aware framework; for example, distributed detection where both the FC and sensors perform sequential tests.
- *Energy-efficient distributed detection/estimation:* In distributed detection, each sensor must sense the observation and transmit its local data to the FC. Since sensors are battery-powered devices, joint design of sensing and transmitting strategies so that sensors will spend energy efficiently is crucial and interesting.
- *Analytical approaches:* The collision-aware framework is an idea originating from the MAC layer's point of view, where the FC observes the time slot states (idle, successful, and collision). The PMF of the FC's observations (to compute the final decision/estimation) is a multinomial distribution. As a result, finding closed forms of the performance measures, such as detection error probability and estimation's variance, which can be used further in parameter optimizing, is still an open problem.

Finally, extending the detection/estimation framework to include multi-hop clusters is a critical subject for future research.

References

[1] Zhong X, and Coyle EJ. eStadium: a wireless "Living Lab" for safety and infotainment applications. In: Communications and Networking in China, 2006. ChinaCom'06. First International Conference on. IEEE; 2006. p. 1–6.

[2] Ault A, Krogmeier JV, Dunlop SR, *et al.* eStadium: the mobile wireless football experience. In: Internet and Web Applications and Services, 2008. ICIW'08. Third International Conference on. IEEE; 2008. p. 644–9.

[3] Coyle EJ, Allebach JP, and Krueger JG. The vertically-integrated projects (VIP) program in ECE at Purdue: fully integrating undergraduate education and graduate research. In: Proceedings of the 2006 ASEE Annual Conference and Exposition; 2006. p. 18–21.
[4] Glotzbach RJ, Coyle EJ, and Bingham N. e-Stadium: wireless football infotainment applications. In: ACM SIGGRAPH 2005 Web program. ACM; 2005. p. 9.
[5] Phanish D, Garver P, Matalkah G, *et al.* A wireless sensor network for monitoring the structural health of a football stadium. In: Internet of Things (WF-IoT), 2015 IEEE 2nd World Forum on. IEEE; 2015. p. 471–7.
[6] Garver PW, Abler R, and Coyle EJ. Theory and development of cross-layer techniques for localization in environments with Extreme Emitter Densities. In: Military Communications Conference, MILCOM 2015-2015 IEEE. IEEE; 2015. p. 471–6.
[7] Bandyopadhyay S, and Coyle EJ. An energy efficient hierarchical clustering algorithm for wireless sensor networks. In: INFOCOM 2003. Twenty-Second Annual Joint Conference of the IEEE Computer and Communications. IEEE Societies. vol. 3. IEEE; 2003. p. 1713–23.
[8] Phanish D, and Coyle EJ. Application-based optimization of multi-level clustering in ad hoc and sensor networks. IEEE Trans Wireless Commun. 2017;16(7):4460–75.
[9] Xu D, and Yao Y. Contention-based transmission for decentralized detection. IEEE Trans Wireless Commun. 2012;11(4):1334–42.
[10] Chang TY, Hsu TC, and Hong YWP. Exploiting data-dependent transmission control and MAC timing information for distributed detection in sensor networks. IEEE Trans Signal Process. 2010;58(3):1369–82.
[11] Laitrakun S, and Coyle EJ. Reliability-based splitting algorithms for time-constrained distributed detection in random-access channels. IEEE Trans Signal Process. 2014;62(21):5536–51.
[12] Whipps GT, Ertin E, and Moses RL. Distributed detection of binary with collisions in a large, random network. IEEE Trans Signal Process. 2015;63(6):1477–89.
[13] Laitrakun S, and Coyle EJ. Collision-aware sequential distributed detection with sensor censoring in random-access WSNs. In: Proc. ECTI-CON. Thailand; 2015.
[14] Laitrakun S, and Coyle EJ. Collision-aware composite hypothesis testing in random-access WSNs with sensor censoring. In: Proc. ICSEC. Thailand; 2015.
[15] Laitrakun S, and Coyle EJ. Collision-aware distributed estimation in WSNs using sensor-censoring random access. In: Proc. MILCOM. USA; 2015.
[16] Bandyopadhyay S, and Coyle EJ. Minimizing communication costs in hierarchically-clustered networks of wireless sensors. Comput Netw. 2004; 44(1):1–16.
[17] Chair Z, and Varshney PK. Optimal data fusion in multiple sensor detection systems. IEEE Trans Aerosp Electr Syst. 1986;(1):98–101.

[18] Pradhan SS, Kusuma J, and Ramchandran K. Distributed compression in a dense microsensor network. IEEE Signal Process Mag. 2002;19(2): 51–60.

[19] Bass T. Intrusion detection systems and multisensor data fusion. Commun ACM. 2000;43(4):99–105.

[20] Bontempi G, and Le Borgne Y. An adaptive modular approach to the mining of sensor network data. In: Proceedings of 1st International Workshop on Data Mining in Sensor Networks as part of the SIAM International Conference on Data Mining. SIAM Press Newport Beach, CA; 2005. p. 3–9.

[21] Baccelli F, Tchoumatchenko K, and Zuyev S. Markov paths on the Poisson–Delaunay graph with applications to routing in mobile networks. Adv Appl Probab. 2000;32(1):1–18.

[22] Foss S, and Zuyev S. On a Voronoi aggregative process related to a bivariate Poisson process. Adv Appl Probab. 1996;28(1):965–81.

[23] Bandyopadhyay S, Tian Q, and Coyle EJ. Spatio-temporal sampling rates and energy efficiency in wireless sensor networks. IEEE/ACM Trans Netw (TON). 2005;13(6):1339–52.

[24] Bertsekas D, and Gallager R. Data Networks. Englewood Cliffs, NJ, USA: Prentice Hall; 1992.

[25] Marano S, Matta V, Willett P, *et al.* Cross-layer design of sequential detectors in sensor networks. IEEE Trans Signal Process. 2006;54(11):4105–17.

[26] Niu R, and Varshney PK. Target location estimation in sensor networks with quantized data. IEEE Trans Signal Process. 2006;54(12):4519–28.

[27] Vempaty A, He H, Chen B, *et al.* On quantizer design for distributed Bayesian estimation in sensor networks. IEEE Trans Signal Process. 2014;62(20): 5359–69.

[28] Gao F, Lili FG, Li H, *et al.* Quantizer design for distributed GLRT detection of weak signals in wireless sensor networks. IEEE Trans Wireless Commun. 2015;14(4):2032–42.

[29] Kay SM. Fundamentals of Statistical Signal Processing: Detection Theory. Englewood Cliffs, NJ, USA: Prentice Hall; 1993.

[30] Ciuonzo D, Romano G, and Salvo Rossi P. Chanel-aware decision fusion in distributed MIMO wireless sensor networks: decode-and-fuse vs. decode-then-fuse. IEEE Trans Wireless Commun. 2012;11(8):2976–85.

[31] Ciuonzo D, Salvo Rossi P, Dey S. Massive MIMO channel-aware decision fusion. IEEE Trans Signal Process. 2015;63(3):604–19.

[32] Wu JY, Wu CW, Wang TY, *et al.* Channel-aware decision fusion with unknown local sensor detection probability. IEEE Trans Signal Process Lett. 2010;58(3):1457–63.

[33] Ciuonzo D, Papa G, Romano G, *et al.* One-bit decentralized detection with a Rao test for multisensor fusion. IEEE Trans Signal Process Lett. 2013;20(9):861–4.

[34] Ciuonzo D, and Rossi PS. Decision fusion with unknown sensor detection probability. IEEE Trans Signal Process Lett. 2014;21(2):208–12.

[35] Ciuonzo D, De Maio A, and Rossi PS. A system framework for composite hypothesis testing of independent Bernoulli trials. IEEE Trans Signal Process Lett. 2015;22(9):1249–53.
[36] Mergen G, Naware V, and Tong L. Asymptotic detection performance of type-based multiple access over multiaccess fading channels. IEEE Trans Signal Process. 2007;55(3):1081–92.
[37] Anandkumar A, and Tong L. Type-based random access for distributed detection over multiaccess fading channels. IEEE Trans Signal Process. 2007; 55(10):5032–43.
[38] Kay SM. Fundamentals of Statistical Signal Processing: Estimation Theory. Englewood Cliffs, NJ, USA: Prentice Hall; 1993.
[39] Ribeiro A, and Giannakis GB. Bandwidth-constrained distributed estimation for wireless sensor networks – Part I: Gaussian case. IEEE Trans Signal Process. 2006;54(3):1131–43.
[40] Msechu M, and Giannakis GB. Sensor-centric data reduction for estimation with WSNs via censoring and quantization. IEEE Trans Signal Process. 2012;60(1):400–14.
[41] Aysal TC, and Barner KE. Constrained decentralized estimation over noisy channels for sensor networks. IEEE Trans Signal Process. 2008;56(4): 1398–410.

Chapter 5

Channel-aware decision fusion in MIMO wireless sensor networks

Domenico Ciuonzo[1] *and Pierluigi Salvo Rossi*[2]

This chapter deals with a distributed version of the binary-hypothesis test which formalizes the case in which a wireless sensor network (WSN) is used for detecting a binary event, and a fusion center (FC) with multiple antennas collects the information for a robust decision. The presence of multiple antennas at both transmit and receive sides resembles a multiple-input–multiple-output (MIMO) system and allows for utilization of array-processing techniques providing spectral efficiency and fading mitigation. The problem is here referred to as MIMO decision fusion. Coherent decision fusion, i.e., the case when instantaneous channel state information (CSI) is available at the FC, is first studied: "Decode-and-Fuse" and "Decode-then-Fuse" approaches are introduced and compared. Successively, noncoherent decision fusion, i.e., the case when statistical CSI is available at the FC, is analyzed: the focus is on the energy test and related optimality characteristics.

5.1 MIMO decision fusion

WSNs may be exploited for centralized or decentralized decision fusion. In the former case, the sensors transmit their local decisions to an FC, which is in charge for a global (and possibly more reliable) decision by appropriately combining the received information [1]. In the latter case, the FC is missing, and each sensor collects the information from (possibly all) the remaining ones in order to reach autonomously a reliable decision [2]. This chapter deals with centralized solutions and focuses on MIMO architectures for decision fusion which will be soon introduced. An algorithm designed for combining the local decisions from the sensors and providing the global decision at the FC is denoted *fusion rule*.

A set of K sensors are responsible for sensing a binary phenomenon of interest and taking autonomously a local decision. The two hypotheses are denoted $\mathcal{H}_0$ and $\mathcal{H}_1$, with corresponding a priori probabilities π_0 and π_1, respectively. We assume that

[1]Network Measurement and Monitoring (NM2), Italy
[2]Department of Advanced Analytics and Machine Learning, Kongsberg Digital AS, Norway

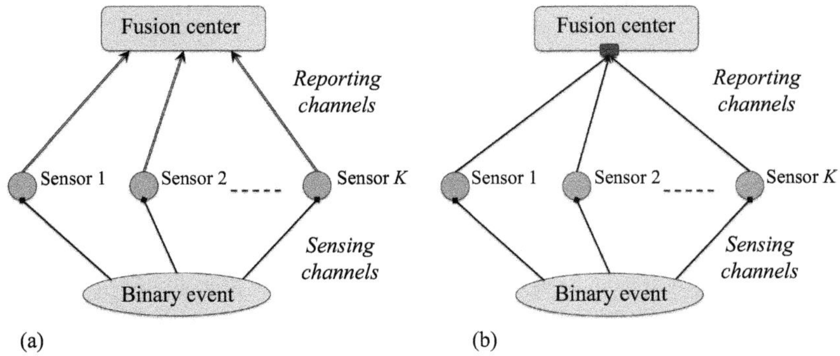

Figure 5.1 The two classical centralized architectures for decision fusion: (a) PAC with orthogonal reporting channels and (b) MAC with interfering reporting channels

the local sensing and decision process at the kth sensor is fully described by the local probability of false alarm ($p_{f,k}$) and the local probability of detection ($p_{d,k}$), with local decision being conditionally independent given the specific hypothesis. A scenario is said *homogeneous* if the probability of false alarm (resp. detection) is the same for each sensor and then simply denoted by p_f (resp. p_d); otherwise, the scenario is named *nonhomogeneous*.

Sensors are equipped each with one single transmit antenna and communicate their decision to the FC using the reporting channel. Information exchange between the sensors and the FC has been commonly based on a parallel-access channel (PAC), i.e., each sensor is provided with a non-interfering dedicated channel to communicate with the FC [3]. Orthogonality among the channels between each sensor and the FC may be achieved via frequency-, time-, or code-division multiple-access techniques. Unfortunately, systems based on PAC architectures require large bandwidths for simultaneous transmissions or large detection delays, which is unfeasible for large-scale sensor networks. A possible alternative is to exploit the intrinsically interfering nature of the wireless medium, i.e., replacing the PAC architecture with a multiple-access channel (MAC) architecture and then increase the bandwidth efficiency [4,5].

The two classical centralized architectures for decision fusion based on PAC and MAC are shown in Figure 5.1. However, recent works have improved even further the achievable system spectral efficiency [6]. Looking at the network as a "virtual" MIMO system and exploiting array-processing techniques at the FC, several fusion rules have been proposed and compared in terms of performance, computational complexity, and system knowledge requirements [7]. Figure 5.2 shows the architecture for a system employing MIMO decision fusion where the FC is equipped with N receive antennas. Recent proposals based on massive MIMO technology [8,9] are not considered here and will be discussed in the next chapter.

In the following, we denote $x_k \in \mathcal{X}$ the symbol transmitted by the kth sensor encoding its local decision, $H_{n,k} \in \mathbb{C}$ the fading channel coefficient on the link

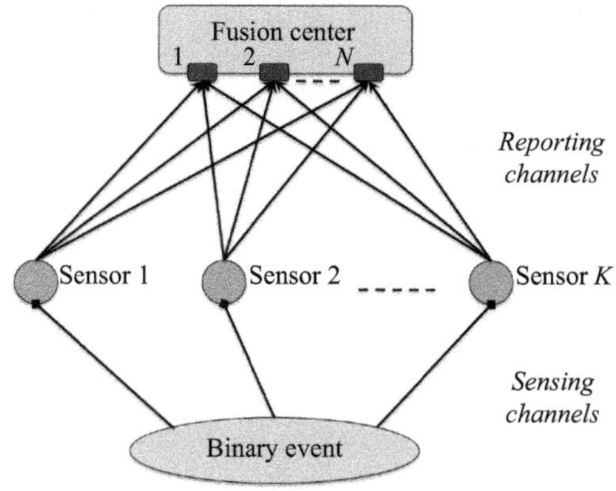

Figure 5.2 Centralized architecture for MIMO decision fusion

between the kth sensor and the nth receive antenna, $y_n \in \mathbb{C}$ the signal received by the nth receive antenna at the FC and $w_n \sim \mathcal{N}_{\mathbb{C}}(0, \sigma_w^2)$ the corresponding additive white Gaussian noise. The binary set $\mathcal{X}$ depends on the modulation employed, e.g., $\mathcal{X} = \{-1, +1\}$ (resp. $\mathcal{X} = \{0, 1\}$) in the case of BPSK (resp. OOK) modulation. The discrete-time model for the received signal at the FC is:

$$\boldsymbol{y} = \boldsymbol{H}\boldsymbol{x} + \boldsymbol{w}, \tag{5.1}$$

where $\boldsymbol{y} = (y_1, y_2, \ldots, y_N)^{\mathrm{T}} \in \mathbb{C}^N$ is the received-signal vector, $\boldsymbol{x} = (x_1, x_2, \ldots, x_K)^{\mathrm{T}} \in \mathcal{X}^K$ is the transmitted vector, $\boldsymbol{w} = (w_1, w_2, \ldots, w_N)^{\mathrm{T}} \in \mathbb{C}^N$ is the noise vector, and

$$\boldsymbol{H} = \begin{pmatrix} H_{1,1} & H_{1,2} & \cdots & H_{1,K} \\ H_{2,1} & H_{2,2} & \cdots & H_{2,K} \\ \vdots & \vdots & \ddots & \vdots \\ H_{N,1} & H_{N,2} & \cdots & H_{N,K} \end{pmatrix} \in \mathbb{C}^{N \times K} \tag{5.2}$$

denotes the reporting-channel matrix. We also denote $\boldsymbol{h}_{(k)} = (H_{1,k}, H_{2,k}, \ldots, H_{N,k})^{\mathrm{T}}$ the kth channel vector, collecting the fading coefficient on the links from the kth sensor to all the antennas at the FC.

The rest of the chapter will go through the design and the analysis of different solutions for MIMO decision fusion. MIMO decision fusion is commonly employed as a threshold test:

$$\lambda(\boldsymbol{y}) \underset{\hat{\mathcal{H}}=\mathcal{H}_0}{\overset{\hat{\mathcal{H}}=\mathcal{H}_1}{\gtrless}} \gamma, \tag{5.3}$$

where $\hat{\mathcal{H}}$ denotes the declared hypothesis, γ is a fixed threshold, and $\lambda(\boldsymbol{y})$ is the decision statistic whose explicit expression depends on the adopted fusion rule. The availability of CSI at the FC, i.e., of the reporting-channel matrix, is crucial for system design based on array processing techniques and both the cases of instantaneous CSI and statistical CSI will be presented. System performance is evaluated in terms of global probability of false alarm (q_f) and global probability of detection (q_d), defined as:

$$q_f = \Pr\left(\hat{\mathcal{H}} = \mathcal{H}_1|\mathcal{H}_0\right) = \Pr(\lambda > \gamma|\mathcal{H}_0), \tag{5.4}$$

$$q_d = \Pr\left(\hat{\mathcal{H}} = \mathcal{H}_1|\mathcal{H}_1\right) = \Pr(\lambda > \gamma|\mathcal{H}_1). \tag{5.5}$$

It is worth noticing that $\Pr(\lambda > \gamma|\mathcal{H}_i)$ generically describes both q_f and q_d (with $i = 0$ and $i = 1$, respectively). The threshold in (5.3) is usually selected in order to keep a fixed probability of false alarm according to the Neyman–Pearson criterion[1] [10].

Possible applications of the framework described in this chapter include spectrum sensing (e.g., see [11]) and underwater monitoring (e.g., see [12]).

5.2 MIMO decision fusion with instantaneous CSI

In this section, it is assumed that the specific realization of the reporting-channel matrix ($\boldsymbol{H}$) is available at the FC without error. In such a case, commonly denoted as *coherent decision fusion*, it is implicitly assumed that the sensors employ BPSK modulation, i.e., $\mathcal{X} = \{-1, +1\}$ with $x_k = -1$ (resp. $x_k = +1$) meaning that the local decision by the kth sensor is $\mathcal{H}_0$ (resp. $\mathcal{H}_1$). Then:

$$\Pr(x_k = +1|\mathcal{H}_0) = p_{f,k}, \tag{5.6}$$

$$\Pr(x_k = +1|\mathcal{H}_1) = p_{d,k}. \tag{5.7}$$

The reporting channels undergo i.i.d. Rayleigh fading with unitary mean power, i.e., $H_{n,k} \sim \mathcal{N}_{\mathbb{C}}(0, 1)$. The channel signal-to-noise ratio (SNR) is defined as the ratio between the average total received energy from the WSN ($E_s = \mathbb{E}\{\|\boldsymbol{H}\boldsymbol{x}\|^2\}$) and the one-sided power spectral density of the continuous process noise (σ_w^2), i.e.,

$$\text{SNR} = \frac{E_s}{\sigma_w^2} = \frac{KN}{\sigma_w^2}. \tag{5.8}$$

Two main approaches have been proposed and analyzed, namely, *decode-and-fuse* and *decode-then-fuse*, which are reviewed in the following [7].

5.2.1 Decode-and-fuse approaches

In the decode-and-fuse approach, the FC provides the declared hypothesis ($\hat{\mathcal{H}}$) directly from the received signal ($\boldsymbol{y}$), as shown in Figure 5.3.

[1]Alternatively, the same test may be used for minimizing the error probability according to the Bayesian criterion.

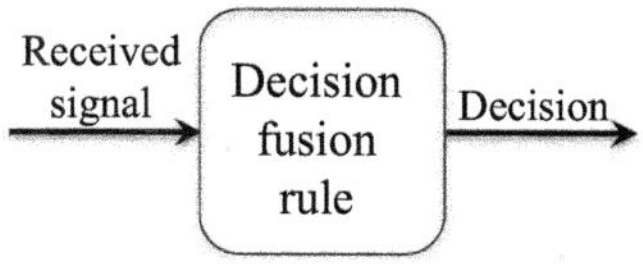

Figure 5.3 Decode-and-fuse approach

5.2.1.1 Log-likelihood ratio

The log-likelihood ratio (LLR) decision statistic is given as:

$$\lambda_{\mathrm{LLR}}(\boldsymbol{y}) = \ln\left(\frac{p(\boldsymbol{y}|\mathcal{H}_1)}{p(\boldsymbol{y}|\mathcal{H}_0)}\right) = \ln\left(\frac{\sum_{\boldsymbol{x}\in\mathcal{X}^K} p(\boldsymbol{y}|\boldsymbol{x})\Pr(\boldsymbol{x}|\mathcal{H}_1)}{\sum_{\boldsymbol{x}\in\mathcal{X}^K} p(\boldsymbol{y}|\boldsymbol{x})\Pr(\boldsymbol{x}|\mathcal{H}_0)}\right) \tag{5.9}$$

$$= \ln\left(\frac{\sum_{\boldsymbol{x}\in\mathcal{X}^K} \exp\left(-(\|\boldsymbol{y}-\boldsymbol{H}\boldsymbol{x}\|^2/\sigma_w^2)\right)\prod_{k=1}^K \Pr(x_k|\mathcal{H}_1)}{\sum_{\boldsymbol{x}\in\mathcal{X}^K} \exp\left(-(\|\boldsymbol{y}-\boldsymbol{H}\boldsymbol{x}\|^2/\sigma_w^2)\right)\prod_{k=1}^K \Pr(x_k|\mathcal{H}_0)}\right), \tag{5.10}$$

where in the last equation, the additional assumption of conditional (given the hypothesis) independent local decisions has been considered, i.e.,

$$\Pr(\boldsymbol{x}|\mathcal{H}_i) = \prod_{k=1}^{K} \Pr(x_k|\mathcal{H}_i). \tag{5.11}$$

Proposition

The LLR decision statistic provides the optimum fusion rule (in terms of performance) for the considered problem.

The practical implementation of the LLR fusion rule is difficult due to exponential functions with large dynamic range (especially for moderate-to-high channel SNR). This becomes a severe issue for fixed-point implementation. Additionally, in order to compute the LLR decision statistic, the FC requires the knowledge of the sensor local performance ($p_{f,k}$ and $p_{d,k}$), the CSI ($\boldsymbol{H}$) and the SNR (i.e., the noise level σ_w^2). Finally, the computational complexity is linear with the number of receive antennas at the FC and exponential with the number of sensors, i.e., $\mathcal{O}(N2^K)$. It is apparent how the computational complexity makes the LLR fusion rule impractical for a moderate number of sensors, thus the need for alternative fusion rules.

5.2.1.2 Maximum ratio combining

Assuming (only for the purpose design of the fusion rule) that sensors are perfect, i.e., $p_{f,k} = 0$ and $p_{d,k} = 1$, only two possible transmitted vectors are allowed: $\boldsymbol{x} = -\boldsymbol{1}_K$ and $\boldsymbol{x} = +\boldsymbol{1}_K$. In such a case, the LLR decision statistic simplifies in the maximum ratio combining (MRC) decision statistic, given as:

$$\lambda_{\mathrm{MRC}}(\boldsymbol{y}) = \ln\left(\frac{\exp\left(-(\|\boldsymbol{y}-\boldsymbol{H}\boldsymbol{1}_K\|^2/\sigma_w^2)\right)}{\exp\left(-(\|\boldsymbol{y}+\boldsymbol{H}\boldsymbol{1}_K\|^2/\sigma_w^2)\right)}\right) \propto \Re\left\{(\boldsymbol{H}\boldsymbol{1}_K)^{\mathrm{H}}\boldsymbol{y}\right\}. \tag{5.12}$$

Proposition

In the case of homogeneous scenario (i.e., $p_{f,k} = p_f$ and $p_{d,k} = p_d$) with realistic sensors (i.e., $p_f < p_d$) and low SNR (i.e., $KN \ll \sigma_w^2$), the MRC fusion rule approaches the optimum fusion rule (i.e., $\lambda_{\text{MRC}} \approx \lambda_{\text{LLR}}$).

In order to compute the MRC decision statistic, the FC needs partial knowledge of the CSI (i.e., $\boldsymbol{H}\boldsymbol{1}_K$) but does not need the knowledge of the sensor local performance and of the SNR. The computational complexity is linear with the number of receive antennas at the FC, i.e., $\mathcal{O}(N)$.

5.2.1.3 Equal gain combining

Based on the analogy with array processing and beamforming techniques, the equal gain combining (EGC) decision statistic is given as:

$$\lambda_{\text{EGC}}(\boldsymbol{y}) = \Re\left\{\exp\left(j\angle(\boldsymbol{H}\boldsymbol{1}_K)\right)^{\text{H}}\boldsymbol{y}\right\}, \tag{5.13}$$

where $\angle(\cdot)$ denotes the phase.

In order to compute the EGC decision statistic, the FC needs partial knowledge of the CSI (i.e. $\angle(\boldsymbol{H}\boldsymbol{1}_K)$) but does not need the knowledge of the sensor local performance and of the SNR. The computational complexity is linear with the number of receive antennas at the FC, i.e., $\mathcal{O}(N)$.

5.2.1.4 Max-log

Exploiting the max-log approximation, common in the turbo-coding literature, i.e.,

$$\ln\left(\sum_{\ell=1}^{L} b_\ell \exp(a_\ell)\right) = \ln\left(\sum_{\ell=1}^{L} \exp(a_\ell + \ln(b_\ell))\right) \approx \max_{\ell}\{a_\ell + \ln(b_\ell)\}, \tag{5.14}$$

where $a_\ell \in \mathbb{R}$ and $b_\ell > 0$, the LLR decision statistic may be approximated as follows:

$$\begin{aligned}\lambda_{\max-\log} = & \min_{\boldsymbol{x}\in\mathcal{X}^K}\left\{\frac{\|\boldsymbol{y} - \boldsymbol{H}\boldsymbol{x}\|^2}{\sigma_w^2} - \sum_{k=1}^{K}\ln(\Pr(x_k|\mathcal{H}_0))\right\} \\ & - \min_{\boldsymbol{x}\in\mathcal{X}^K}\left\{\frac{\|\boldsymbol{y} - \boldsymbol{H}\boldsymbol{x}\|^2}{\sigma_w^2} - \sum_{k=1}^{K}\ln(\Pr(x_k|\mathcal{H}_1))\right\},\end{aligned} \tag{5.15}$$

which can be interpreted as the difference between hypothesis prior-weighted minimum distance searches.

Proposition

In the case of high SNR (i.e., $KN \gg \sigma_w^2$), the max-log fusion rule approaches the optimum fusion rule (i.e., $\lambda_{\max-\log} \approx \lambda_{\text{LLR}}$).

Proposition

In the case of homogeneous scenario (i.e., $p_{f,k} = p_f$ and $p_{d,k} = p_d$) with good sensors (i.e., $p_f < 1/2$ and $p_d > 1/2$) and low SNR (i.e., $KN \ll \sigma_w^2$), the max-log fusion rule approaches the optimum fusion rule (i.e., $\lambda_{\text{max-log}} \approx \lambda_{\text{LLR}}$).

Proposition

In the case of good sensors (i.e., $p_{f,k} < 1/2$ and $p_{d,k} > 1/2$) and low SNR (i.e., $KN \ll \sigma_w^2$), the max-log fusion rule approaches the MRC fusion rule (i.e., $\lambda_{\text{max-log}} \approx \lambda_{\text{MRC}}$).

In order to compute the max-log decision statistic, the FC needs knowledge of the sensor local performance ($p_{f,k}$ and $p_{d,k}$), the CSI ($\boldsymbol{H}$) and the SNR (i.e., the noise level σ_w^2). The computational complexity is linear with the number of receive antennas at the FC and exponential with the number of sensors (but lower than in the case of LLR if implemented using computationally efficient techniques like sphere decoding [13]), i.e., $\mathcal{O}(N2^{K-c_1})$ with $c_1 > 0$.

5.2.2 Decode-then-fuse approaches

In the decode-and-fuse approach, the FC provides the declared hypothesis ($\hat{\mathcal{H}}$) after estimating the transmitted vector ($\boldsymbol{x}$) from the received signal ($\boldsymbol{y}$) by exploiting tools from communication theory, as shown in Figure 5.4. The declared hypothesis is given using the Chair–Varshney (CV) rule, i.e., the optimum fusion rule in the case of noiseless channels, while different solutions have been proposed for decoding the transmitted vector.

The CV fusion rule is given as:

$$\lambda_{\text{CV}}(\hat{\boldsymbol{x}}) = \sum_{k=1}^{K} \left[\left(\frac{\hat{x}_k + 1}{2} \right) \ln \left(\frac{p_{d,k}}{p_{f,k}} \right) + \left(1 - \frac{\hat{x}_k + 1}{2} \right) \ln \left(\frac{1 - p_{d,k}}{1 - p_{f,k}} \right) \right], \quad (5.16)$$

where $\hat{\boldsymbol{x}}$ is the decoded transmitted vector. Also, it is worth noticing that in the case of homogeneous scenarios, the CV rule simplifies in the well-known counting rule, i.e.,

$$\lambda_{\text{C}}(\hat{\boldsymbol{x}}) = \sum_{k=1}^{K} \left(\frac{\hat{x}_k + 1}{2} \right). \quad (5.17)$$

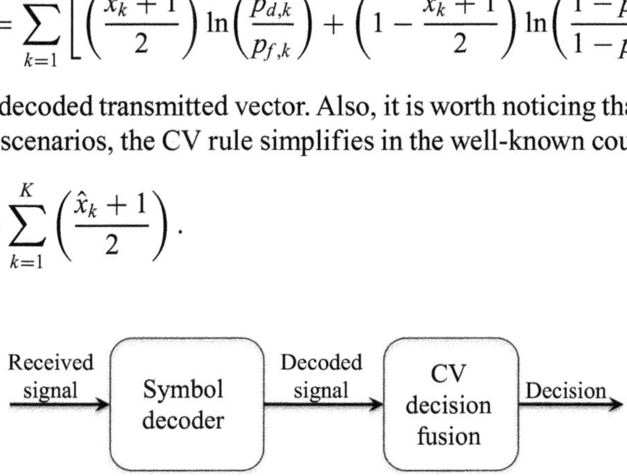

Figure 5.4 Decode-then-fuse approach

5.2.2.1 CV with maximum likelihood (ML)

One possibility to decode the transmitted vector is to use the maximum likelihood (ML) rule, i.e.,

$$\hat{\boldsymbol{x}}_{\mathrm{ML}} = \arg \min_{\boldsymbol{x} \in \mathscr{X}^K} \|\boldsymbol{y} - \boldsymbol{H}\boldsymbol{x}\|^2. \tag{5.18}$$

Proposition

In the case of high SNR (i.e., $KN \gg \sigma_w^2$), the CV-ML fusion rule approaches the optimum fusion rule (i.e., $\lambda_{\mathrm{CV}}(\hat{\boldsymbol{x}}_{\mathrm{ML}}) \approx \lambda_{\mathrm{LLR}}$).

In order to compute the CV-ML decision statistic, the FC needs knowledge of the sensor local performance ($p_{f,k}$ and $p_{d,k}$) and the CSI ($\boldsymbol{H}$). The computational complexity is linear with the number of receive antennas at the FC and exponential with the number of sensors (but lower than in the case of max-log if implemented using computationally efficient techniques like sphere decoding [13]), i.e., $\mathcal{O}(N2^{K-c_2})$ with $c_2 > c_1 > 0$.

5.2.2.2 CV with minimum mean square error (MMSE)

One possibility to decode the transmitted vector with reduced complexity is to use the minimum mean square error (MMSE) rule, which should be properly modified in order to take into account the fact that symbols x_k are correlated differently than the standard scenarios found in communication systems. The MMSE rule provides the following estimated vector:

$$\hat{\boldsymbol{x}}_{\mathrm{MMSE}} = \mathrm{sign}\big(\boldsymbol{\mu}_x + \boldsymbol{C}_x\boldsymbol{H}^{\mathrm{H}}(\boldsymbol{H}\boldsymbol{C}_x\boldsymbol{H}^{\mathrm{H}} + \sigma_w^2\boldsymbol{I}_N)^{-1}(\boldsymbol{y} - \boldsymbol{H}\boldsymbol{\mu}_x)\big), \tag{5.19}$$

where $\boldsymbol{\mu}_x = \mathbb{E}\{\boldsymbol{x}\}$ and $\boldsymbol{C}_x = \mathbb{E}\{(\boldsymbol{x} - \boldsymbol{\mu}_x)(\boldsymbol{x} - \boldsymbol{\mu}_x)^{\mathrm{H}}\}$ are the mean vector and the covariance matrix of the transmitted vector ($\boldsymbol{x}$), respectively.

In order to compute the CV-MMSE decision statistic, the FC needs knowledge of the sensor local performance ($p_{f,k}$ and $p_{d,k}$), the CSI ($\boldsymbol{H}$) and the SNR (i.e., the noise level σ_w^2). The computational complexity is polynomial both with the number of receive antennas at the FC and with the number of sensors, i.e., $\mathcal{O}(NK^2 + N^2K)$.

5.2.3 Performance comparison

Table 5.1 provides a complete comparison of the presented fusion rules in terms of computational complexity (both with respect to the number of sensors and the number of receive antennas at the FC) and required system knowledge. It is worth noticing that the dependence of the complexity with respect to N is moderate for all the considered fusion rules (polynomial in the worst case), thus the deployment of multiple receive antennas at the FC is an effective solution for performance increase.

Table 5.1 Comparison of the fusion rules in terms of computational complexity and system knowledge

Fusion rule	**Complexity**	**Local performance**	**CSI**	**SNR**
LLR	$\mathcal{O}(N2^K)$	Required	Required	Required
MRC	$\mathcal{O}(N)$	Not required	Partially required	Not required
EGC	$\mathcal{O}(N)$	Not required	Partially required	Not required
Max-log	$\mathcal{O}(N2^{K-c_1})$	Required	Required	Required
CV-ML	$\mathcal{O}(N2^{K-c_2})$	Required	Required	Not required
CV-MMSE	$\mathcal{O}(NK^2 + N^2K)$	Required	Required	Required

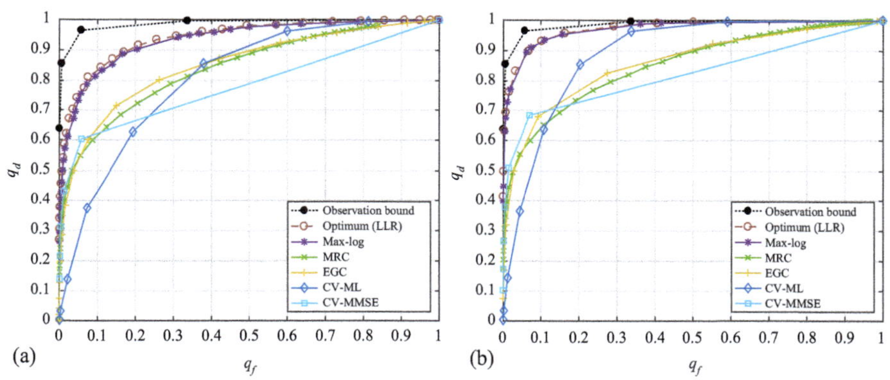

Figure 5.5 ROC for different fusion rules in the case of WSN with $K = 8$ sensors (homogeneous scenario with $p_f = 0.05$ and $p_d = 0.5$) and $N = 2$ receive antennas at the FC: (a) SNR = 10 dB and (b) SNR = 15 dB

As a benchmark for comparison purposes, we consider the (upper) "observation bound" [4], i.e., the optimum performance over a noiseless channel, given by:

$$q_{\mathrm{OB},f} = \sum_{k=g}^{K} \binom{K}{k} p_f^k (1 - p_f)^{K-k}, \tag{5.20}$$

$$q_{\mathrm{OB},d} = \sum_{k=g}^{K} \binom{K}{k} p_d^k (1 - p_d)^{K-k}, \tag{5.21}$$

where g is a discrete threshold.

Figure 5.5 shows the receiver operating characteristics (ROCs), i.e., the q_d-vs.-q_f curves, for the presented rules in a homogeneous WSN with $K = 8$ sensors and $N = 2$ receive antennas at the FC. The local performance of the sensors are $p_f = 0.05$ and $p_d = 0.5$, while the reporting phase undergoes SNR = 10 dB and SNR = 15 dB. It is apparent that the LLR and max-log fusion rules exhibit similar performance, although

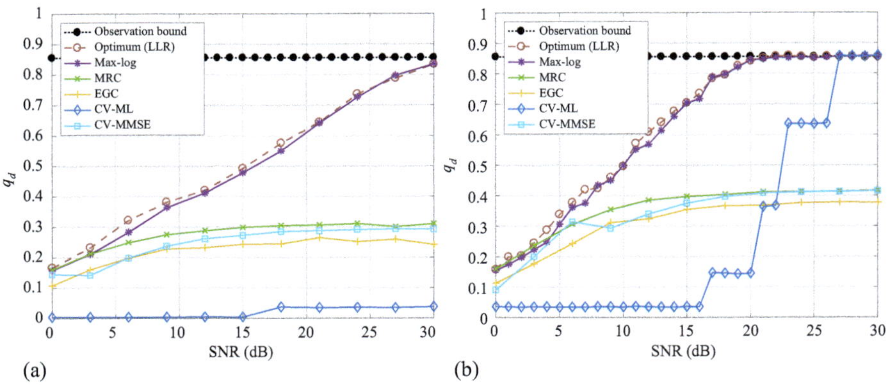

Figure 5.6 *q_d vs. SNR for different fusion rules in the case of WSN with $K = 8$ sensors (homogeneous scenario with $p_f = 0.05$ and $p_d = 0.5$) and $q_f \leq 0.01$: (a) $N = 1$ receive antenna at the FC and (b) $N = 2$ receive antennas at the FC*

a significant gap from the unrealistic observation bound exists. Also, the ROCs of the MRC and EGC fusion rules present a crossing point, and the same applies to the ROCs of the CV-ML and CV-MMSE fusion rules. However, in the first case, the crossing point appears to be SNR-independent, while this is not true in the second case.

Figure 5.6 shows the probability of detection vs. the channel SNR for the presented rules in a homogeneous WSN with $K = 8$ sensors assuming a upper bound on the probability of false alarm ($q_f \leq 0.01$). The local performance of the sensors are $p_f = 0.05$ and $p_d = 0.5$, while the number of receive antennas at the FC is $N = 1$ and $N = 2$. It is apparent how the MRC (resp. CV-ML) fusion rule approaches the LLR fusion rule at low (resp. high) SNR. The max-log fusion rule approaches the performance of the LLR fusion rule over the whole SNR range. It is worth noticing that the curves corresponding to the decode-then-fuse approach exhibit jumpy-step (CV-ML fusion rule) and non-monotonic (CV-MMSE) behaviors. Such phenomena are due to the discrete nature of CV decision statistics.

5.3 MIMO decision fusion with statistical CSI

In this section, it is assumed that the statistical distribution of the reporting-channel matrix ($\boldsymbol{H}$) is available at the FC, but the specific realization is not. In such a case, commonly denoted as *noncoherent decision fusion*, it is implicitly assumed that the sensors employ OOK modulation [14–16], i.e., $\mathcal{X} = \{0, 1\}$ with $x_k = 0$ (resp. $x_k = 1$) meaning that the local decision by the kth sensor is $\mathcal{H}_0$ (resp. $\mathcal{H}_1$) and the sensor itself is considered *inactive* (resp. *active*). Then:

$$\Pr(x_k = 1|\mathcal{H}_0) = p_{f,k}, \tag{5.22}$$

$$\Pr(x_k = 1|\mathcal{H}_1) = p_{d,k}. \tag{5.23}$$

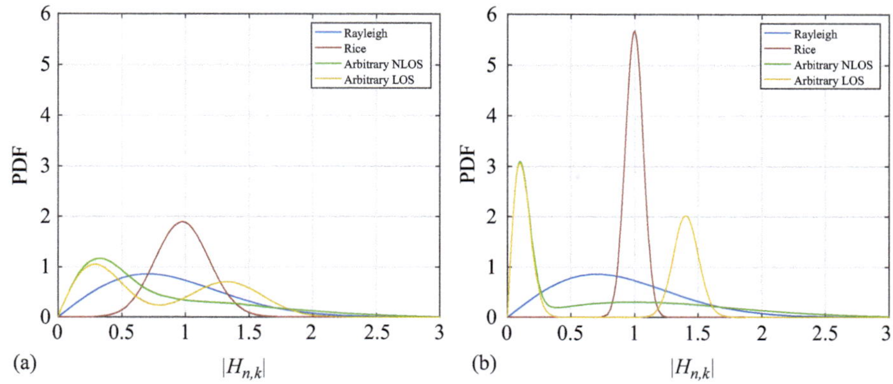

Figure 5.7 Fading statistics of the single link: (a) smooth modes and (b) sharp modes

The reporting channels undergo small-scale fading, and four cases will be considered in detail[2]:

- Rayleigh fading, with channel vectors assumed i.i.d. according to a zero-mean complex Gaussian distribution, i.e., $\boldsymbol{h}_{(k)} \sim \mathcal{N}_{\mathbb{C}}(\boldsymbol{0}_N, \sigma_H^2 \boldsymbol{I}_N)$;
- Rice fading, with channel coefficients assumed i.i.d. according to a complex Gaussian distribution, i.e., $\boldsymbol{h}_{(k)} \sim \mathcal{N}_{\mathbb{C}}(\boldsymbol{\mu}_H, \sigma_H^2 \boldsymbol{I}_N)$;
- Arbitrary non-line-of-sight (NLOS) fading, with channel vectors assumed i.i.d. according to a complex Gaussian mixture distribution with zero-mean components, i.e., $\boldsymbol{h}_{(k)} \sim \sum_{m=1}^{M} \rho_{H,m} \mathcal{N}_{\mathbb{C}}(\boldsymbol{0}_N, \sigma_{H,m}^2 \boldsymbol{I}_N)$; and
- Arbitrary fading, with channel vectors assumed i.i.d. according to a complex Gaussian mixture distribution, i.e., $\boldsymbol{h}_{(k)} \sim \sum_{m=1}^{M} \rho_{H,m} \mathcal{N}_{\mathbb{C}}(\boldsymbol{\mu}_{H,m}, \sigma_{H,m}^2 \boldsymbol{I}_N)$.

Figure 5.7 shows the PDF of the generic (modulus) channel gain for the four considered cases, with parameters providing smooth or sharp modes. Assuming unitary mean-power fading, i.e., $\mathbb{E}\{|H_{n,k}|^2\} = 1$, for sake of simplicity, the channel SNR is defined as the inverse of the one-sided power spectral density of the continuous process noise (σ_w^2), i.e., a per-link SNR:

$$\mathrm{SNR} = \frac{1}{\sigma_w^2}. \tag{5.24}$$

In the following, the optimum LLR test will be introduced as a benchmark for performance comparison and the simple energy test will be considered as a practical fusion rule.

[2]Mixture distributions have commonly been assumed as effective approximations of real-world measured channels: Gaussian (resp. Gamma) mixtures have been considered in [17] (resp. [18]).

5.3.1 LLR test

The LLR decision statistic is given as:

$$\lambda_{\mathrm{LLR}}(\boldsymbol{y}) = \ln\left(\frac{p(\boldsymbol{y}|\mathcal{H}_1)}{p(\boldsymbol{y}|\mathcal{H}_0)}\right) = \ln\left(\frac{\sum_{\ell=0}^{K} p(\boldsymbol{y}|\ell)\Pr(\ell|\mathcal{H}_1)}{\sum_{\ell=0}^{K} p(\boldsymbol{y}|\ell)\Pr(\ell|\mathcal{H}_0)}\right) \tag{5.25}$$

where ℓ is a shorthand notation for $\ell(\boldsymbol{x}) = \boldsymbol{x}^{\mathrm{T}}\mathbf{1}_K$, i.e., the number of active sensors, and the conditional (given ℓ) independence between $\boldsymbol{y}$ and $\mathcal{H}_i$ has been exploited. More specifically, in the case of homogeneous scenarios, $\ell|\mathcal{H}_i$ is a binomial RV, i.e.,

$$\Pr(\ell|\mathcal{H}_i) = \binom{K}{\ell} p_i^{\ell}(1-p_i)^{K-\ell}, \tag{5.26}$$

where $p_i = p_f$ (resp. $p_i = p_d$) under $\mathcal{H}_0$ (resp. $\mathcal{H}_1$), while in the case of nonhomogeneous scenarios, $\ell|\mathcal{H}_i$ is a Poisson-binomial RV, i.e.,

$$\Pr(\ell|\mathcal{H}_i) = \sum_{\boldsymbol{x}:\boldsymbol{x}^{\mathrm{T}}\mathbf{1}_K=\ell} \left(\prod_{k=1}^{K} p_{i,k}^{x_k}\right)\left(\prod_{k=1}^{K}(1-p_{i,k})^{1-x_k}\right), \tag{5.27}$$

where $p_{i,k} = p_{f,k}$ (resp. $p_{i,k} = p_{d,k}$) under $\mathcal{H}_0$ (resp. $\mathcal{H}_1$). More details on (5.27) and effective evaluation are found in [19].

As for the conditional (given the number of active sensors) PDF of the received signal, in the case of Rayleigh fading, it is given as:

$$p(\boldsymbol{y}|\ell) = \frac{1}{\pi^N\left(\sigma_w^2 + \ell\sigma_H^2\right)^N}\exp\left(-\frac{\|\boldsymbol{y}\|^2}{\sigma_w^2 + \ell\sigma_H^2}\right), \tag{5.28}$$

i.e., $\boldsymbol{y}|\ell$ is Rayleigh distributed; in the case of Rice fading, it is given as:

$$p(\boldsymbol{y}|\ell) = \frac{1}{\pi^N\left(\sigma_w^2 + \ell\sigma_H^2\right)^N}\exp\left(-\frac{\|\boldsymbol{y} - \ell\boldsymbol{\mu}_H\|^2}{\sigma_w^2 + \ell\sigma_H^2}\right), \tag{5.29}$$

i.e., $\boldsymbol{y}|\ell$ is Rice distributed; in the case of arbitrary NLOS fading, it is given as:

$$p(\boldsymbol{y}|\ell) = \sum_{\boldsymbol{m}:\boldsymbol{m}^{\mathrm{T}}\mathbf{1}_M=\ell} \frac{\theta_{\boldsymbol{m}}(\ell)}{\left(\pi\omega_{\boldsymbol{m}}^2\right)^N}\exp\left(-\frac{\|\boldsymbol{y}\|^2}{\omega_{\boldsymbol{m}}^2}\right), \tag{5.30}$$

i.e., $\boldsymbol{y}|\ell$ is Gaussian-mixture distributed with zero-mean components; in the case of arbitrary fading, it is given as:

$$p(\boldsymbol{y}|\ell) = \sum_{\boldsymbol{m}:\boldsymbol{m}^{\mathrm{T}}\mathbf{1}_M=\ell} \frac{\theta_{\boldsymbol{m}}(\ell)}{\left(\pi\omega_{\boldsymbol{m}}^2\right)^N}\exp\left(-\frac{\|\boldsymbol{y} - \boldsymbol{\nu}_m\|^2}{\omega_{\boldsymbol{m}}^2}\right), \tag{5.31}$$

i.e., $\boldsymbol{y}|\ell$ is Gaussian-mixture distributed. In the last two equations, $\boldsymbol{m} = (m_1, m_2, \ldots, m_M)^{\mathrm{T}}$ is a vector of integers (with m_u representing the number of links undergoing the uth component of the reporting-channel complex Gaussian mixture) denoting the aggregate component of the new resulting Gaussian mixture. With reference to the new resulting Gaussian mixture, the probability of the single (aggregate) component is:

$$\theta_{\boldsymbol{m}}(\ell) = \binom{\ell}{\boldsymbol{m}} \exp(\boldsymbol{m}^{\mathrm{T}} \ln(\boldsymbol{\rho}_H)), \tag{5.32}$$

with $\boldsymbol{\rho}_H = (\rho_{H,1}, \rho_{H,2}, \ldots, \rho_{H,M})^{\mathrm{T}}$, and the corresponding mean and variance are:

$$\boldsymbol{\nu}_{\boldsymbol{m}} = \sum_{u=1}^{M} m_u \boldsymbol{\mu}_{H,u}, \tag{5.33}$$

$$\omega_{\boldsymbol{m}}^2 = \boldsymbol{m}^{\mathrm{T}} \boldsymbol{\sigma}_H^2 + \sigma_w^2, \tag{5.34}$$

respectively, where (with a slight abuse of notation) $\boldsymbol{\sigma}_H^2 = (\sigma_{H,1}, \sigma_{H,2}, \ldots, \sigma_{H,M})^{\mathrm{T}}$. For sake of simplicity, as the conditional received signal $\boldsymbol{y}|\ell$ is in general a Gaussian mixture with $L(\ell) = \binom{M+\ell-1}{\ell}$ components, the following notation will be used:

$$\boldsymbol{y}|\ell \sim \sum_{m=1}^{L(\ell)} \theta_m(\ell) \mathcal{N}_{\mathbb{C}} \left(\boldsymbol{\nu}_m(\ell); \omega_m^2(\ell) \boldsymbol{I}_N \right), \tag{5.35}$$

with apparent meaning for the symbols $\boldsymbol{\nu}_m(\ell)$ and $\omega_m^2(\ell)$.

The performance of the LLR test is not available in closed form. However, theoretical results and numerical results based on computer simulations have shown that, in many relevant scenarios of practical interest, the energy test (which theoretical performance analysis is available for) is equivalent to the LLR test or exhibit negligible performance degradation.

5.3.2 Energy test

The energy decision statistic is given as:

$$\lambda_{\mathrm{E}}(\boldsymbol{y}) = \|\boldsymbol{y}\|^2, \tag{5.36}$$

i.e., the energy of the received signal, which apparently requires little computational complexity and also has the advantage that neither local sensor performance nor CSI nor SNR are needed. Another interesting aspect of the energy test is the (near-)optimal behavior in many scenarios of practical interest.

Proposition

In the case of Rayleigh fading, the energy test is equivalent to the LLR test and then is the optimum test.

Proposition

In the case of arbitrary NLOS fading, the inequality $\max_{m=1,2,\ldots,M} \sigma_{H,m}^2 / \min_{m=1,2,\ldots,M} \sigma_{H,m}^2 < K/(K-1)$ represents a sufficient condition for the energy test being equivalent to the LLR test, i.e., the optimum test.

Proposition

In most cases of arbitrary NLOS fading not matching the sufficient condition of the previous proposition, the energy test is near-optimum, i.e., exhibiting a negligible gap in performance when compared with the LLR test.

The performance of the energy test is available in closed form. More specifically, in the case of Rayleigh fading, it is given as:

$$\Pr(\lambda_E > \gamma|\mathcal{H}_i) = \sum_{\ell=0}^{K} \Pr(\ell|\mathcal{H}_i)\frac{1}{\Gamma(N)}\Gamma\left(N; \frac{\gamma}{\sigma_w^2 + \ell\sigma_H^2}\right); \tag{5.37}$$

in the case of Rice fading, it is given as:

$$\Pr(\lambda_E > \gamma|\mathcal{H}_i) = \sum_{\ell=0}^{K} \Pr(\ell|\mathcal{H}_i)Q_N\left(\frac{\sqrt{2\ell}\|\boldsymbol{\mu}_H\|}{\sqrt{\sigma_w^2 + \ell\sigma_H^2}}, \sqrt{\frac{2\gamma}{\sigma_w^2 + \ell\sigma_H^2}}\right); \tag{5.38}$$

in the case of arbitrary NLOS fading, it is given as:

$$\Pr(\lambda_E > \gamma|\mathcal{H}_i) = \sum_{\ell=0}^{K}\sum_{m=1}^{L(\ell)} \Pr(\ell|\mathcal{H}_i)\theta_m(\ell)\frac{1}{\Gamma(N)}\Gamma\left(N; \frac{\gamma}{\omega_m^2(\ell)}\right); \tag{5.39}$$

and in the case of arbitrary fading, it is given as:

$$\Pr(\lambda_E > \gamma|\mathcal{H}_i) = \sum_{\ell=0}^{K}\sum_{m=1}^{L(\ell)} \Pr(\ell|\mathcal{H}_i)\theta_m(\ell)Q_N\left(\frac{\sqrt{2}\|\boldsymbol{v}_m(\ell)\|}{\omega_m(\ell)}, \frac{\sqrt{2\gamma}}{\omega_m(\ell)}\right), \tag{5.40}$$

where $Q_m(\cdot;\cdot)$ and $\Gamma(\cdot;\cdot)$ denote the generalized Marcum Q-function and the upper incomplete Gamma function defined respectively as:

$$Q_m(a;b) = \int_b^\infty \frac{\xi^m}{a^{m-1}} \exp\left(-\frac{\xi^2 + a^2}{2}\right) I_{m-1}(a\xi)\, d\xi, \tag{5.41}$$

and

$$\Gamma(a,b) = \int_b^\infty \xi^{a-1} \exp(-\xi) d\xi. \tag{5.42}$$

being $I_n(\cdot)$ and $\Gamma(\cdot)$ the modified Bessel function of the first kind and the Gamma function, defined as $I_n(x) = (1/\pi)\int_0^\pi \exp(x\cos(\theta))\cos(n\theta)d\theta$ and $\Gamma(a) = \Gamma(a, 0)$.

5.3.3 Asymptotic performance of LLR and energy tests

In the case of large-size system, i.e., $K \gg 1$, both the LLR statistic and the energy statistic have a compact closed form for the characteristic function. The characteristic function allows low-complexity numerical performance evaluation by means of the Gauss–Chebyshev quadrature formulas [20,21]. The analysis relies on the multivariate versions of the central limit theorem for complex-valued improper Gaussian random vectors in order to compute the limit distribution of the scaled-version received signal ($\tilde{\boldsymbol{y}}$). Two scenarios are of interest:

- Individual power constraint (IPC) on each single sensor with the scaled received signal defined as $\tilde{\boldsymbol{y}} = \boldsymbol{y}/\sqrt{K}$.
- Total power constraint (TPC) on the whole set of sensors with the scaled received signal defined as $\tilde{\boldsymbol{y}} = \boldsymbol{H}\boldsymbol{x}/\sqrt{K} + \boldsymbol{w}$.

The IPC case is relevant for systems using predefined sensors with given performance, while the TPC case is relevant when it is possible to design the sensors in advance.

In this subsection, the analysis is limited to the general case of arbitrary fading channel distribution, as the first three cases are special cases of the fourth case. The first-order and second-order moments of the channel vectors are denoted as:

$$\boldsymbol{\mu}_H = \mathbb{E}\left\{\boldsymbol{h}_{(k)}\right\} = \sum_{m=1}^{M} \rho_{H,m}\boldsymbol{\mu}_{H,m}, \tag{5.43}$$

$$\boldsymbol{R}_H = \mathbb{E}\left\{\boldsymbol{h}_{(k)}\boldsymbol{h}_{(k)}^{\mathrm{H}}\right\} = \sum_{m=1}^{M} \rho_{H,m}(\sigma_{H,m}^2\boldsymbol{I}_N + \boldsymbol{\mu}_{H,m}\boldsymbol{\mu}_{H,m}^{\mathrm{H}}), \tag{5.44}$$

$$\boldsymbol{Q}_H = \mathbb{E}\left\{\boldsymbol{h}_{(k)}\boldsymbol{h}_{(k)}^{\mathrm{T}}\right\} = \sum_{m=1}^{M} \rho_{H,m}\boldsymbol{\mu}_{H,m}\boldsymbol{\mu}_{H,m}^{\mathrm{T}}. \tag{5.45}$$

The limit distribution of the scaled received signal is found as [16]:

$$\tilde{\boldsymbol{y}}|\mathcal{H}_i - \tilde{\boldsymbol{\mu}}_i \overset{(a)}{\sim} \mathcal{N}_{\mathbb{C}}\left(\boldsymbol{0}_N; \boldsymbol{\Omega}_i, \boldsymbol{\Upsilon}_i\right), \tag{5.46}$$

where, referring to a nonhomogeneous scenario and denoting $S_{p_i} = \left(\sum_{k=1}^{K} p_{i,k}\right)$ and $S_{p_i^2} = \left(\sum_{k=1}^{K} p_{i,k}^2\right)$, in the IPC case:

$$\tilde{\boldsymbol{\mu}}_i = \frac{S_{p_i}}{\sqrt{K}}\boldsymbol{\mu}_H, \tag{5.47}$$

$$\boldsymbol{\Omega}_i = \lim_{K\to\infty}\left(\frac{S_{p_i}}{K}\boldsymbol{R}_H - \frac{S_{p_i^2}}{K}\boldsymbol{\mu}_H\boldsymbol{\mu}_H^{\mathrm{H}}\right), \tag{5.48}$$

$$\boldsymbol{\Upsilon}_i = \lim_{K\to\infty}\left(\frac{S_{p_i}}{K}\boldsymbol{Q}_H - \frac{S_{p_i^2}}{K}\boldsymbol{\mu}_H\boldsymbol{\mu}_H^{\mathrm{T}}\right). \tag{5.49}$$

while in the TPC case $\boldsymbol{\Omega}_i$ is replaced with:

$$\boldsymbol{\Omega}_i = \lim_{K\to\infty}\left(\frac{S_{p_i}}{K}\boldsymbol{R}_H - \frac{S_{p_i^2}}{K}\boldsymbol{\mu}_H\boldsymbol{\mu}_H^{\mathrm{H}}\right) + \sigma_w^2\boldsymbol{I}_N. \tag{5.50}$$

Referring to a homogeneous scenario, in the IPC case, the moments simplify as:

$$\tilde{\boldsymbol{\mu}}_i = \sqrt{K} p_i \boldsymbol{\mu}_H, \tag{5.51}$$

$$\boldsymbol{\Omega}_i = p_i \big(\boldsymbol{R}_H - p_i \boldsymbol{\mu}_H \boldsymbol{\mu}_H^{\mathrm{H}}\big), \tag{5.52}$$

$$\boldsymbol{\Upsilon}_i = p_i \big(\boldsymbol{Q}_H - p_i \boldsymbol{\mu}_H \boldsymbol{\mu}_H^{\mathrm{T}}\big), \tag{5.53}$$

while in the TPC case, $\boldsymbol{\Omega}_i$ is replaced with:

$$\boldsymbol{\Omega}_i = p_i \left(\boldsymbol{R}_H - p_i \boldsymbol{\mu}_H \boldsymbol{\mu}_H^{\mathrm{H}}\right) + \sigma_w^2 \boldsymbol{I}_N. \tag{5.54}$$

As the received signal in large-size system results in an improper Gaussian distribution, the augmented representation of vectors and matrices [22] is more effective for analysis. More specifically, the augmented received vector and related moments are defined as:

$$\underline{\boldsymbol{y}} = \left(\tilde{\boldsymbol{y}}^{\mathrm{T}}, \tilde{\boldsymbol{y}}^{\mathrm{H}}\right)^{\mathrm{T}}, \qquad \underline{\boldsymbol{\mu}}_i = \left(\tilde{\boldsymbol{\mu}}_i^{\mathrm{T}}, \tilde{\boldsymbol{\mu}}_i^{\mathrm{H}}\right)^{\mathrm{T}}, \qquad \underline{\boldsymbol{\Omega}}_i = \begin{pmatrix} \boldsymbol{\Omega}_i & \boldsymbol{\Upsilon}_i \\ \boldsymbol{\Upsilon}_i^* & \boldsymbol{\Omega}_i^* \end{pmatrix}. \tag{5.55}$$

The characteristic functions of the LLR and energy statistics are:

$$\phi_{\lambda_{\mathrm{LLR}}|\mathscr{H}_i}(\tau) = \frac{\exp\left(-(1/2)\underline{\boldsymbol{u}}_i^{\mathrm{H}} \underline{\boldsymbol{\Omega}}_i^{-1} \left(\boldsymbol{I}_{2N} - \left(\boldsymbol{I}_{2N} - j\tau \underline{\boldsymbol{\Omega}}_i \underline{\boldsymbol{G}}\right)^{-1}\right) \underline{\boldsymbol{u}}_i\right)}{\det^{1/2}\left(\boldsymbol{I}_{2N} - j\tau \underline{\boldsymbol{\Omega}}_i \underline{\boldsymbol{G}}\right)}, \tag{5.56}$$

and

$$\phi_{\lambda_{\mathrm{E}}|\mathscr{H}_i}(\tau) = \frac{\exp\left(-(1/2)\underline{\boldsymbol{\mu}}_i^{\dagger} \underline{\boldsymbol{\Omega}}_i^{-1} \left(\boldsymbol{I}_{2N} - \left(\boldsymbol{I}_{2N} - j\tau \underline{\boldsymbol{\Omega}}_i\right)^{-1}\right) \underline{\boldsymbol{\mu}}_i\right)}{\det^{1/2}\left(\boldsymbol{I}_{2N} - j\tau \underline{\boldsymbol{\Omega}}_i\right)}, \tag{5.57}$$

respectively, where $\underline{\boldsymbol{G}} = \underline{\boldsymbol{\Omega}}_0^{-1} - \underline{\boldsymbol{\Omega}}_1^{-1}$ and $\underline{\boldsymbol{u}}_i = \underline{\boldsymbol{\mu}}_i + \underline{\boldsymbol{G}}^{-1}\left(\underline{\boldsymbol{\Omega}}_1^{-1}\underline{\boldsymbol{\mu}}_1 - \underline{\boldsymbol{\Omega}}_0^{-1}\underline{\boldsymbol{\mu}}_0\right)$.

The computational complexity for evaluating the exact performance, based on (5.40) is approximately $\mathscr{O}(K^M)$, while the computational complexity for evaluating the asymptotic performance, based on (5.57), is approximately $\mathscr{O}(N^3)$. The saving in the case of large-size systems is apparent.

5.3.4 Performance analysis

Figure 5.8 shows the ROCs for the energy test in a homogeneous WSN with $K = 8$ sensors and $N = 2$ receive antennas at the FC under different fading statistics for the reporting channels. The local performances of the sensors are $p_f = 0.05$ and $p_d = 0.5$, while the reporting phase undergoes SNR = 10 dB and SNR = 15 dB. It is apparent how the fading statistics impact the performance, e.g., the presence of a LOS component is beneficial while multimodal statistical behavior is detrimental.

Figure 5.9 shows the asymptotic ROCs for the energy test considering a homogeneous WSN with $K = 50$ sensors under different fading statistics for the reporting channels. FCs with $N = 2$ and $N = 4$ receive antennas are considered. The local performance of the sensors are $p_f = 0.25$ and $p_d = 0.5$. Again, it is apparent how the fading statistics impact the performance, e.g., the presence of a LOS component is

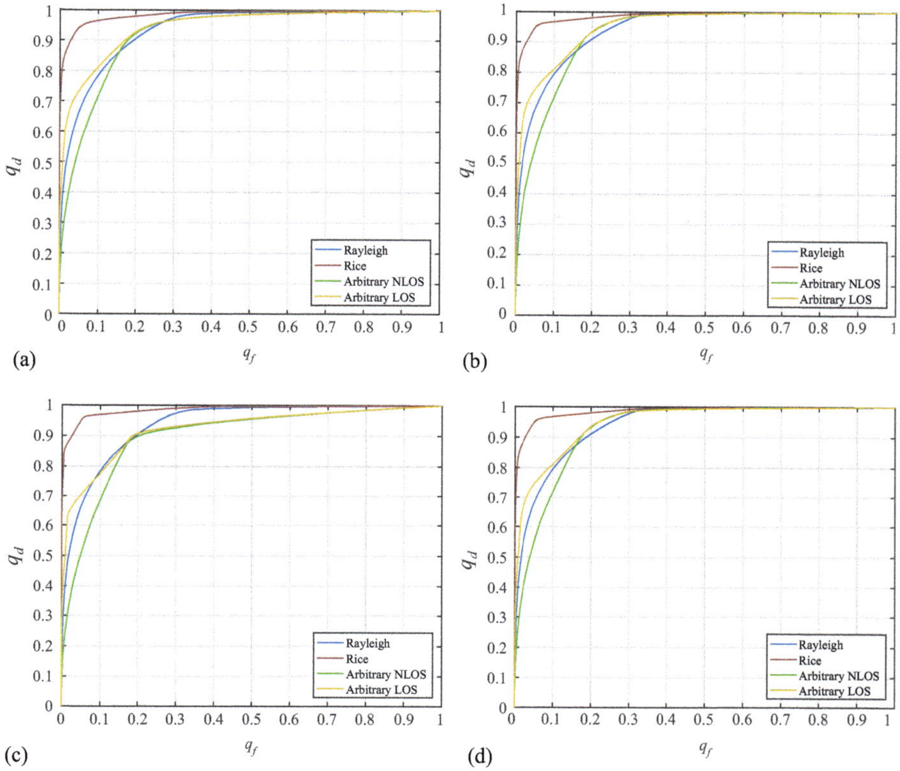

Figure 5.8 *ROC for energy test in the case of WSN with $K = 8$ sensors (homogeneous scenario with $p_f = 0.05$ and $p_d = 0.5$) and $N = 2$ receive antennas at the FC: (a) smooth modes and SNR = 10 dB, (b) smooth modes and SNR = 15 dB, (c) sharp modes and SNR = 10 dB, and (d) sharp modes and SNR = 15 dB*

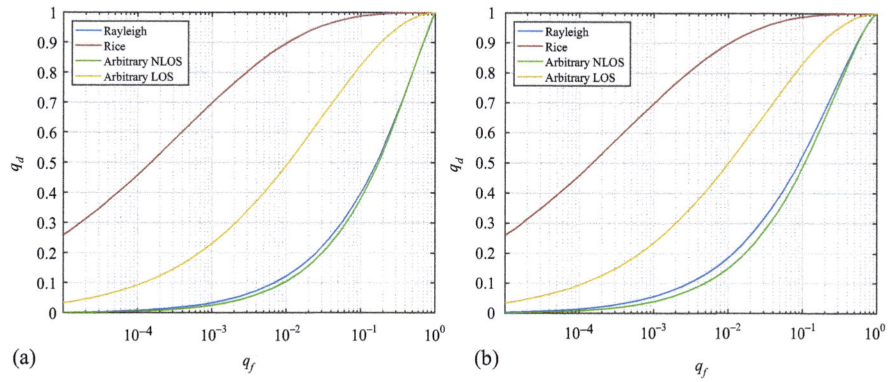

Figure 5.9 *ROC for energy test in the case of WSN with $K = 50$ sensors (homogeneous scenario with $p_f = 0.25$ and $p_d = 0.5$): (a) $N = 2$ receive antennas at the FC and (b) $N = 4$ receive antennas at the FC*

beneficial while multimodal statistical behavior is detrimental. Also, the number of receive antennas at the FC has negligible impact on the performance.

5.4 Conclusion

This chapter focused on the design of various fusion rules for MIMO decision fusion when instantaneous or statistical CSI is available at the FC. The problem arises when multiple antennas at the FC are considered in order to obtain a relevant performance improvements improvement paired with reduced energy budget.

In the former case (instantaneous CSI), several suboptimal fusion rules have been reviewed as practical alternative solutions to the exact LLR-based fusion rule. More specifically, the presented rules have been classified into two different families: (i) decode-and-fuse approaches, including MRC, EGC, and max-log fusion rules and (ii) decode-than-fuse approaches, including CV-ML and CV-MMSE fusion rules.

In the latter case (statistical CSI), the focus has been on the energy test, which reveals to be (near-)optimal in many significant scenarios of practical interest. Asymptotic analysis is also shown.

References

[1] Chen B, Tong L, and Varshney PK. Channel-aware distributed detection in wireless sensor networks. IEEE Signal Process. Mag. 2006; 23(4):16–26.
[2] Dimakis AG, Kar S, Moura JMF, Rabbat MG, and Scaglione A. Gossip algorithms for distributed signal processing. Proc. IEEE. 2010; 98(11):1847–1864.
[3] Chen B, Jiang R, Kasetkasem T, and Varshney PK. Channel aware decision fusion in wireless sensor networks. IEEE Trans. Signal Process. 2004; 52(12):3454–3458.
[4] Berger CR, Guerriero M, Zhou S, and Willett P. PAC vs. MAC for decentralized detection using noncoherent modulation. IEEE Trans. Signal Process. 2009; 57(9):3562–3575.
[5] Li F, Evans JS, and Dey S. Decision fusion over noncoherent fading multiaccess channels. IEEE Trans. Signal Process. 2011; 59(9):4367–4380.
[6] Banavar MK, Smith AD, Tepedelenlioglu C, and Spanias A. On the effectiveness of multiple antennas in distributed detection over fading MACs. IEEE Trans. Wireless Commun. 2012; 11(5):1744–1752.
[7] Ciuonzo D, Romano G, and Salvo Rossi P. Channel-aware decision fusion in distributed MIMO wireless sensor networks: decode-and-fuse vs. decode-then-fuse. IEEE Trans. Wireless Commun. 2012; 11(8):2976–2985.
[8] Ciuonzo D, Salvo Rossi P, and Dey S. Massive MIMO channel-aware decision fusion. IEEE Trans. Signal Process. 2015; 63(3):604–619.
[9] Feng J, Chen J, Swindlehurst AL, and Lopez-Salcedo JA. Massive MIMO for wireless sensing with a coherent multiple access channel. IEEE Trans. Signal Process. 2015; 63(12):3005–3017.

[10] Kay SM. Fundamentals of Statistical Signal Processing, Volume 2: Detection Theory. Englewood Cliffs, NJ: Prentice Hall, 1998.

[11] Salvo Rossi P, Ciuonzo D, and Romano G. Orthogonality and cooperation in collaborative spectrum sensing through MIMO decision fusion. IEEE Trans. Wireless Commun. 2013; 12(11):5826–5836.

[12] Salvo Rossi P, Ciuonzo D, Ekman T, and Dong H. Energy detection for MIMO decision fusion in underwater sensor networks. IEEE Sensors J. 2015; 15(3): 1630–1640.

[13] Cui T, and Tellambura C. An efficient generalized sphere decoder for rank-deficient MIMO systems. IEEE Commun. Lett. 2005; 9(5):423âŁ"–425.

[14] Ciuonzo D, Romano G, and Salvo Rossi P. Optimality of received energy in decision fusion over a Rayleigh fading diversity MAC with non-identical sensors. IEEE Trans. Signal Process. 2013; 61(1):22–27.

[15] Salvo Rossi P, Ciuonzo D, Kansanen K, and Ekman T. On energy detection for MIMO decision fusion in wireless sensor networks over NLOS fading. IEEE Commun. Lett. 2015; 19(2):303–306.

[16] Salvo Rossi P, Ciuonzo D, Kansanen K, and Ekman T. Performance analysis of energy detection for MIMO decision fusion in wireless sensor networks over arbitrary fading channels. IEEE Trans. Wireless Commun. 2016; 15(11):7794–7806.

[17] Selim B, Alhussein O, Muhaidat S, Karagiannidis G, and Liang J. Modeling and analysis of wireless channels via the mixture of Gaussians distributions. IEEE Trans. Veh. Tech. 2016; 65(10):8309–8321.

[18] Olofsson T, Ahlén A, and Gidlund M. Modeling of the fading statistics of wireless sensor network channels in industrial environments. IEEE Trans. Signal Process. 2016; 64(12):3021–3034.

[19] Fernandez M, and Williams S. Closed-form expression for the Poisson-binomial probability density function. IEEE Trans. Aerosp. Electron. Syst. 2010; 46(2):803–817.

[20] Lei A, and Schober R. Coherent max-log decision fusion in wireless sensor networks. IEEE Trans. Commun. 2010; 58(5):1327–1332.

[21] Ciuonzo D, Romano G, and Salvo Rossi P. Performance analysis and design of maximum ratio combining in channel-aware MIMO decision fusion. IEEE Trans. Wireless Commun. 2013; 12(9):4716–4728.

[22] Schreier PJ, and Scharf LL. Statistical Signal Processing of Complex-Valued Data: The Theory of Improper and Noncircular Signals. Cambridge: Cambridge University Press, 2010.

Chapter 6

Channel-aware detection and estimation in the massive MIMO regime

Domenico Ciuonzo[1]*, Pierluigi Salvo Rossi*[2]*, and Subhrakanti Dey*[3]

This chapter investigates channel-aware distributed detection (viz. binary hypothesis testing, HT) and estimation (EST) over a "virtual" and "massive" multiple-input–multiple-output (MIMO) channel at the fusion center (FC), underlining analogies and differences with uplink communication in a multiuser (massive) MIMO setup. The considered scenario takes into account channel estimation and inhomogeneous large-scale fading between the sensors and the FC. In the former case, the aim is the development of (widely) linear fusion rules, as opposed to the unsuitable (optimum) log-likelihood ratio (LLR). In the latter case, the aim is the power allocation design for decentralized estimation of a correlated random source vector with amplify-and-forward sensors and an FC adopting a minimum mean square error (MMSE) approach. In both cases, the well-known favorable propagation condition achieved in massive MIMO is exploited. In the HT problem, this greatly simplifies the development of suboptimal rules, whereas for EST problem this allows to obtain an asymptotic MSE approximation, which is then used with convex optimization techniques to solve the optimal sensor power allocation problem in an efficient fashion.

Specifically, in the HT problem, the performance of the proposed fusion rules is evaluated in two different scenarios: (a) the energy spent in the training phase cannot be assumed to be negligible with respect to the energy spent in the transmitting phase and (b) the energy used for training is negligible, and thus transmit energy can be scaled down aggressively. On the other hand, in the EST problem, the amplification factor (or equivalently transmission power) is optimized at each sensor node under two different rationales: (a) with the objective of total power minimization subject to MSE of source estimation constraint and (b) with the objective of minimizing MSE subject to total power constraint.

[1]Network Measurement and Monitoring (NM2), Italy
[2]Department of Advanced Analytics and Machine Learning, Kongsberg Digital AS, Norway
[3]School of Information Technology and Mathematical Sciences, University of South Australia, Australia

6.1 Introduction

Wireless sensor networks (WSNs) have recently attracted much research interest due to their practical popularity in accomplishing decentralized tasks, such as monitoring, sensing, computation and communication. Among these, two key tasks typically studied in the literature are *decentralized detection* and *decentralized estimation*.

Typically, most of the studies have been focused on a parallel access channels (PACs) with instantaneous or statistical channel-state information (CSI), although decentralized inference over multiple access channels (MACs) has become increasingly attractive, since it exploits both the interfering nature of the broadcast wireless medium (for spectral-efficiency purposes) and the correlated nature of reported information, all regarding the same unknown event being observed. Indeed, transmissions over coherent MAC normally provide a better performance (the improvement is achieved at the price of precise phase synchronization at sensor nodes) compared to PAC that requires more resources such as time and bandwidth. Also, deep fading scenarios are typically overcome with the use of multiple antennas at the FC (via spatial multiplexing), thus determining a "virtual" MIMO channel, and only demanding further complexity at the FC side (with respect to a single-antenna MAC) while not affecting simplicity of sensors implementation. Accordingly, the main difficulty consists in the design of *efficient* (in terms of complexity and performance) *fusion rules* at the FC and FC-coordinated optimization of sensors parameters.

In this respect, massive MIMO is an emerging technology for communication applications where antenna arrays with a few hundred elements are placed at base stations, serving many tens of terminals simultaneously [1,2]. Its potential benefits consist in (a) significant increase of the capacity and the radiated energy efficiency, (b) reduced latency through deep-fading mitigation, (c) simplification of the multiple-access layer and (d) increased robustness to unintended man-made interference and to intentional jamming [2]. From a mathematical point of view, there are *three main advantages* in using a large antenna array. First, the effect of small-scale fading is averaged out. Second, the random channel vectors between the users and the base station become pairwise orthogonal as the array size grows. Finally, massive MIMO systems allow for reduction of transmitted energy/power. Therefore, the advantages offered by massive MIMO can be surely beneficial in the context of WSNs, especially for decentralized inference over MACs. The FC may be equipped with a large-antenna array, thus giving rise to a "virtual" massive MIMO, similarly to the uplink of a multiuser MIMO communication, e.g., with a WSN made of many sensors reporting their statistics to an FC equipped with much more antennas. Alternatively, this setup may be also verified in the case of massively deployed sensor nodes, through the use of appropriate sensor subset selection approaches [3]. Accordingly, there has been a recent interest to equip the FC with large arrays of antennas, also known as the massive MIMO framework [4–6].

Accordingly, the main focus of this chapter is on decentralized *detection* and *estimation* capitalizing a massive MIMO framework. Specifically, within such framework, *decentralized detection* is accomplished via decision fusion (DF), consisting in transmitting local decisions about an observed phenomenon from sensors to an FC for a global decision, with the intent of surveillance and/or situation awareness purposes [7]. On the other hand, for *decentralized estimation*, the general case of a vector source comprising random and correlated components is considered. The two most popular forwarding techniques, *decode-and-forward* and *amplify-and-forward*, are here considered for decentralized detection and estimation, respectively.

The main stepping stone of the present discussion is the asymptotic approximation arising from the so-called favorable propagation conditions in the massive MIMO literature. Such conditions allow the design of efficient (widely)linear fusion rules for the detection problem and to simplify the optimal MSE expression for the estimation problem when the number of receive antennas largely exceeds the number of sensors, i.e., $N \gg K$ [8]. It is also seen that within the massive MIMO approximation, the asymptotic performance (and the optimal sensor power allocation in the estimation problem, as an immediate consequence) depends only on the distance-based attenuation components and not on the randomly varying fading gains.

Specifically, in the case of *detection*, the aforementioned results are capitalized to characterize asymptotically the performance of the considered fusion rules under two relevant energy-efficiency scenarios: (a) the energy spent in the training phase is not negligible and (b) the energy used for training is negligible and thus transmit energy can be scaled down aggressively. On the other hand, in the *estimation* problem, the asymptotic MSE expression allows to optimally design the sensors' amplification factors with respect to the following two complementary objectives: (a) total power minimization subject to a maximum (optimal) MSE constraint and (b) MSE minimization subject to a total power constraint. It is shown that the both power allocation (optimization) problems, mentioned earlier, are *convex* and can be thus solved numerically using well-known optimization routines in polynomial time, and under some conditions, closed-form solutions are available.

The remainder of the chapter is organized as follows. Section 6.2 introduces the considered system model for both the considered inference tasks; in Section 6.3, decentralized inference techniques for massive MIMO regime are developed an discussed; Section 6.4 presents a corresponding theoretical analysis (and corresponding asymptotic power allocation for decentralized estimation). Final considerations and suggested further reading in Section 6.5 end the chapter.

6.2 System model

6.2.1 WSN model

In this section, the system model, sketched in Figure 6.1, is briefly described. Specifically, the following two *decentralized* inference problems are considered for a WSN with K sensors:

- **HT**: A binary hypothesis test, where the hypotheses $\mathcal{H}_0$ and $\mathcal{H}_1$ may represent the absence and the presence of a specific target of interest, respectively.
- **EST**: A vector source estimation problem, characterized by $\boldsymbol{\theta} = [\theta_1, \ldots, \theta_K]^T \in \mathbb{C}^K$, where $\boldsymbol{\theta} \sim \mathcal{N}_{\mathbb{C}}(\mathbf{0}_K, \boldsymbol{\Sigma}_\theta)$, where the source (positive-definite) covariance matrix $\boldsymbol{\Sigma}_\theta \in \mathbb{H}_+^{K\times K}$ is not necessarily diagonal. In this chapter, the general *correlated case* is tackled.

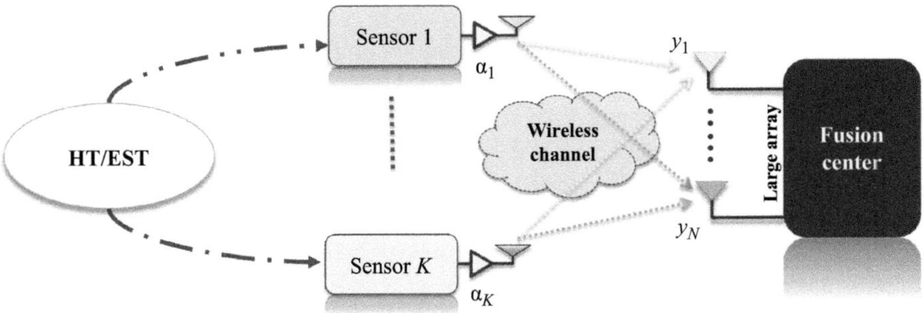

Figure 6.1 Decentralized inference in a virtual massive MIMO channel

In HT, the kth sensor, $k \in \mathcal{K} \triangleq \{1, 2, \ldots, K\}$, takes a binary-valued local decision b_k about the observed phenomenon on the basis of its own measurements. Unless otherwise stated, here it is not made any conditional (given $\mathcal{H}_i$) mutual independence assumption on b_k. Each decision b_k is mapped to a symbol $x_k \in \mathcal{X} = \{-1, +1\}$ representing a binary phase-shift keying (BPSK) modulation[1]: without loss of generality it is assumed that $b_k = \mathcal{H}_i$ maps into $x_k = (2i - 1)$, $i \in \{0, 1\}$. The quality of the kth sensor decisions is characterized by the conditional probabilities $P(x_k|\mathcal{H}_i)$. Specifically, $P_{D,k} \triangleq P\,(x_k = 1|\mathcal{H}_1)$ and $P_{F,k} \triangleq P\,(x_k = 1|\mathcal{H}_0)$ are used to denote the (local) probability of detection and false alarm of the kth sensor, respectively.

In EST, the kth sensor, $k \in \mathcal{K} \triangleq \{1, 2, \ldots, K\}$, forwards its analog measurement $x_k = \theta_k + n_k$, where n_k denotes the kth source-to-sensor noise component, which are stacked compactly in $\boldsymbol{n} \triangleq \begin{bmatrix} n_1 & \cdots & n_K \end{bmatrix}^T$, where $\boldsymbol{n} \sim \mathcal{N}_{\mathbb{C}}(\mathbf{0}_K, \boldsymbol{\Sigma}_n)$, and $\boldsymbol{\Sigma}_n \triangleq \mathrm{diag}(\sigma^2_{n_1}, \ldots, \sigma^2_{n_K})$ is the source-to-sensor noise covariance matrix.

In *both* HT and EST, it is assumed that the kth sensor multiplies its transmission by the complex amplification gain α_k, all collected in the (diagonal) matrix $\boldsymbol{A} \triangleq \mathrm{diag}(\alpha_1, \ldots, \alpha_K)$. In the following, the corresponding energy matrix $\boldsymbol{P} \triangleq \mathrm{diag}(p_1, \ldots, p_K)$, where $p_k \triangleq |\alpha_k|^2$, will be also employed.

Sensors communicate with an FC equipped with N receive antennas over a wireless flat-fading MAC to exploit diversity and combat signal attenuation due to small-scale fading; this setup determines a distributed or "virtual" MIMO channel [9,10]. It is worth remarking that the results shown in this chapter rely on the assumption that the *number of antennas at the FC is much larger than the number of transmitting sensors*. In other terms, this condition is formalized as $N \gg K$; nonetheless, the results which still apply in the generic virtual MIMO setup will be emphasized. Also, perfect synchronization, as in [9–12], is assumed at the FC. It is worth highlighting that multiple antennas at the FC do not make these assumptions harder to verify with respect to a single-antenna MAC.

[1]In the case of an absence/presence task, where $\mathcal{H}_0$ is much more probable, on-off keying (OOK) can be employed for energy-efficiency purposes.

The vector model at the FC is for *both* inference tasks:

$$\boldsymbol{y} = \boldsymbol{G}\boldsymbol{A}\boldsymbol{x} + \boldsymbol{w} \tag{6.1}$$

where $\boldsymbol{y} \in \mathbb{C}^N$, $\boldsymbol{G} \in \mathbb{C}^{N\times K}$ and $\boldsymbol{w} \sim \mathcal{N}_{\mathbb{C}}(\boldsymbol{0}_N, \boldsymbol{\Sigma}_w)$ (with $\boldsymbol{\Sigma}_w \triangleq \sigma_w^2 \boldsymbol{I}_N$) are the received-signal vector, the channel matrix and the noise vector, respectively. Also, $\boldsymbol{x} \in \mathcal{X}^K$ (resp. $\boldsymbol{x} \in \mathbb{C}^K$) holds in HT (EST) for the transmitted-signal vector. Additionally, for EST, the vector model at the FC can be rewritten explicitly as:

$$\boldsymbol{y} = \boldsymbol{G}\boldsymbol{A}\boldsymbol{\theta} + \boldsymbol{w}_{\text{est}}\,, \tag{6.2}$$

where $\boldsymbol{w}_{\text{est}} \triangleq (\boldsymbol{G}\boldsymbol{A}\boldsymbol{n} + \boldsymbol{w})$, which allows to highlight the separation between useful-signal and noise contributions.

In detail, the matrix $\boldsymbol{G}$ models independent fast fading, geometric attenuation and log-normal shadow fading. The coefficient $g_{n,k}$ is then expressed as:

$$g_{n,k} = \sqrt{\beta_k}\, h_{n,k}, \qquad n = 1, \ldots, N \tag{6.3}$$

where the known (since it is assumed constant over many coherence intervals) term $\sqrt{\beta_k}$ models the geometric attenuation and shadow fading.[2] On the other hand, the term $h_{n,k}$ denotes the fast fading coefficient between the kth sensor and the nth receive antenna, implying that there are non-line of sight wireless transmissions between sensors and the FC, which arise due to obstacles (fixed or mobile) resulting in multipath fading propagation and a rich scattering environment (application examples include outdoor or residential WSNs [13–15]). All the coefficients $h_{n,k}$ are assumed independent and identically distributed (i.i.d.) random variables, with $h_{n,k} \sim \mathcal{N}_{\mathbb{C}}(0, 1)$. Based on these assumptions, the overall channel matrix can be expressed as $\boldsymbol{G} = \boldsymbol{H}\boldsymbol{D}^{1/2}$, with $\boldsymbol{H} \in \mathbb{C}^{N\times K}$ denoting the matrix of fast-fading coefficients and $\boldsymbol{D} \triangleq \text{diag}(\beta_1, \ldots, \beta_K) \in \mathbb{R}_+^{K\times K}$ the slow-fading matrix.

Finally, another central design criterion that needs to be considered is the average total power consumed by the WSN for transmission. For the inference stage (i.e., not including channel estimation contribution), this can be simply expressed as:

$$P_{tot} \triangleq \mathbb{E}\left\{\|\boldsymbol{A}\boldsymbol{x}\|^2\right\} = \sum_{k=1}^{K} |\alpha_k|^2 \, [\boldsymbol{\Sigma}_{\boldsymbol{x}}]_{k,k} \tag{6.4}$$

In HT, the kth diagonal term of the covariance matrix $\boldsymbol{\Sigma}_{\boldsymbol{x}}$ is simply equal to $[\boldsymbol{\Sigma}_{\boldsymbol{x}}]_{k,k} = 1$, whereas in EST simple manipulations show that $[\boldsymbol{\Sigma}_{\boldsymbol{x}}]_{k,k} = [\boldsymbol{\Sigma}_{\boldsymbol{\theta}} + \boldsymbol{\Sigma}_{\boldsymbol{n}}]_{k,k}$

[2]It is assumed that β_k is constant over n, as this assumption is justified since the sensor-DFC distance is typically much higher w.r.t. the inter-antenna distance.

(as a consequence of statistical independence of $\boldsymbol{\theta}$ and $\boldsymbol{n}$). Therefore, P_{tot} in the two inference tasks considered equals:

$$P_{tot}^{\mathrm{HT}} = \sum_{k=1}^{K} |\alpha_k|^2 = \sum_{k=1}^{K} \frac{\lambda_k}{\beta_k} \tag{6.5}$$

$$P_{tot}^{\mathrm{EST}} = \sum_{k=1}^{K} |\alpha_k|^2 \left[\boldsymbol{\Sigma}_{\theta} + \boldsymbol{\Sigma}_{n}\right]_{k,k} = \sum_{k=1}^{K} \frac{\lambda_k}{\beta_k} \left[\boldsymbol{\Sigma}_{\theta} + \boldsymbol{\Sigma}_{n}\right]_{k,k} \tag{6.6}$$

where in the second expression of each line, the definition $\lambda_k \triangleq \beta_k |\alpha_k|^2, \forall k$ has been used.

6.2.2 Channel training and estimation

With reference to HT problem, the general case where the FC has available estimated instantaneous CSI $\hat{\boldsymbol{G}}$ is considered here. On the other hand, for EST problem, a simplified analysis with ideal channel estimation $\hat{\boldsymbol{G}} \to \boldsymbol{G}$ is considered for simplicity, while leaving the much more involved imperfect-channel analysis to the interested reader.

This may be achieved by pilot transmissions from the sensors at the beginning of each fading block. Specifically, it is assumed that a part of the coherence interval is used for channel training. Let be $\tau \in \mathbb{N}$ the number of symbols used as pilots and $(\tau_c - \tau)$ symbols are used for the generic inference task, where τ_c is the number of symbols within the channel coherence interval. During the training phase, all sensors simultaneously transmit mutually orthogonal pilot sequences of length τ. The pilot sequences are collected in a matrix $(\sqrt{\tau\rho_{\mathrm{P}}}\, \boldsymbol{\Phi}) \in \mathbb{C}^{\tau \times K}$ (with $\tau \geq K$), where ρ_{P} controls pilot energy and satisfies the condition $\boldsymbol{\Phi}^{\dagger}\boldsymbol{\Phi} = \boldsymbol{I}_K$. Then, the received signal matrix model of the training phase is:

$$\boldsymbol{Y}_{\mathrm{P}} = \sqrt{\tau\rho_{\mathrm{P}}}\, \boldsymbol{G}\boldsymbol{\Phi}^T + \boldsymbol{W}_{\mathrm{P}} \tag{6.7}$$

where $\boldsymbol{Y}_{\mathrm{P}} \in \mathbb{C}^{N \times \tau}$ and $\boldsymbol{W}_{\mathrm{P}} \in \mathbb{C}^{N \times \tau}$ has i.i.d. elements $w_{\mathrm{P},i,j} \sim \mathcal{N}_{\mathbb{C}}(0, \sigma_w^2)$. Given the model in (6.7), the MMSE estimator of $\boldsymbol{G}$ given $\boldsymbol{Y}_{\mathrm{P}}$ is [16]:

$$\hat{\boldsymbol{G}} = \boldsymbol{Y}_{\mathrm{P}}\boldsymbol{\Phi}^{*} \frac{1}{\sqrt{\tau\rho_{\mathrm{P}}}} \tilde{\boldsymbol{D}} \tag{6.8}$$

where $\tilde{\boldsymbol{D}} \triangleq ((\sigma_w^2/\tau\rho_{\mathrm{P}})\boldsymbol{D}^{-1} + \boldsymbol{I}_K)^{-1}$ and can be expressed, after some standard manipulations, in the alternative form:

$$\hat{\boldsymbol{G}} = \left(\boldsymbol{G} + \frac{1}{\sqrt{\tau\rho_{\mathrm{P}}}} \boldsymbol{W}_{\mathrm{P}}\boldsymbol{\Phi}^{*}\right) \tilde{\boldsymbol{D}}. \tag{6.9}$$

Also, for completeness, it is defined the error matrix as $\boldsymbol{\Xi} \triangleq (\hat{\boldsymbol{G}} - \boldsymbol{G})$. It can be shown that $\hat{\boldsymbol{G}}$ is independent on $\boldsymbol{\Xi}$ (also the columns of both $\hat{\boldsymbol{G}}$ and $\boldsymbol{\Xi}$ are mutually independent) and that both kth columns of the aforementioned matrices (denoted with $\hat{\boldsymbol{g}}_k$ and $\boldsymbol{\xi}_k$, respectively) are proper complex-valued Gaussian vectors with the following moments:

$$\mathbb{E}\{\hat{\boldsymbol{g}}_k\} = \mathbb{E}\{\boldsymbol{\xi}_k\} = \boldsymbol{0}_N; \tag{6.10}$$

$$\mathbb{E}\{\hat{\boldsymbol{g}}_k\,\hat{\boldsymbol{g}}_k^\dagger\} = d_{\hat{g},k}\,\boldsymbol{I}_N, \quad d_{\hat{g},k} \triangleq \frac{(\tau\rho_{\mathrm{P}})\beta_k^2}{(\tau\rho_{\mathrm{P}})\beta_k + \sigma_w^2}; \tag{6.11}$$

$$\mathbb{E}\{\boldsymbol{\xi}_k\,\boldsymbol{\xi}_k^\dagger\} = d_{e,k}\,\boldsymbol{I}_N, \quad d_{e,k} \triangleq \frac{\sigma_w^2\beta_k}{(\tau\rho_{\mathrm{P}})\beta_k + \sigma_w^2}. \tag{6.12}$$

Clearly, as $\tau \to +\infty$, the perfect instantaneous CSI case is approached. It can be also highlighted that the channel-error matrix can be written as $\boldsymbol{\Xi} = \left[\sqrt{d_{e,1}}\,\boldsymbol{e}_1 \;\cdots\; \sqrt{d_{e,K}}\,\boldsymbol{e}_K\right]$ (i.e., highlighting the variance of the error components), where $\boldsymbol{e}_k \in \mathbb{C}^{1\times K}$ $(k = 1,\ldots,K)$ is a vector accounts for the standardized estimation error from the channels, i.e., $\boldsymbol{e}_k \sim \mathcal{N}_{\mathbb{C}}(\boldsymbol{0}_N, \boldsymbol{I}_N)$.

Finally, it is worth noticing that in HT (where channel-estimation is also considered), the overall energy spent from the whole WSN is given by ($\boldsymbol{\phi}_k$ denotes the kth column of $\boldsymbol{\Phi}$):

$$\mathscr{E}_{\mathrm{ht}} \triangleq (\tau_c - \tau)\sum_{k=1}^{K}|\alpha_k|^2 + \tau\rho_{\mathrm{P}}\sum_{k=1}^{K}\mathbb{E}\{\|\boldsymbol{\phi}_k\|^2\} = (\tau_c - \tau)\,P_{tot}^{\mathrm{HT}} + \tau\rho_{\mathrm{P}}\,K \tag{6.13}$$

which accounts for both training and transmitting phases.

6.2.3 Favorable propagation

From Section 6.2.2 it is known that $\hat{\boldsymbol{g}}_k \sim \mathcal{N}_{\mathbb{C}}(\boldsymbol{0}_N, d_{\hat{g},k}\,\boldsymbol{I}_N)$. Also, the vectors $\hat{\boldsymbol{g}}_k$, $k \in \mathcal{K}$, are mutually independent. Thus, the so-called favorable propagation conditions hold [17]:

$$\left(\frac{1}{N}\right)\boldsymbol{R}_{\hat{g}} \approx \boldsymbol{P}\,\boldsymbol{D}_{\hat{g}}, \qquad N \gg K; \tag{6.14}$$

where $\boldsymbol{R}_{\hat{g}} \triangleq [(\hat{\boldsymbol{G}}\boldsymbol{A})^\dagger\,\hat{\boldsymbol{G}}\boldsymbol{A}]$ and $\boldsymbol{D}_{\hat{g}}$ is a diagonal matrix whose kth element equals $d_{\hat{g},k}$ (cf. (6.11)). Furthermore, it is worth noticing that, in the case of perfect channel estimation, the above condition reduces to:

$$\left(\frac{1}{N}\right)\boldsymbol{R} \approx \boldsymbol{P}\,\boldsymbol{D}, \qquad N \gg K; \tag{6.15}$$

where $\boldsymbol{R} \triangleq [(\boldsymbol{G}\boldsymbol{A})^\dagger\,\boldsymbol{G}\boldsymbol{A}]$. Experimental results [1] have shown that practical large arrays approximate well the aforementioned propagation assumption. The above condition will be capitalized in what follows to *simplify the design stage* in both the considered inference tasks.

6.3 Inference in massive MIMO regime

The FC is in charge of providing a decision $\hat{\mathcal{H}}$ (resp. an estimate $\hat{\boldsymbol{\theta}}$ of the source vector $\boldsymbol{\theta}$) from the received signal vector $\boldsymbol{y}$ given by the model in (6.1). Henceforth, it is assumed that the FC has perfect knowledge of the statistical characterization of the received noise vector $\boldsymbol{w}$, and of source $\boldsymbol{\theta}$ an the sensing noise $\boldsymbol{n}$ vectors in EST.

6.3.1 *Fusion techniques for HT*

The optimal test [18] for this problem, given the estimate $\hat{\boldsymbol{G}}$, is formulated as:

$$\left\{\Lambda_{\text{opt}} \triangleq \ln\left[\frac{p(\boldsymbol{y}|\hat{\boldsymbol{G}},\mathcal{H}_1)}{p(\boldsymbol{y}|\hat{\boldsymbol{G}},\mathcal{H}_0)}\right]\right\} \underset{\hat{\mathcal{H}}=\mathcal{H}_0}{\overset{\hat{\mathcal{H}}=\mathcal{H}_1}{\gtrless}} \gamma \tag{6.16}$$

where $\hat{\mathcal{H}}$, Λ_{opt} and γ denote the estimated hypothesis, the LLR and the threshold which the LLR is compared to. The threshold γ can be determined to assure a fixed system false-alarm rate (Neyman–Pearson approach) or can be chosen to minimize the probability of error (Bayesian approach) [18]. Exploiting the independence of $\boldsymbol{y}$ from $\mathcal{H}_i$, given $\boldsymbol{x}$, an explicit expression of the LLR in (6.16) is obtained as:

$$\begin{aligned}\Lambda_{\text{opt}} &= \ln\left[\frac{\sum_{\boldsymbol{x}\in\mathcal{X}^K} p(\boldsymbol{y}|\hat{\boldsymbol{G}},\boldsymbol{x})P(\boldsymbol{x}|\mathcal{H}_1)}{\sum_{\boldsymbol{x}\in\mathcal{X}^K} p(\boldsymbol{y}|\hat{\boldsymbol{G}},\boldsymbol{x})P(\boldsymbol{x}|\mathcal{H}_0)}\right] \\ &= \ln\left[\frac{\sum_{\boldsymbol{x}\in\mathcal{X}^K} \exp\left(-(\|\boldsymbol{y}-\hat{\boldsymbol{G}}\boldsymbol{A}\boldsymbol{x}\|^2)/\sigma_e^2\right)P(\boldsymbol{x}|\mathcal{H}_1)}{\sum_{\boldsymbol{x}\in\mathcal{X}^K} \exp\left(-(\|\boldsymbol{y}-\hat{\boldsymbol{G}}\boldsymbol{A}\boldsymbol{x}\|^2)/\sigma_e^2\right)P(\boldsymbol{x}|\mathcal{H}_0)}\right]\end{aligned} \tag{6.17}$$

where $\boldsymbol{G} = (\hat{\boldsymbol{G}} - \boldsymbol{\Xi})$ has been replaced in (6.1), and the definition $\sigma_e^2 \triangleq (\sum_{k=1}^{K} d_{e,k}\,p_k + \sigma_w^2)$ has been employed. Unfortunately, the optimal rule in (6.17) presents several difficulties in its implementation, such as (a) availability of $\hat{\boldsymbol{G}}$, $P(\boldsymbol{x}|\mathcal{H}_i)$ and σ_w^2 and (b) numerical instability of the expression, due to the presence of exponential functions with large dynamics [10,19]. More importantly, the exponential growth of the complexity with K is prohibitive for a practical design. Design of suboptimal fusion rules with simpler implementation and (possibly) reduced system knowledge is then *extremely desirable*. Accordingly, in what follows, *three different philosophies* to reduced-complexity DF are introduced, able to capitalize massive MIMO gains.

6.3.1.1 Linear filters plus decision fusion

The *first approach* exploits the observation that Λ_{opt} can be well-approximated by a simple two-step architecture when N grows large. More specifically, the *first step* consists in a linear filter which recovers the soft decision from each sensor, while the *second step* is in charge of fusing them so as to obtain a final decision. First, consider the model in (6.1) and the linear processing $z \triangleq F^{\dagger}y$, closely related to matched filter (MF), zero forcing (ZF) and MMSE processing, that is,

$$F \triangleq \begin{cases} \hat{G}A & \text{MF} \\ \hat{G}A(R_{\hat{g}})^{-1} & \text{ZF} \\ \hat{G}A(R_{\hat{g}} + \sigma_e^2 I_K)^{-1} & \text{MMSE} \end{cases} \tag{6.18}$$

It is worth observing that the compression given by z (since it is here assumed $N \gg K$) does not entail loss of detection power for the considered HT, since z is a *sufficient statistic* for the considered HT in the cases of MF, ZF and MMSE processing [5]. It is also straightforward to show that $z|\hat{G}, x$ is Gaussian-distributed [5] in all the three cases (as an immediate consequence of the linear processing involved).

Furthermore, exploiting favorable propagation conditions in (6.14), the vector pdf $z|\hat{G}, x$ is approximately distributed ($N \gg K$) as follows:

$$z|\hat{G}, x \overset{a}{\sim} \begin{cases} \mathcal{N}_{\mathbb{C}}(NPD_{\hat{g}}x,\ N\sigma_e^2 PD_{\hat{g}}) & \text{MF} \\ \mathcal{N}_{\mathbb{C}}(x,\ \frac{\sigma_e^2}{N}(PD_{\hat{g}})^{-1}) & \text{ZF/MMSE} \end{cases} \tag{6.19}$$

It is apparent from (6.19) that all the filters show the desirable *independence* property $p(z|\hat{G}, x) \approx \prod_{k=1}^{K} \varphi(z_k|x_k)$ (recall that P and $D_{\hat{g}}$ are *both diagonal*). In what follows, a fusion step will be considered which is based (a) either on the complex-valued filter output (i.e., z) or (b) the estimated (BPSK) transmitted vector, i.e., $z_q \triangleq \text{sign}(\Re\{z\})$ ($z_q \in \mathcal{X}^K$), where:

$$\bar{\varphi}(z_{q,k}|x_k) \triangleq \begin{cases} P_{e,k} & z_{q,k} \neq x_k \\ 1 - P_{e,k} & z_{q,k} = x_k \end{cases} \tag{6.20}$$

In the above equation, the term $P_{e,k}$ denotes the expression of the channel bit-error probability between kth sensor and the FC, namely, $P_{e,k} \triangleq \mathcal{Q}\left((\sqrt{2Nd_{\hat{g},k}}\,|\alpha_k|)/\sigma_e\right)$. Obviously as N grows large, also $P(z_q|\hat{G}, x) \approx \prod_{k=1}^{K} \bar{\varphi}(z_{q,k}|x_k)$ holds. It is worth noticing that while the latter approach clearly belongs to a *decode-then-fuse* philosophy [10], the former approach represents a hybrid solution in between *decode-then-fuse and decode-and-fuse*.

Once processing of y to obtain z is performed, asymptotic large-array pdfs in (6.19) can be then effectively exploited as follows, with the help of assumption of *conditionally independent decisions*,[3] i.e., $P(x|\mathcal{H}_i) = \prod_{k=1}^{K} P(x_k|\mathcal{H}_i)$. Indeed, the following two asymptotic expressions will hold with the two considered approaches:

$$p(z|\hat{G}, \mathcal{H}_i) \approx \prod_{k=1}^{K} \varphi(z_k|x_k) P(x_k|\mathcal{H}_i) \tag{6.21}$$

$$P(z_q|\hat{G}, \mathcal{H}_i) \approx \prod_{k=1}^{K} \bar{\varphi}(z_{q,k}|x_k) P(x_k|\mathcal{H}_i) \tag{6.22}$$

[3]This assumption is frequently used in DF and is used either as a simplifying assumption or a requirement to be satisfied by placing appropriate requirements on the WSN [20]. The extension to conditionally dependent decisions, although tedious, is however straightforward.

Specifically, in case (a), the *max-log approximation* [21] is adopted in (6.21), which leads to the simple rule $\Lambda_{\mathrm{u}} \triangleq \sum_{k=1}^{K} \lambda_{\mathrm{u},k}$, where $\lambda_{\mathrm{u},k}$ is obtained with the following linear structure with saturation:

$$\lambda_{\mathrm{u},k} \triangleq \begin{cases} \ln\left[\frac{1-P_{D,k}}{1-P_{F,k}}\right], & f_k < \frac{1}{2}\ln\left[\frac{1-P_{D,k}}{P_{D,k}}\right] \\ 2f_k + \ln\left[\frac{P_{D,k}}{1-P_{F,k}}\right], & \frac{1}{2}\ln\left[\frac{1-P_{D,k}}{P_{D,k}}\right] \le f_k < \frac{1}{2}\ln\left[\frac{1-P_{F,k}}{P_{F,k}}\right], \\ \ln\left[\frac{P_{D,k}}{P_{F,k}}\right], & f_k \ge \frac{1}{2}\ln\left[\frac{1-P_{F,k}}{P_{F,k}}\right] \end{cases} \tag{6.23}$$

where f_k is easily obtained from z_k as follows:

$$f_k \triangleq \begin{cases} \frac{2\,\Re\{z_k\}}{\sigma_e^2} & \text{MF} \\ \frac{2p_k\, d_{\hat{g},k}\, N\, \Re\{z_k\}}{\sigma_e^2} & \text{ZF/MMSE} \end{cases} \tag{6.24}$$

On the other hand, in the case (b) the *Chair–Varshney rule* [22] is exploited (i.e., the explicit LLR $\ln\left[P(z_{\mathrm{q}}|\hat{\boldsymbol{G}}, \mathcal{H}_1)/P(z_{\mathrm{q}}|\hat{\boldsymbol{G}}, \mathcal{H}_0)\right]$ based on (6.22), *without further approximations*), in order to obtain the fusion statistic $\Lambda_{\mathrm{q}} \triangleq \sum_{k=1}^{K} \lambda_{\mathrm{q},k}$, where:

$$\lambda_{\mathrm{q},k} \triangleq \hat{b}_k \ln\left[\frac{\alpha_k(P_{D,k})}{\alpha_k(P_{F,k})}\right] + (1-\hat{b}_k)\ln\left[\frac{\eta_k(P_{D,k})}{\eta_k(P_{F,k})}\right], \tag{6.25}$$

with $\hat{b}_k \triangleq (1 + z_{\mathrm{q,k}})/2$, $\alpha_k(P) \triangleq (P\cdot(1-P_{e,k}) + (1-P)\cdot P_{e,k})$ and $\eta_k(P) \triangleq (1-\alpha_k(P))$.

6.3.1.2 WL rules

The *second approach* consists in adopting a WL statistic, that is, designing a fusion rule $\Lambda_{\mathrm{WL}} \triangleq \underline{\boldsymbol{a}}^{\dagger}\underline{\boldsymbol{y}}$, where $\underline{\boldsymbol{x}} \triangleq \left[\boldsymbol{x}^T \; \boldsymbol{x}^{\dagger}\right]^T$ denotes the *augmented form* of the complex-valued vector $\boldsymbol{x}$. Such approach is motivated by reduced complexity and $\boldsymbol{y}|\hat{\boldsymbol{G}}, \mathcal{H}_i$ being an *improper* complex-valued random vector, that is, the pseudocovariance $\bar{\boldsymbol{\Sigma}}_{\boldsymbol{y}|\hat{\boldsymbol{G}},\mathcal{H}_i} \neq \boldsymbol{O}_{N\times N}$. Additionally, several optimization metrics may be considered for obtaining $\underline{\boldsymbol{a}}$. In the following discussion, for simplicity, the augmented vector $\underline{\boldsymbol{a}}$ is chosen as the maximizer of the so-called deflection measure [23,24], that is,

$$\underline{\boldsymbol{a}}_{\mathrm{WL},i} \triangleq \arg \max_{\underline{\boldsymbol{a}}:\, \|\underline{\boldsymbol{a}}\|^2=1} \left\{ D_i\left(\underline{\boldsymbol{a}}\right) \triangleq \frac{(\mathbb{E}\{\Lambda_{\mathrm{WL}}|\mathcal{H}_1\} - \mathbb{E}\{\Lambda_{\mathrm{WL}}|\mathcal{H}_0\})^2}{\mathrm{var}\{\Lambda_{\mathrm{WL}}|\mathcal{H}_i\}} \right\} \tag{6.26}$$

$D_0(\underline{\boldsymbol{a}})$ and $D_1(\underline{\boldsymbol{a}})$ correspond to the *normal* [23] and *modified* [24] deflections, respectively. Hereinafter $\Lambda_{\mathrm{WL},i} \triangleq (\underline{\boldsymbol{a}}_{\mathrm{WL},i}^{\dagger}\underline{\boldsymbol{y}})$ will be used to denote the WL statistic maximizing $D_i(\underline{\boldsymbol{a}})$. The explicit form of $\underline{\boldsymbol{a}}_{\mathrm{WL},i}$ is obtained as detailed hereinafter.

WL rules closed-form

The vector $\underline{\boldsymbol{a}}_{\mathrm{WL},i}$, being the optimal solution of the optimization problem in (6.26), is given by:

$$\underline{\boldsymbol{a}}_{\mathrm{WL},i} = (\boldsymbol{\Sigma}^{-1}_{\underline{\boldsymbol{y}}|\hat{\boldsymbol{G}},\mathcal{H}_i}\, \underline{\hat{\boldsymbol{G}}\boldsymbol{A}}\, \boldsymbol{\mu}_{1,0}) \times \|\boldsymbol{\Sigma}^{-1}_{\underline{\boldsymbol{y}}|\hat{\boldsymbol{G}},\mathcal{H}_i}\, \underline{\hat{\boldsymbol{G}}\boldsymbol{A}}\, \boldsymbol{\mu}_{1,0}\|^{-1} \tag{6.27}$$

where $\mu_{1,0} \triangleq 2\left[(P_{D,1} - P_{F,1}) \cdots (P_{D,K} - P_{F,K}) \right]^T$ and $\boldsymbol{\Sigma}_{\underline{y}|\hat{G},\mathcal{H}_i} = (\underline{\hat{\boldsymbol{G}}\boldsymbol{A}}\, \boldsymbol{\Sigma}_{x|\mathcal{H}_i}\, (\underline{\hat{\boldsymbol{G}}\boldsymbol{A}})^\dagger + \sigma_e^2 \boldsymbol{I}_{2N})$, respectively.

It is worth noticing that WL rules take into account the general case of *conditionally dependent* decisions, while requiring only the first and second order moments of $\boldsymbol{x}|\mathcal{H}_i$.

6.3.1.3 (Modified) MRC

The *third approach* relies on LLR simplification in (6.17) under the (design) assumption of perfect sensors [25,26], i.e., $P(\boldsymbol{x} = \boldsymbol{1}_K|\mathcal{H}_1) = P(\boldsymbol{x} = -\boldsymbol{1}_K|\mathcal{H}_0) = 1$. In this case, $\boldsymbol{x} \in \{\boldsymbol{1}_K, -\boldsymbol{1}_K\}$ and (6.17) reduces to [10]:

$$\ln\left[\frac{\exp\left(-(\|\boldsymbol{y} - \sqrt{\rho}\,\hat{\boldsymbol{G}}\boldsymbol{A}\boldsymbol{1}_K\|^2)/\sigma_e^2\right)}{\exp\left(-(\|\boldsymbol{y} + \sqrt{\rho}\,\hat{\boldsymbol{G}}\boldsymbol{A}\boldsymbol{1}_K\|^2)/\sigma_e^2\right)}\right] \propto \Re\left(\boldsymbol{a}_{\text{MRC}}^\dagger \boldsymbol{y}\right) \triangleq \Lambda_{\text{MRC}} \tag{6.28}$$

where $\boldsymbol{a}_{\text{MRC}} \triangleq (\hat{\boldsymbol{G}}\boldsymbol{A}\,\boldsymbol{1}_K)$ and terms independent from $\boldsymbol{y}$ have been incorporated in γ as in (6.16). It is worth noticing that the MRC is a suboptimal rule since, in the practice, the sensor local decisions are far from being perfect. Nonetheless, it is interesting to note that the MRC in the case of estimated CSI is simply obtained by replacing $\boldsymbol{G}$ in the MRC formula with perfect CSI [26] with its estimate $\hat{\boldsymbol{G}}$, i.e., knowledge of error statistics is not needed.

Additionally, in order to exploit the linear SNR increase with N, which would inevitably make the fusion process mainly dependent on the "sensing" errors (and consequently MRC rule becomes clearly inappropriate, since its design is unaware of sensing errors), a modified form of MRC (mMRC) can be also considered:

$$\Lambda_{\text{mMRC}} \triangleq \Re\left(\boldsymbol{a}_{\text{mMRC}}^\dagger \boldsymbol{y}\right) \tag{6.29}$$

where $\boldsymbol{a}_{\text{mMRC}} \triangleq (\hat{\boldsymbol{G}}\boldsymbol{A}^{-1}\boldsymbol{D}_{\hat{g}}^{-1}\,\boldsymbol{1}_K)$. It can be observed that mMRC applies a sort of "static" ZF (via $\boldsymbol{A}^{-1}\boldsymbol{D}_{\hat{g}}^{-1}$) in order to remove rule dependence on large-scale fading coefficients (which may become detrimental as the SNR grows). Also, it can be shown that (6.29) asymptotically approaches (as N grows large) the performance of the well-known *counting rule*, which has been shown to be robust even with heterogeneous sensors, since as N grows, it holds:

$$\left(\frac{\Lambda_{\text{mMRC}}}{N}\right) \approx \sum_{k=1}^{K} x_k + \Re\{w_{\text{m}}\}, \tag{6.30}$$

where $w_{\text{m}} \sim \mathcal{N}_{\mathbb{C}}(0, (\sigma_e^2/N)\sum_{k=1}^{K} d_{\hat{g},k}^{-1} p_k^{-1})$.

6.3.2 *Fusion techniques for EST*

Given the model under investigation, the MMSE estimator can be applied in order to give the lowest possible MSE. Herein, it is shown the explicit fusion formula and the resulting MSE . Exploiting the MMSE estimator at the FC, the estimated vector $\hat{\boldsymbol{\theta}} \triangleq [\hat{\theta}_1, \ldots, \hat{\theta}_K]^T$ is obtained as the *conditional posterior mean* [16, Chap. 15]:

$$\hat{\boldsymbol{\theta}} = \mathbb{E}\{\boldsymbol{\theta}|\boldsymbol{y}, \boldsymbol{G}\} = \left(\boldsymbol{\Sigma}_{\boldsymbol{\theta}}^{-1} + \boldsymbol{A}^{\dagger}\boldsymbol{G}^{\dagger}\boldsymbol{\Sigma}_{w_{\text{est}}}^{-1}\boldsymbol{G}\boldsymbol{A}\right)^{-1}(\boldsymbol{G}\boldsymbol{A})^{\dagger}\,\boldsymbol{\Sigma}_{w_{\text{est}}}^{-1}\boldsymbol{y} \tag{6.31}$$

which gives the following optimal MSE:

$$\text{MSE} \triangleq \mathbb{E}\left\{\|\boldsymbol{\theta} - \hat{\boldsymbol{\theta}}\|^2|\boldsymbol{G}\right\} = \text{Tr}\left\{\left(\boldsymbol{\Sigma}_{\boldsymbol{\theta}}^{-1} + \boldsymbol{A}^{\dagger}\boldsymbol{G}^{\dagger}\boldsymbol{\Sigma}_{w_{\text{est}}}^{-1}\boldsymbol{G}\boldsymbol{A}\right)^{-1}\right\}, \tag{6.32}$$

where $\boldsymbol{\Sigma}_{w_{\text{est}}} = (\boldsymbol{G}\boldsymbol{A})\,\boldsymbol{\Sigma}_{\boldsymbol{n}}(\boldsymbol{G}\boldsymbol{A})^{\dagger} + \boldsymbol{\Sigma}_{\boldsymbol{w}}$. By using the matrix inversion lemma on $\boldsymbol{\Sigma}_{w_{\text{est}}}^{-1}$, (6.32) can be rewritten as:

$$\text{MSE} = \text{Tr}\left\{\left[\boldsymbol{\Sigma}_{\boldsymbol{\theta}}^{-1} + \sigma_w^{-2}\boldsymbol{A}^{\dagger}\boldsymbol{G}^{\dagger}\boldsymbol{G}\boldsymbol{A} - \sigma_w^{-4}\boldsymbol{A}^{\dagger}\boldsymbol{G}^{\dagger}\boldsymbol{G}\boldsymbol{A}\left(\boldsymbol{\Sigma}_{\boldsymbol{n}}^{-1} + \sigma_w^{-2}\boldsymbol{A}^{\dagger}\boldsymbol{G}^{\dagger}\boldsymbol{G}\boldsymbol{A}\right)^{-1}\boldsymbol{A}^{\dagger}\boldsymbol{G}^{\dagger}\boldsymbol{G}\boldsymbol{A}\right]^{-1}\right\} \tag{6.33}$$

The derivation of the asymptotic form of the optimal MSE in (6.45) is later obtained in Section 6.4.2, along with its use in *power-allocation problems*.

6.4 Asymptotic performance analysis and optimization

6.4.1 Performance measures for HT

The performance of the considered fusion rules for HT can be, in principle, either analyzed in terms of performance conditioned to the instantaneous channel (IC) experienced or in terms of its long-term channel-averaged (CA) counterpart. More specifically, IC system false alarm and detection probabilities are defined as:

$$P_{F_0}(\gamma, \hat{\boldsymbol{G}}) \triangleq \Pr\{\Lambda > \gamma|\hat{\boldsymbol{G}}, \mathcal{H}_0\}, \quad P_{D_0}(\gamma, \hat{\boldsymbol{G}}) \triangleq \Pr\{\Lambda > \gamma|\hat{\boldsymbol{G}}, \mathcal{H}_1\}, \tag{6.34}$$

whereas the corresponding CA counterparts are obtained as:

$$P_{F_0}(\gamma) \triangleq \mathbb{E}_{\hat{\boldsymbol{G}}}\{P_{F_0}(\gamma, \hat{\boldsymbol{G}})\}, \quad P_{D_0}(\gamma) \triangleq \mathbb{E}_{\hat{\boldsymbol{G}}}\{P_{D_0}(\gamma, \hat{\boldsymbol{G}})\} \tag{6.35}$$

where Λ is the generic (decision) statistic employed at the FC.

Hereinafter, these metrics are investigated in the massive MIMO setup, namely, under the limit $N \to +\infty$. The analysis of massive MIMO performance IC metrics will highlight that the IC and CA system probabilities approach the same value and thus a unified analysis can be pursued. Within the massive MIMO setup, the two relevant scenarios are discussed [8] and consist of (a) either taking the limit $N \to +\infty$ and considering an energy cut in both training and reporting phases as $\rho_{\mathrm{P}} = \bar{\rho}_{\mathrm{P}}/\sqrt{N}$ and $\boldsymbol{P} = \bar{\boldsymbol{P}}/\sqrt{N}$ or (b) taking the limit $N \to +\infty$ and considering an energy cut only in the reporting phase as $\boldsymbol{P} = \bar{\boldsymbol{P}}/N$. Scenario (a) refers to a setup with a fast-fading channel, where the energy spent in the training phase is not negligible, while scenario (b) refers to a case where the energy used for training is negligible and thus transmit energy can be scaled down *aggressively* [8].

Clearly, both scenarios are of high significance for WSN applications, since the results presented hereinafter are achieved in a regime where each sensor keeps reducing the energy spent as the number of antennas at the DFC grows. Therefore each sensor behaves "*efficiently*" (in terms of consumed energy) and consequently prolongs its expected battery life.

From now on, performance characterization will be limited to WL rules and (modified) MRC, due to *their tractability*, while leaving the theoretical analysis of linear filter plus DF aside to avoid a cumbersome discussion. Indeed, the IC system probabilities for WL rules and (modified) MRC can be obtained in a *unified form* as follows, since they both apply a *widely linear processing* to the received data $\boldsymbol{y}$. Then, the two asymptotic scenarios previously described are considered. First, it is easily shown that $\boldsymbol{y}|\hat{\boldsymbol{G}},\boldsymbol{x} \sim \mathcal{N}_{\mathbb{C}}(\hat{\boldsymbol{G}}\boldsymbol{A}\boldsymbol{x}, \sigma_e^2 \boldsymbol{I}_N)$ (cf. (6.17)). As an immediate consequence, and $\Lambda_{\mathrm{WL},i}|\hat{\boldsymbol{G}},\boldsymbol{x} \sim \mathcal{N}(\mathbb{E}\{\Lambda_{\mathrm{WL},i}|\hat{\boldsymbol{G}},\boldsymbol{x}\}, \mathrm{var}\{\Lambda_{\mathrm{WL},i}|\hat{\boldsymbol{G}},\boldsymbol{x}\})$ and $\Lambda_{\mathrm{MRC}}|\hat{\boldsymbol{G}},\boldsymbol{x} \sim \mathcal{N}\left(\mathbb{E}\{\Lambda_{\mathrm{MRC}}|\hat{\boldsymbol{G}},\boldsymbol{x}\}, \mathrm{var}\{\Lambda_{\mathrm{MRC}}|\hat{\boldsymbol{G}},\boldsymbol{x}\}\right)$, where:

$$\mathbb{E}\{\Lambda_{\mathrm{WL},i}|\hat{\boldsymbol{G}},\boldsymbol{x}\} = \frac{\boldsymbol{\mu}_{1,0}^T\, (\underline{\hat{\boldsymbol{G}}\boldsymbol{A}})^\dagger\, \boldsymbol{\Sigma}^{-1}_{\underline{\boldsymbol{y}}|\hat{\boldsymbol{G}},\mathcal{H}_i}\, \underline{\hat{\boldsymbol{G}}\boldsymbol{A}}\boldsymbol{x}}{\left\| \boldsymbol{\Sigma}^{-1}_{\underline{\boldsymbol{y}}|\hat{\boldsymbol{G}},\mathcal{H}_i}\, \underline{\hat{\boldsymbol{G}}\boldsymbol{A}}\, \boldsymbol{\mu}_{1,0} \right\|}; \quad \mathrm{var}\{\Lambda_{\mathrm{WL},i}|\hat{\boldsymbol{G}},\boldsymbol{x}\} = \sigma_e^2. \tag{6.36}$$

and

$$\mathbb{E}\{\Lambda_{\mathrm{MRC}}|\hat{\boldsymbol{G}},\boldsymbol{x}\} = \Re\{\boldsymbol{a}_{\mathrm{MRC}}^\dagger \hat{\boldsymbol{G}}\boldsymbol{x}\}; \quad \mathrm{var}\{\Lambda_{\mathrm{MRC}}|\hat{\boldsymbol{G}},\boldsymbol{x}\} = \frac{1}{2}\sigma_e^2\, \|\boldsymbol{a}_{\mathrm{MRC}}\|^2\,; \tag{6.37}$$

It is worth noticing that both $\mathrm{var}\{\Lambda_{\mathrm{WL},i}|\hat{\boldsymbol{G}},\boldsymbol{x}\}$ and $\mathrm{var}\{\Lambda_{\mathrm{MRC}}|\hat{\boldsymbol{G}},\boldsymbol{x}\}$ do not depend on $\boldsymbol{x}$. Thus, without loss of generality (because of statistical equivalence to a positive scaling factor), performance can be evaluated in terms of the decision statistic $\Lambda_{\mathrm{WL},i}^\circ \triangleq \Lambda_{\mathrm{WL},i}/\sigma_e$. A similar reasoning applies to (modified) MRC when considering $\Lambda_{\mathrm{MRC}}^\circ \triangleq \sqrt{2}\Lambda_{\mathrm{MRC}}/\sigma_e\|\boldsymbol{a}_{\mathrm{MRC}}\|$. In the following, these unit-variance decision statistics will be generically denoted with Λ° for simplicity. Then, it is readily shown that $\Lambda^\circ|\hat{\boldsymbol{G}},\mathcal{H}_j$ is distributed as:

$$\Lambda^\circ|\hat{\boldsymbol{G}},\mathcal{H}_j \sim \sum_{\boldsymbol{x}\in\mathcal{X}^K} P(\boldsymbol{x}|\mathcal{H}_j)\,\mathcal{N}(\mathbb{E}\{\Lambda^\circ|\hat{\boldsymbol{G}},\boldsymbol{x}\},1) \tag{6.38}$$

where $\mathbb{E}\{\Lambda_{\mathrm{WL},i}^\circ|\hat{\boldsymbol{G}},\boldsymbol{x}\} = (\boldsymbol{\mu}_{1,0}^T\,(\underline{\hat{\boldsymbol{G}}\boldsymbol{A}})^\dagger\,\boldsymbol{\Sigma}^{-1}_{\underline{\boldsymbol{y}}|\hat{\boldsymbol{G}},\mathcal{H}_i}\,\underline{\hat{\boldsymbol{G}}\boldsymbol{A}}\boldsymbol{x})/\left(\sigma_e\left\|\boldsymbol{\Sigma}^{-1}_{\underline{\boldsymbol{y}}|\hat{\boldsymbol{G}},\mathcal{H}_i}\,\underline{\hat{\boldsymbol{G}}\boldsymbol{A}}\,\boldsymbol{\mu}_{1,0}\right\|\right)$ and $\mathbb{E}\{\Lambda_{\mathrm{MRC}}^\circ|\hat{\boldsymbol{G}},\boldsymbol{x}\} = (\sqrt{2}\,\Re\{\boldsymbol{1}_K^T(\hat{\boldsymbol{G}}\boldsymbol{A})^\dagger(\hat{\boldsymbol{G}}\boldsymbol{A})\boldsymbol{x}\})/(\sigma_e\sqrt{\boldsymbol{1}_K^T(\hat{\boldsymbol{G}}\boldsymbol{A})^\dagger\hat{\boldsymbol{G}}\boldsymbol{A}\boldsymbol{1}_K})$, respectively. Equation (6.38) denotes a mixture of 2^K one-dimensional real-valued Gaussians, all depending on $\hat{\boldsymbol{G}}$ (which is random) through their mean. A direct application of (6.38) into (6.34) leads to the following exact expression for $P_{F_0}(\gamma,\hat{\boldsymbol{G}})$) and $P_{D_0}(\gamma,\hat{\boldsymbol{G}})$ of Λ°:

$$P_{F_0}(\gamma,\hat{\boldsymbol{G}}) = \sum_{\boldsymbol{x}\in\mathcal{X}^K} P(\boldsymbol{x}|\mathcal{H}_0)\,\mathcal{Q}(\gamma - \mathbb{E}\{\Lambda^\circ|\hat{\boldsymbol{G}},\boldsymbol{x}\}) \tag{6.39}$$

$$P_{D_0}(\gamma,\hat{\boldsymbol{G}}) = \sum_{\boldsymbol{x}\in\mathcal{X}^K} P(\boldsymbol{x}|\mathcal{H}_1)\,\mathcal{Q}(\gamma - \mathbb{E}\{\Lambda^\circ|\hat{\boldsymbol{G}},\boldsymbol{x}\}) \tag{6.40}$$

Such expression can now be evaluated in the asymptotic scenarios (a) and (b) previously described. This is accomplished by taking the limit $N \to +\infty$ of $\mathbb{E}\{\Lambda^\circ|\hat{\boldsymbol{G}},\boldsymbol{x}\}$ under the assumptions determining scenarios (a) and (b), respectively. The reason is the dependence of $P_{F_0}(\gamma, \hat{\boldsymbol{G}})$ and $P_{D_0}(\gamma, \hat{\boldsymbol{G}})$ on N is *only through the means* of the mixture components.

Hence, for WL rules, scenario (a) leads to:

$$\lim_{N\to+\infty} (\mathbb{E}\{\Lambda^{\circ}_{\mathrm{WL},i}|\hat{\boldsymbol{G}},\boldsymbol{x}\})_{\mathrm{a}} = \frac{\sqrt{2\tau\,\bar{\rho}_{\mathrm{P}}}\,\boldsymbol{\mu}_{1,0}^T\,\bar{\boldsymbol{V}}_i\,\boldsymbol{D}^2\,\bar{\boldsymbol{P}}\boldsymbol{x}}{\sigma_w^2\sqrt{\boldsymbol{\mu}_{1,0}^T\,\bar{\boldsymbol{V}}_i\,\boldsymbol{D}^2\,\bar{\boldsymbol{V}}_i^T\,\boldsymbol{\mu}_{1,0}}} \tag{6.41}$$

where $\bar{\boldsymbol{S}}_i \triangleq (\sigma_w^2\,\boldsymbol{\Sigma}^{-1}_{\boldsymbol{x}|\mathscr{H}_i} + (2\tau\bar{\rho}_{\mathrm{P}}/\sigma_w^2)\boldsymbol{D}^2\bar{\boldsymbol{P}})$ and $\bar{\boldsymbol{V}}_i \triangleq (\boldsymbol{I}_K - (2\tau\bar{\rho}_{\mathrm{P}}/\sigma_w^2)\boldsymbol{D}^2\,\bar{\boldsymbol{P}}\,\bar{\boldsymbol{S}}_i^{-1})$. On the other hand, in the asymptotic scenario (b), $\mathbb{E}\{\Lambda^{\circ}_{\mathrm{WL},i}|\hat{\boldsymbol{G}},\boldsymbol{x}\}$ converges to:

$$\lim_{N\to+\infty} (\mathbb{E}\{\Lambda^{\circ}_{\mathrm{WL},i}|\hat{\boldsymbol{G}},\boldsymbol{x}\})_{\mathrm{b}} = \frac{\sqrt{2}\,\boldsymbol{\mu}_{1,0}^T\,\breve{\boldsymbol{V}}_i\,\boldsymbol{D}_{\hat{g}}\,\bar{\boldsymbol{P}}\boldsymbol{x}}{\sigma_w\sqrt{\boldsymbol{\mu}_{1,0}^T\,\breve{\boldsymbol{V}}_i\,\boldsymbol{D}_{\hat{g}}\,\bar{\boldsymbol{P}}\,\breve{\boldsymbol{V}}_i^T\,\boldsymbol{\mu}_{1,0}}} \tag{6.42}$$

where $\breve{\boldsymbol{S}}_i \triangleq (\sigma_w^2\,\boldsymbol{\Sigma}^{-1}_{\boldsymbol{x}|\mathscr{H}_i} + 2\,\boldsymbol{D}_{\hat{g}}\,\bar{\boldsymbol{P}})$ and $\breve{\boldsymbol{V}}_i \triangleq (\boldsymbol{I}_K - 2\,\boldsymbol{D}_{\hat{g}}\,\bar{\boldsymbol{P}}\,\breve{\boldsymbol{S}}_i^{-1})$. Derivation of both (6.41) and (6.42) can be found in [5].

Analogously, referring to MRC, the following asymptotic performance are obtained (by similar reasoning, the performance of mMRC in both the asymptotic scenarios can be obtained). Specifically, considering scenario (a) gives:

$$\lim_{N\to+\infty} (\mathbb{E}\{\Lambda^{\circ}_{\mathrm{MRC}}|\hat{\boldsymbol{G}},\boldsymbol{x}\})_{\mathrm{a}} = \frac{\sqrt{2\tau\bar{\rho}_{\mathrm{P}}}\sum_{k=1}^{K}\beta_k^2\,\bar{p}_k\,x_k}{\sigma_w^2\sqrt{\sum_{k=1}^{K}\beta_k^2\,\bar{p}_k}} \tag{6.43}$$

Differently, in the case of scenario (b), it is obtained:

$$\lim_{N\to+\infty} (\mathbb{E}\{\Lambda^{\circ}_{\mathrm{MRC}}|\hat{\boldsymbol{G}},\boldsymbol{x}\})_{\mathrm{b}} = \frac{\sqrt{2}}{\sigma_w}\frac{\sum_{k=1}^{K}\bar{p}_k\,d_{\hat{g},k}\,x_k}{\sqrt{\sum_{k=1}^{K}\bar{p}_k\,d_{\hat{g},k}}} \tag{6.44}$$

Based on the reported results, it is worth remarking that, in both asymptotic scenarios considered, all the above decision statistics correspond to a real-valued Gaussian mixture with 2^K components that *can be evaluated offline*. Finally, it is worth remarking that further simplification of these expressions can be obtained in the case of conditionally uncorrelated decisions [5].

6.4.2 Performance measures and power-allocation for EST

The optimal MSE in (6.33) can be asymptotically *approximated* (and *simplified*), using favorable propagation conditions in (6.15), as:

$$\mathrm{MSE} \stackrel{a}{=} \mathrm{Tr}\left\{\left[\boldsymbol{\Sigma}_{\boldsymbol{\theta}}^{-1} + N\sigma_w^{-2}\boldsymbol{\Lambda} - N^2\sigma_w^{-4}\boldsymbol{\Lambda}\left(\boldsymbol{\Sigma}_{\boldsymbol{n}}^{-1} + N\,\sigma_w^{-2}\boldsymbol{\Lambda}\right)^{-1}\boldsymbol{\Lambda}\right]^{-1}\right\} \tag{6.45}$$

In (6.45), the definition $\boldsymbol{\Lambda} \triangleq (\boldsymbol{P}\boldsymbol{D}) = \mathrm{diag}(\lambda_1, \ldots, \lambda_K)$ has been exploited. It is worth noticing that, under this approximation, the optimal design of the sensor amplification factors α_k reduces to optimizing only the (squared) modulus of α_k, and not the complex-valued factor α_k. Also, as a byproduct, the obtained power allocation problems become mathematically more tractable and *easier to solve* given the simplifications arising from the massive MIMO approximation.

In what follows, *two optimization problems* dealing with sensor power allocation for EST are posed. Specifically, the *first problem* minimizes the total power consumption by the sensors subject to a reconstruction MSE constraint (therefore, this optimization problem is desirable when power is limited, due to sensors' battery limitations, and a worst case MSE should be guaranteed), i.e.,

$$\mathrm{P}_1 \triangleq \begin{cases} \underset{\{\lambda_k \geq 0\}_{k=1}^K}{\text{minimize}} \;\; P_{tot}^{\mathrm{EST}} \\ \text{subject to} \;\; \mathrm{MSE} \leq \bar{d}, \end{cases} \tag{6.46}$$

where P_{tot}^{EST} and (optimal) MSE are specified by (6.6) and (6.45), respectively, whereas $\bar{d}$ is a maximum user-defined distortion (viz. optimal MSE) threshold and has to be chosen such that $\mathrm{Tr}\{(\mathbf{C}_\theta^{-1} + \mathbf{C}_n^{-1})^{-1}\} < \bar{d} \leq \mathrm{Tr}\{\mathbf{C}_\theta\}$. Note that the lower bound on $\bar{d}$ corresponds to the case of transmission over an ideal noiseless channels (the uncertainty arising from measurement noise is however still present), i.e., all measurements are directly available at the FC. Therefore, this optimization problem is desirable when power is limited, due to sensors' battery limitations, and guarantee on worst case (optimal) MSE is imposed.

In the *second problem* (which corresponds to the case where the estimation accuracy, in terms of MSE, is more demanding rather than power consumption of the sensors), the MSE is a minimized subject to a total power constraint, i.e.,

$$\mathrm{P}_2 \triangleq \begin{cases} \underset{\{\lambda_k \geq 0\}_{k=1}^K}{\text{minimize}} \;\; \mathrm{MSE} \\ \text{subject to} \;\; P_{tot}^{\mathrm{EST}} \leq \bar{P}. \end{cases} \tag{6.47}$$

Although it is more desirable to address the convex optimization problems (6.46) and (6.47) analytically, or provide closed-form solutions, this is not, however, feasible in some cases. Therefore, in such situations, it is more convenient to rewrite the optimization problems in a way to be solved in a computationally more efficient manner. A typical approach is to use the semidefinite programming (SDP) method that can be solved via the low, polynomial-complexity interior point method [27].

The optimizations (6.46) and (6.47) can be expressed as SDPs by introducing slack random variables and by using Schur's complement [28,29]. More precisely, optimization problem (6.46) can be equivalently solved as:

$$\mathrm{P}_1 = \begin{cases} \underset{\{\lambda_k \geq 0\}_{k=1}^K, \boldsymbol{X}, \boldsymbol{Y}}{\text{minimize}} & \sum_{k=1}^K \lambda_k \frac{1}{\beta_k} [\boldsymbol{\Sigma}_\theta + \boldsymbol{\Sigma}_n]_{k,k} \\ \text{subject to} & \mathrm{Tr}[\boldsymbol{Y}] \leq \bar{d} \\ & \begin{bmatrix} [\boldsymbol{\Sigma}_\theta^{-1} + N\sigma_w^{-2}\boldsymbol{\Lambda} - N^2\sigma_w^{-4}\boldsymbol{X}] & \boldsymbol{I}_K \\ \boldsymbol{I}_K & \boldsymbol{Y} \end{bmatrix} \succeq \mathbf{0} \\ & \begin{bmatrix} \boldsymbol{X} & \boldsymbol{\Lambda} \\ \boldsymbol{\Lambda} & \boldsymbol{\Sigma}_n^{-1} + N\sigma_w^{-2}\boldsymbol{\Lambda} \end{bmatrix} \succeq \mathbf{0}, \end{cases} \tag{6.48}$$

where $\boldsymbol{X} \succeq \mathbf{0}$ is a $K \times K$ diagonal matrix, and $\boldsymbol{Y} \succeq \mathbf{0}$ is a Hermitian $K \times K$ matrix (the notation "$\succeq$" is here used to refer to positive semidefinite matrices). Similarly, the optimization problem 6.47 can be equivalently solved as the following SDP:

$$\mathrm{P}_2 = \begin{cases} \underset{\{\lambda_k \geq 0\}_{k=1}^K, \widetilde{\boldsymbol{X}}, \widetilde{\boldsymbol{Y}}}{\text{minimize}} & \mathrm{Tr}[\widetilde{\boldsymbol{X}}] \\ \text{subject to} & \sum_{l=1}^{L} \lambda_k \frac{1}{\beta_k} [\boldsymbol{\Sigma}_{\boldsymbol{\theta}} + \boldsymbol{\Sigma}_{\boldsymbol{n}}]_{k,k} \leq \bar{P} \\ & \begin{bmatrix} [\boldsymbol{\Sigma}_\theta^{-1} + N\sigma_w^{-2}\boldsymbol{\Lambda} - N^2\sigma_w^{-4}\widetilde{\boldsymbol{Y}}] & \boldsymbol{I}_K \\ \boldsymbol{I}_K & \widetilde{\boldsymbol{X}} \end{bmatrix} \succeq \mathbf{0} \\ & \begin{bmatrix} \widetilde{\mathbf{Y}} & \boldsymbol{\Lambda} \\ \boldsymbol{\Lambda} & \boldsymbol{\Sigma}_{\boldsymbol{n}}^{-1} + N\sigma_w^{-2}\boldsymbol{\Lambda} \end{bmatrix} \succeq \mathbf{0} \end{cases} \tag{6.49}$$

where $\widetilde{\boldsymbol{X}} \succeq \mathbf{0}$ is a Hermitian $K \times K$ matrix and $\widetilde{\boldsymbol{Y}} \succeq \mathbf{0}$ is a $K \times K$ diagonal matrix. The optimization problems (6.46) and (6.49) are now analyzed, and approximate approaches are discussed to obtain the optimized $\{\lambda_k^\star\}_{k=1}^K$ when N is sufficiently large. First, using the matrix inversion lemma, the MSE in (6.45) can be rewritten as:

$$\mathrm{MSE} \overset{a}{=} \mathrm{Tr}\left[\left(\boldsymbol{\Sigma}_\theta^{-1} + \left(\boldsymbol{\Sigma}_{\boldsymbol{n}} + \frac{1}{N}\sigma_w^2 \boldsymbol{\Lambda}^{-1}\right)^{-1}\right)^{-1}\right] \tag{6.50}$$

Next, the above expression is expanded as $\mathrm{MSE}(1/N)$ via Taylor series around the point $(1/N) = 0$, which yields:

$$\mathrm{MSE} \overset{a}{=} \underbrace{\mathrm{Tr}\left[(\Sigma_\theta^{-1} + \Sigma_{\boldsymbol{n}}^{-1})^{-1}\right]}_{\mathrm{MSE}(1/N=0)} + \frac{1}{N} \underbrace{\mathrm{Tr}\left[(\Sigma_\theta^{-1} + \Sigma_{\boldsymbol{n}}^{-1})^{-2}\, \mathrm{diag}\left(\ldots, \frac{\sigma_w^2}{\sigma_{n_k}^4 \lambda_k}, \ldots\right)\right]}_{\text{MSE first derivative evaluated at } (1/N)=0} + \mathcal{O}\left(\frac{1}{N^2}\right), \tag{6.51}$$

where the first term represents the MSE expression at $(1/N) = 0$, and the coefficient of $1/N$ in the second term represents the first-order partial derivative of the (optimal) MSE, with respect to $1/N$, evaluated at $(1/N) = 0$. Note it has been implicitly assumed that sensors always amplify their observations with positive (non-zero) gains, i.e., $\lambda_k > 0$. The reason for expanding the Taylor series around $(1/N) = 0$ is due to the massive MIMO framework with a large number of antennas N, and when N is sufficiently large, the second- and higher order terms in the series *can be safely neglected*.

In what follows, the diagonal structure of $\boldsymbol{\Sigma}_n = \mathrm{diag}(\sigma_{n1}^2, \ldots, \sigma_{nK}^2)$ and the auxiliary notation $\boldsymbol{Q} \triangleq (\boldsymbol{\Sigma}_\theta^{-1} + \boldsymbol{\Sigma}_n^{-1})^{-1}$ will be employed for compactness. Then, by neglecting second- and higher order moments in (6.51) onward, the *approximated MSE* is denoted by:

$$\widetilde{\mathrm{MSE}} \triangleq \mathrm{Tr}[\boldsymbol{Q}] + \frac{1}{N}\mathrm{Tr}\left[\boldsymbol{Q}^2 \,\mathrm{diag}\left(\ldots, \frac{\sigma_w^2}{\sigma_{n_k}^4 \lambda_k}, \ldots\right)\right] \tag{6.52}$$

Accordingly, plugging back the asymptotic MSE into the constraint in (6.46), the optimization problem P_1 can be approximated as:

$$\bar{\mathrm{P}}_1 \triangleq \begin{cases} \underset{\{\lambda_k>0\}_{k=1}^K}{\text{minimize}} & \sum_{k=1}^K \lambda_k \frac{1}{\beta_k}[\boldsymbol{\Sigma}_\theta + \boldsymbol{\Sigma}_n]_{k,k} \\ \text{subject to} & \widetilde{\mathrm{MSE}} \le \bar{d}. \end{cases} \tag{6.53}$$

The above problem, as opposed to P_1 in (6.46), admits a closed-form solution, as stated henceforth.

Power allocation based on $\bar{\mathrm{P}}_1$:
The optimal solution to $\bar{\mathrm{P}}_1$ has the closed-form expression:

$$\lambda_k^\star = \frac{1}{N}\left[\frac{([\boldsymbol{Q}^2]_{k,k}^{1/2}\sigma_w^2\sqrt{\gamma_l}/\sigma_{n_k}^2[\boldsymbol{\Sigma}_\theta + \boldsymbol{\Sigma}_n]_{k,k}^{1/2})\sum_{m=1}^K (d_m^\beta[\boldsymbol{\Sigma}_\theta + \boldsymbol{\Sigma}_n]_{m,m}^{1/2}[\boldsymbol{Q}^2]_{m,m}^{1/2}/\sigma_{n_m}^2)}{\bar{d} - \sum_{m=1}^K \boldsymbol{Q}_{m,m}}\right] \tag{6.54}$$

By studying (6.54), it can be observed that the total power consumed by sensors decays with a factor proportional to $1/N$. It should be noted that since an approximation of MSE, i.e, $\widetilde{\mathrm{MSE}}$, is used in the constraint of $\bar{\mathrm{P}}_1$, λ_k, $\forall k$, might not satisfy the exact MSE by equality. In this case, one can easily rescale the resulting $\lambda_k^\star$ in (6.54) in order to satisfy the original MSE constraint by equality.

Similar to the above analysis, an approximate closed-form solution for problem P_2 in (6.47) can be derived when N is sufficiently large. To do so, the MSE approximation in (6.51) is again exploited. Hence, the problem can be simplified into:

$$\bar{\mathrm{P}}_2 \triangleq \begin{cases} \underset{\{\lambda_k>0\}_{k=1}^K}{\text{minimize}} & \mathrm{Tr}\left\{\boldsymbol{Q}\,\mathrm{diag}\left(\frac{\sigma_w^2}{\sigma_{n_k}^2\lambda_k}\right)\right\} \\ \text{subject to} & \sum_{k=1}^K \lambda_k \frac{1}{\beta_k}[\boldsymbol{\Sigma}_\theta + \boldsymbol{\Sigma}_n]_{k,k} \le \bar{P} \end{cases} \tag{6.55}$$

where the term $\mathrm{Tr}\{\boldsymbol{Q}\}$ has been discarded from $\widetilde{\mathrm{MSE}}$, as it does not depend on λ_k. A closed-form solution exists also for the approximate problem $\bar{\mathrm{P}}_2$, as follows.

Power allocation based on $\bar{\mathrm{P}}_2$:
The optimal solution to $\bar{\mathrm{P}}_2$ has the closed-form expression:

$$\lambda_k^\star = \frac{\bar{P}\,\left[\boldsymbol{Q}^2\right]_{k,k}^{1/2}}{(\sigma_{n_k}^2/\sqrt{\gamma_k})\,[\boldsymbol{\Sigma}_\theta + \boldsymbol{\Sigma}_n]_{k,k}^{1/2}\sum_m([\boldsymbol{Q}^2]_{m,m}^{1/2} d_m^\beta [\boldsymbol{\Sigma}_\theta + \boldsymbol{\Sigma}_n]_{m,m}/\sigma_{n_m}^2)} \tag{6.56}$$

As $N \to \infty$, the MSE floor incurred by using the optimal power allocation in (6.56), derived for $\bar{\mathrm{P}}_2$, becomes:

$$\lim_{N\to\infty} \mathrm{MSE} = \mathrm{Tr}\left\{\left(\boldsymbol{\Sigma}_\theta^{-1} + \boldsymbol{\Sigma}_n^{-1}\right)^{-1}\right\} \tag{6.57}$$

Clearly, the MSE cannot decay to zero as $N \to \infty$. This is indeed expected since using the favorable propagation condition, there exist K orthogonal channels between the sensors and the FC, and it is well-known (see, e.g., [29]) that the MSE for an orthogonal MAC cannot decay to zero as $N \to \infty$.

6.5 Conclusions and further reading

This chapter focused on decentralized detection (HT) and estimation (EST) over (virtual) MIMO channels, in the presence of a large antenna-array at the FC, which allowed to capitalize the analytical advantages of favorable propagation condition. In the former case (namely, HT), a wide choice of low-complexity suboptimal fusion rules was discussed, efficiently exploiting large-array benefits, and a corresponding theoretical analysis of these rules was provided under energy-reduction laws commonly considered in massive MIMO literature. In the latter case (namely, EST), MMSE estimation was considered and the power allocation (to improve naive uniform power allocation) problem tackled at each sensor node, with the objectives of minimizing total power consumption or MSE for source estimation, under a maximum distortion or total power constraint, respectively. The analysis revealed that as N increases, the optimal total power consumption of the sensors decay with a factor proportional to $1/N$ in order to satisfy a targeted MSE.

We now conclude the section with suggestions for further reading. The pioneering work of Marzetta [8] showed that the use of unlimited number of antennas at the base station provides novel interesting phenomena in the context of cellular systems: the detrimental effects of uncorrelated noise and fast fading vanish, the throughput is independent of the size of the cells, the spectral efficiency is independent of the bandwidth, the required transmitted energy per bit decreases and the simplest sort of precoders and detectors are permitted. In the context of very large size MIMO systems, unified performance analysis for both uplink and downlink of noncooperative multicell systems with linear precoders and detectors is proposed in [30], where the loss of realistic systems due to finite number of antennas w.r.t. asymptotic theoretical performance is evaluated. Performance of linear detectors for the uplink in presence of imperfect CSI is analyzed in [17] focusing on the trade-off between energy and spectral efficiency.

DF over a (virtual) MIMO was first studied in [9], with focus on power-allocation design based on instantaneous CSI, under the framework of J-divergence. The same model was adopted to study data fusion over MIMO channels with amplify-and-forward sensors in [31,32]. Design of several suboptimal fusion rules for the MIMO scenario was proposed in [10], and a theoretical study on MRC was presented in [26]. In WSNs, the massive MIMO framework was employed in [33] for maximum likelihood (ML) estimation of a scalar deterministic source. A further application of massive MIMO is sought in [4], considering decentralized detection of a random scalar source and the corresponding (conceptually related) decentralized estimation problem. A detailed analysis of DF over massive MIMO is considered in [5], whereas the estimation problem of a correlated vector source investigated in the same setup is analyzed in [6].

References

[1] Rusek F, Persson D, Lau BK, *et al.* Scaling up MIMO: opportunities and challenges with very large arrays. IEEE Signal Processing Magazine. 2013;30(1):40–60.

[2] Larsson E, Edfors O, Tufvesson F, *et al.* Massive MIMO for next generation wireless systems. IEEE Communications Magazine. 2014;52(2):186–195.

[3] Joshi S, and Boyd S. Sensor selection via convex optimization. IEEE Transactions on Signal Processing. 2009;57(2):451–462.

[4] Jiang F, Chen J, Swindlehurst AL, *et al.* Massive MIMO for wireless sensing with a coherent multiple access channel. IEEE Transactions on Signal Processing. 2015;63(12):3005–3017.

[5] Ciuonzo D, Salvo Rossi P, and Dey S. Massive MIMO channel-aware decision fusion. IEEE Transactions on Signal Processing. 2015;63(3):604–619.

[6] Shirazinia A, Dey S, Ciuonzo D, *et al.* Massive MIMO for decentralized estimation of a correlated source. IEEE Transactions on Signal Processing. 2016;64(10):2499–2512.

[7] Chen B, Tong L, and Varshney PK. Channel-aware distributed detection in wireless sensor networks. IEEE Signal Processing Magazine. 2006;23(4):16–26.

[8] Marzetta TL. Noncooperative cellular wireless with unlimited numbers of base station antennas. IEEE Transactions on Wireless Communications. 2010;9(11):3590–3600.

[9] Zhang X, Poor HV, and Chiang M. Optimal power allocation for distributed detection over MIMO channels in wireless sensor networks. IEEE Transactions on Signal Processing. 2008;56(9):4124–4140.

[10] Ciuonzo D, Romano G, and P Salvo Rossi. Channel-aware decision fusion in distributed MIMO wireless sensor networks: decode-and-fuse vs. decode-then-fuse. IEEE Transactions on Wireless Communications. 2012;11(8): 2976–2985.

[11] Li W, and Dai H. Distributed detection in wireless sensor networks using a multiple access channel. IEEE Transactions on Signal Processing. 2007;55(3): 822–833.
[12] Ciuonzo D, Romano G, and Salvo Rossi P. Optimality of received energy in decision fusion over Rayleigh fading diversity MAC with non-identical sensors. IEEE Transactions on Signal Processing. 2013;61(1): 22–27.
[13] Savvides A, Han CC, and Strivastava MB. Dynamic fine-grained localization in ad-hoc networks of sensors. In: 7th ACM Annual International Conference on Mobile Computing and Networking; 2001. p. 166–179.
[14] Cheng X, Thaeler A, Xue G, *et al.* TPS: a time-based positioning scheme for outdoor wireless sensor networks. In: 23rd Annual Joint Conference of the IEEE Computer and Communications Societies (INFOCOM). vol. 4; 2004. p. 2685–2696.
[15] Chen H, Wang G, Wang Z, *et al.* Non-line-of-sight node localization based on semi-definite programming in wireless sensor networks. IEEE Transactions on Wireless Communications. 2012;11(1):108–116.
[16] Kay SM. Fundamentals of Statistical Signal Processing, Volume 1: Estimation Theory. Upper Saddle River, NJ: Prentice Hall PTR; 1993.
[17] Ngo HQ, Larsson EG, and Marzetta TL. Energy and spectral efficiency of very large multiuser MIMO systems. IEEE Transactions on Communications. 2013;61(4):1436–1449.
[18] Kay SM. Fundamentals of Statistical Signal Processing, Volume 2: Detection Theory. Upper Saddle River, NJ: Prentice Hall PTR; 1998.
[19] Lei A, and Schober R. Coherent max–log decision fusion in wireless sensor networks. IEEE Transactions on Communications. 2010;58(5):1327–1332.
[20] Kotecha JH, Ramachandran V, and Sayeed AM. Distributed multi-target classification in wireless sensor networks. IEEE Journal of Selected Areas in Communications. 2005;23(4):703–713.
[21] Robertson P, Villebrun E, and Hoeher P. A comparison of optimal and sub-optimal MAP decoding algorithms operating in the log domain. In: IEEE International Conference on Communications (ICC '95). vol. 2. Seattle; 1995. p. 1009–1013.
[22] Chen B, Jiang R, Kasetkasem T, *et al.* Channel aware decision fusion in wireless sensor networks. IEEE Transactions on Signal Processing. 2004;52(12): 3454–3458.
[23] Picinbono B. On deflection as a performance criterion in detection. IEEE Transactions on Aerospace and Electronic Systems. 1995;31(3):1072–1081.
[24] Quan Z, Cui S, and Sayed AH. Optimal linear cooperation for spectrum sensing in cognitive radio networks. IEEE Journal of Selected Topics in Signal Processing. 2008;2(1):28–40.
[25] Yiu S, and Schober R. Nonorthogonal transmission and noncoherent fusion of censored decisions. IEEE Transactions on Vehicular Technology. 2009;58(1): 263–273.

[26] Ciuonzo D, Romano G, and Salvo Rossi P. Performance analysis of maximum ratio combining in channel-aware MIMO decision fusion. IEEE Transactions on Wireless Communications. 2013;12(9):4716–4728.
[27] Vandenberghe L, and Boyd S. Semidefinite programming. SIAM Review. 1996;38(1):49–95.
[28] Boyd S, and Vandenberghe L. Convex Optimization. New York, NY, USA: Cambridge University Press; 2004.
[29] Xiao JJ, Cui S, Luo ZQ, *et al.* Linear coherent decentralized estimation. IEEE Transactions on Signal Processing. 2008;56(2):757–770.
[30] Hoydis J, ten Brink S, and Debbah M. Massive MIMO in the UL/DL of cellular networks: how many antennas do we need? IEEE Journal of Selected Areas in Communications. 2013;31(2):160–171.
[31] Banavar MK, Smith AD, Tepedelenlioglu C, *et al.* On the effectiveness of multiple antennas in distributed detection over fading MACs. IEEE Transactions on Wireless Communications. 2012;11(5):1744–1752.
[32] Nevat I, Peters GW, and Collings IB. Distributed detection in sensor networks over fading channels with multiple antennas at the fusion centre. IEEE Transactions on Signal Processing. 2014;62(3):671–683.
[33] Jiang F, Chen J, and Swindlehurst AL. Estimation in phase-shift and forward wireless sensor networks. IEEE Transactions on Signal Processing. 2013; 61(15):3840–3851.

Part III

Distributed inference over graphs

Chapter 7

Decentralized detection via running consensus

Paolo Braca[1], Stefano Marano[2], Vincenzo Matta[2], and Peter Willett[3]

Consensus by sensor gossip, which ensures information retrieval from any subset of sensors at an arbitrary instant of time, is a popular paradigm for modern sensor networks designed for inference purposes. In realistic applications, the network continuously senses the surrounding environment, while consensus among its nodes is simultaneously enforced. The basic consensus equation is coherently modified allowing the sensor state update to include both the consensus contribution of neighboring nodes and new measurements. This new paradigm is often referred as running consensus. We review the state-of-the-art of running consensus techniques and discuss their applications with special emphasis to detection problems.

The running consensus, when compared with the ideal centralized detection statistic, is affected by an error term that, under suitable conditions, can be negligible. We study such conditions exploiting the theory of locally optimum statistics. We prove the asymptotic equivalence of the running consensus with the ideal centralized system, in terms of detection capabilities. Specifically, such asymptotic optimality is demonstrated for two cases: the fixed sample size (FSS) test and the sequential test.

7.1 Introduction

One of the first distributed solutions to averaging S values with local exchanges of information was a procedure proposed by DeGroot [1] to solve the opinion pool problem. DeGroot titled his paper "Reaching a Consensus," and consensus is now the term universally adopted in computer science, control, signal processing, and information theory communities (see entry points in [2–4]) for referring to numerous applications including flocking, multi-agent collaboration, vehicle formation, and distributed inference [2].

Consensus protocols have the attractive property that a routing strategy is not required because at each iteration, information is exchanged among a subset of nodes

[1]Research Department NATO STO Centre for Maritime Research and Experimentation (CMRE), La Spezia, Italy
[2]Dipartimento di Ingegneria dell'Informazione ed Elettrica e Matematica applicata (DIEM), University of Salerno, Italy
[3]Electrical and Computer Engineering (ECE), University of Connecticut, USA

(often referred as agents or sensors), and then this information is processed by the receiving nodes to compute a local update. In addition, as further detailed below, consensus algorithms allow to design the network with a completely flat architecture, meaning that there is no central unit to coordinate, process, or gather data, and all the information processing and data exchange take place locally at the remote nodes in a completely distributed fashion. At the same time, single agents perform as a centralized unit able to fully exploit the whole network information, at the cost of a performance degradation (error) that under suitable conditions can be made negligible. In the case of the averaging consensus, often referred as gossip [3], the error decreases exponentially with the number of exchanges among the agents. The rate of error decrease depends on the network topology and the consensus weight matrices, and minimizing this rate has been the focus of some literature, see e.g., [2,3].

The basic idea of the averaging consensus has been widely studied and extended in several directions. For instance, in [5], the consensus procedure was generalized to operate only with quantized values at each node, while the multiple voting problem is proposed and analyzed in [6]. In [7], the distributed optimization is performed exploiting consensus iterations. Noisy and imperfect communication among network nodes has been also widely studied, see e.g. [8,9]. Target tracking problems have been addressed in a distributed fashion exploiting the consensus paradigm [10–12], including the extensions to the case of multiple targets with data association, see e.g. [13]. The concept of privacy preservation (see, e.g., [14] and reference therein) has been recently extended to averaging consensus networks, see, e.g., [15,16].

7.1.1 The estimation problem

Because of its nature, the averaging consensus was naturally proposed for distributed inference approaches. Existing distributed inference schemes in the literature can be broadly divided into two classes. The first class is the long-standing architecture proposed for wireless sensor networks and is characterized by the presence of a fusion center (e.g., see [17–22]) that receives local measurements, or local statistics, from the sensors and then estimate the state of the nature. The second class, to which we shall refer in the following, involves a completely flat architecture (without fusion center) in which node cooperation is based on the consensus paradigm.

Consensus-based architectures can be further divided in two groups. The first one is the classic consensus scheme which involves a two-step approach: (i) sensing phase, i.e., data collection, followed by (ii) distributed consensus or optimization protocols to fuse the initial estimates, see, e.g., [8,23]. In contrast to this classic consensus approach, the second group involves sequential observations collected by sensors over time, while the local exchange of information occurs at the same rate as sensing, see, e.g., [10,24–32]. The main representative schemes from this class are the *running consensus* [25,26] (with its generalization *consensus* + *innovations* [33]) and the diffusion adaptation [29,32,34].

The key difference between running consensus and diffusion adaptation schemes is the nature of the step-size or innovation gain, see, e.g., [31,34] and references therein. In the diffusion framework, the step size is taken to be constant, whereas in the running consensus schemes, this is made to decay over time in a controlled fashion. The constant step size in the diffusion approaches facilitates adaptation in dynamic parameter environments, but at the same time leads to nonzero residual estimation error of the sensor estimates, whereas the time-varying step size in the running consensus approach ensures consistent parameter estimates at the agents. Because of its constant step size (decaying memory of the gathered information), the diffusion adaptation network scheme is responsive in tracking drifting state, while the running consensus, which integrates all the gathered information, exhibits poor performance in adaptation. On the other hand, in stationary environments, the running consensus scheme could be preferred. This is a manifestation of the inherent trade-off between adaptation properties and inference performance. In the case of diffusion adaptation, the mean square error (MSE) in estimation problems is proportional to the step size μ [32]. In detection problems, it was established, exploiting a detailed large deviations analysis [35], that error probabilities of all agents vanish exponentially as functions of $1/\mu$ [34], and that all agents share the same error exponent derived in closed form. The differences among the agents are discovered by conducting a refined asymptotic analysis based on the mathematical framework of exact asymptotics [36].

Running consensus schemes were initially developed to sequentially estimate a static parameter [24,25]. This estimation scheme was then generalized in many directions including nonlinear observation models, noisy intersensor communication and intermittent sensor, and link failures [31,33,37]. Comparison is made with the ideal centralized system that collects at each time all the data of the sensor network. Specifically, analytical bounds on the MSE and correlation among the sensors are provided in [25], showing that the ratio between the ideal centralized MSE and that of running consensus converges to unity. The asymptotic properties of the consensus + innovation statistic is established in [31,33,37], where conditions for the asymptotic efficiency are provided.

In contrast to the running consensus estimation schemes (estimation of a static parameter), the Bayesian filtering formulation assumes that the state of nature to be estimated is a stochastic process, usually modeled as hidden Markov. In this case, asymptotic convergence makes no sense because the quantity to be estimated is time-varying. In this framework, assuming linear state and observation equations with Gaussian processes, Olfati-Saber proposed the consensus-based Kalman filter [10]. General nonlinear, and non-Gaussian distributed estimation problems have been addressed in [11,12] in which the authors proposed the likelihood consensus approach to design a fully distributed particle filter. The case of multiple moving targets with data association was addressed in [13].

7.1.2 The detection problem

The other major application context of the consensus scheme is that of detection, in which the state of nature is discrete. A recent overview of detection strategies over sensor networks is provided in [38]. Several architectures have been proposed during the past decades. Again two main classes can be identified: parallel architectures (sensors transmit local decision statistics, including decisions) to a fusion center which makes the final decision, see, e.g., [18,20,39–42], and flat architecture without fusion center, exploiting consensus-based update rules.

Running consensus statistic

Let us consider a networked system composed by S sensors (or agents). The running consensus statistic is recursively defined by [25,26]:

$$\mathbf{s}_n = W_n\left(\alpha_n \mathbf{s}_{n-1} + \beta_n t(x_n)\right), \quad n = 1, 2, \ldots, \quad \mathbf{s}_0 = \beta_0 t(x_0), \tag{7.1}$$

where entries $s_{n,i}$ of the vector $\mathbf{s}_n$ of size S represent the state of the nodes at the nth time slot, the measurements $x_n = \left[x_{n,1}, \ldots, x_{n,S}\right]^T$ are modeled as iid (independent, identically distributed) random variables, $t(x) = [t(x_1), \ldots, t(x_S)]^T$, $t(x)$ is a real function, and α_n, β_n are deterministic weights. For each n, the agents exchange data according to (random) consensus matrices W_n, $n = 1, 2, \ldots,$ [3,26].

Given the conditional independence of measurements, several inference statistics are based on cumulative sum (CUSUM) or random walks (e.g., see details in [43]). Let us consider the following ideal centralized CUSUM:

$$s_n^{(c)} = \eta_n \sum_{i=1}^{n} \sum_{s=1}^{S} t(x_{i,s}), \tag{7.2}$$

where usually $t(x)$ is a likelihood function, and η_n is a deterministic sequence of weights. The running consensus state of each sensors is an approximate version of the ideal centralized CUSUM (7.2):

$$\mathbf{s}_n = s_n^{(c)} \mathbf{1} + e_n, \tag{7.3}$$

where $\mathbf{1}$ is a vector made of all unit entries. The error term e_n under technical conditions can be asymptotically neglected ($n \to \infty$), see details in [25,26]. The qualitative behavior of this approximation is provided in Figure 7.1 with Gaussian observations, $t(x) = x$, and randomized pairwise consensus matrices. In panel (b), running consensus is approximating a sequential average designed to estimate the average of observations [25], while in panel (c) running consensus is approximating a CUSUM designed for sequential detection [26].

The consensus architectures for detection problems, as already discussed in the previous section, can be divided in two classes. In the first class sensing and communication are separate (see, e.g., [23]), while the second one is based on the running consensus paradigm [26–30,34,44].

The assumption in the first class, in which the consensus process starts after all data had been collected, is in contrast to sequential and real time detection problems. The running consensus paradigm is instead a natural candidate for designing distributed sequential decision schemes [26,27,31]. Sequential detection, introduced

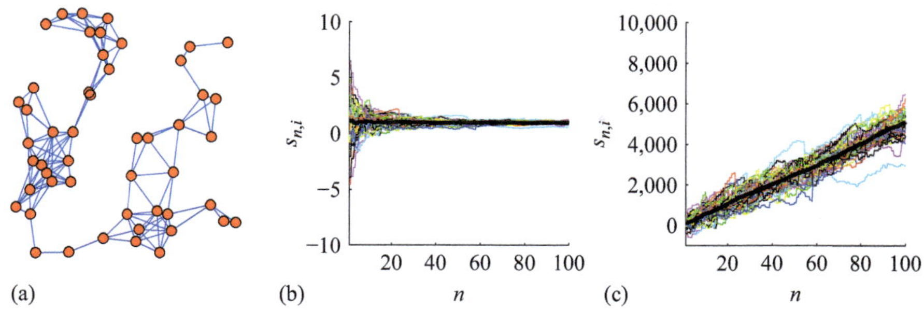

Figure 7.1 Realizations of running consensus (lines in color for each sensors); ideal statistic (thick line in black). Running consensus (7.1) is generated with randomized pairwise matrices and Gaussian observations with mean $\mu = 1$ and variance $\sigma^2 = 5$. In each time, five agents are selected to perform pairwise gossip exchanges. Panel (a): network topology with $N = 50$; panel (b): $\alpha_n = ((n-1)/n)$, $\beta_n = \eta_n = (1/nS)$; panel (c): $\alpha_n = 1, \beta_n = S, \eta_n = 1$

in [45,46] (see as entry-point [43]), allows to learn from the data when is the most suitable time to stop the data observation and to take the decision. Thus, instead of minimizing the decision error probabilities using a fixed sample size, the sequential decision objective is to distinguish between the two hypotheses with guaranteed accuracy (given in terms of both false alarm and miss detection probabilities) in the quickest time or using the minimal amount of sensed data samples at the agents.

Sequential change detection focuses on the problem in which the transition time among statistical hypotheses is unknown. In this framework, the procedure decides to stop collecting data when enough evidence is observed to declare a hypothesis change. Distributed change detection procedures based on consensus have been recently proposed, see, e.g., [27,47,48].

In the following section, we briefly review asymptotic properties and optimality criteria of running consensus detection schemes.

7.1.3 Asymptotic optimality criteria of running consensus in hypothesis testing problems

In this subsection, an overview of the optimality criteria of running consensus detection schemes is provided, with special emphasis on the comparison with the ideal centralized detector that processes at each time n all the data of the sensor network.

In detection problems, the error probability is seldom available in closed-form expressions. Then in the classic detection literature, the error behavior was studied in the asymptotic regime. Two different asymptotic frameworks have been proposed. One relies on large deviation techniques in which it is established that typically the error probabilities vanish exponentially [35], and the rate of error probability is analytically computed. The other framework is usually considered when the signal to be detected is small compare to the noise, often referred as low signal-to-noise ratio (SNR) regime, but at the same time, the number of observations is large. Exploiting the weak convergences and the central limit theorem, the detection problem is studied when the hypotheses get closer and closer as the number of observation grows, leading to nonzero asymptotic error probabilities [49,50]. Typically, the decision statistic converges to a normal random variable and the asymptotic performance is expressed in terms of a closed-form receiver operating characteristic (ROC) curve, equivalent to a Gaussian shift-in-mean problem [49,50].

Running consensus detection problems are studied under both the asymptotic frameworks.

The work [26] establishes the asymptotic optimality, or *local optimality*, of running consensus for a binary hypothesis testing problem. The optimality is in the sense that when the hypotheses get closer and closer as the time n grows, the running consensus statistics converge to normal random variables, and the asymptotic error probabilities are the same as those of the (optimal) ideal centralized detector. Asymptotically, the number of samples required to achieve a given power for the hypothesis test is exactly the same of the number of samples required by the ideal system, i.e., the asymptotic relative efficiency (ARE, see [50,51]) converges to one [26].

In contrast to the scenario considered in [26], when the hypotheses are fixed, an exponential decay rate of error probabilities is observed, see, e.g., [28,52]. Interestingly, with Gaussian observations, in this case, running consensus exhibits a phase transition, when the network connectivity is above a certain threshold, then the distributed detector is asymptotically optimal, i.e., asymptotically equivalent to the centralized detector that optimally processes the observations of all sensors [28]. In other words, the exponential decay rate of the error probability for running consensus equals the Chernoff information [28]. When the rate of information flow is below the threshold, running consensus achieves only a fraction of the Chernoff information rate [28]. This achievable rate is quantified as a function of the network rate of information flow. The network information flow does not affect the asymptotic optimality in the framework described in [26] but only the convergence speed, as it has been shown by numerical simulations.

The authors of [52] observe a similar phase transition for the running consensus over random networks for generic non-Gaussian sensor observations. The optimality threshold is a function of the log-moment generating function of the sensors' observations and of the number of sensors S. This reveals an interesting interplay between the distribution of the sensor observations and the network connectivity. Furthermore, they determine the detection threshold for which each sensor achieves an exponential decay of the error probability.

In [30], the running consensus scheme is properly modified to accommodate noisy communication among the agents. Large deviations are exploited, and it is established that all agents can still achieve an exponential error rate, even when certain (or most) agents cannot detect the event of interest in isolation. A communication payoff threshold on the communication noise power is observed and characterized [30]. There is a critical noise power below which cooperation among neighbors improves detection performance and above which the noise in the communication among agents overwhelms the distributed detector performance.

Composite hypothesis testing problems are investigated in [44]. The network objective is to test a simple null hypothesis against a composite alternative concerning the state of the field, modeled as a vector of (continuous) unknown parameters. Specifically, under the alternative hypothesis, each agent sequentially observes an iid time-series consisting of a function of the true but unknown parameter corrupted by Gaussian noise, whereas, under the null, they obtain noise only. The running consensus is generalized in such a way that the agents estimate the underlying parameter and in parallel also update their test decision statistics by simultaneously processing the latest local sensed information and information obtained from neighboring agents [44]. Conditions to ensure asymptotically decaying probabilities of errors are characterized [44].

Sequential detection schemes via running consensus are proposed in [26,31] and analyzed in both asymptotic regimes, local optimality [26] and large deviations [31]. Specifically, in [26], both the ideal centralized sequential statistic and the sequential running consensus converge asymptotically to the Wiener process with the same positive drift $d/2$ under the alternative hypothesis and the same negative drift $-d/2$ under the null hypothesis. The asymptotic properties of the stopping time are ruled by the well-known Wiener stopping time [26]. By defining the ARE in the sequential case as the ratio of the expected samples sizes of two detection strategies for the same error probabilities, in [26], it is shown that the running consensus is asymptotically optimal, i.e., the ARE converges to one. The drift d is the *efficacy* of the test and is proportional to the square root of the number of sensors S. The larger the d is, the better is the detection performance and, as expected, by increasing the number of sensors, the asymptotic performance increases as well.

In [31], large deviation exponents for the tail probabilities of the agent stopping time distributions are obtained and it is shown that asymptotically (in the number of agents or in the high SNR regime) the exponents associated with the distributed algorithm approach that of the optimal centralized detector. The expected stopping time for the proposed algorithm at each network agent is evaluated and is benchmarked with respect to the optimal centralized detector. The efficiency of the proposed algorithm, measured by the expected stopping times, is characterized in terms of network connectivity.

The main theme of this contribution is the local optimality of running consensus in both sequential and fixed sample size (FSS) detection schemes. The reader is referred to [26] for further details. The remainder of this chapter is structured as follows. In Section 7.2, we describe the running consensus algorithm and the addressed problem is precisely formalized. The local optimality of running consensus is established in Section 7.3. Computer experiments are given in Section 7.4. Finally, Section 7.5 draws the conclusions.

7.2 Hypothesis testing in sensor networks with running consensus

Let us consider a network of S sensors monitoring a phenomenon of physical interest, modeled as a binary state of the nature. Each sensor collects n iid samples. By denoting with $x_{i,s}$ the ith sample collected by the sth sensor, the goal of network is to discriminate between the two hypotheses:

$$\begin{aligned} \mathscr{H}_0 &: x_{i,s} \sim f_{\theta_0}(x), \\ \mathscr{H}_1 &: x_{i,s} \sim f_\theta(x), \end{aligned} \tag{7.4}$$

where $i = 1, 2, \ldots, n$, $s = 1, 2, \ldots, S$, the symbol $\sim$ stands for "is distributed according to," and $f_{\theta_0}(x)$ and $f_\theta(x)$ are the marginal probability density functions of the data, parametrized by θ.

In this work, we focus on detectors that are asymptotically equivalent when the hypotheses come close to each other, as the number of samples increases. We consider, as is standard in these contexts [49,50], detection statistics of the form:

$$T_n^{(c)} = \sum_{i=1}^{n} \sum_{s=1}^{S} t(x_{i,s}), \tag{7.5}$$

where the suffix c emphasizes that this is the ideal, centralized, detection statistic. Since data are iid, the first two moments of the ideal centralized statistic are the sum of the moments of $t(x_{i,s})$: $\mathrm{E}_\theta[T_n^{(c)}] = nS\mu(\theta)$ and $\mathrm{VAR}_\theta[T_n^{(c)}] = nS\sigma^2(\theta)$, where $\mathrm{E}_\theta[\cdot]$ stands for expectation under $f_\theta(x)$, including as particular case $\theta = \theta_0$.

We assume that sensors measure and exchange data according to the running consensus strategy [25,26]. Specifically, it is assumed that (i) data are acquired by sensors at discrete time instants $n = 1, 2, \ldots$ and (ii) for each n, a randomly chosen subset of nodes is allowed to exchange data; according to the so-called gossip algorithm [3], the original measurements made at a given node at each time instant are averaged with those owned by some of its neighbors.

Let us denote with $T_{n,s}$ the state (detection statistic) computed at the nth epoch by the sth node. The updating rule for the states of the nodes according to the running consensus scheme is [26]:

$$T_n = W_n \left[T_{n-1} + S\, t(x_n) \right] \tag{7.6}$$

where the $S \times S$ matrices W_n, $n = 1, 2, \ldots$ are *iid* and doubly stochastic. For later use, we also define $\widetilde{W}_n = W_n - \mathbf{1}\mathbf{1}^T/S$, where $\mathbf{1}$ is a vector of all unit entries. The subscript n will be omitted when inessential.

For our purposes, it is convenient to make explicit the relationships between the statistic available at the S sensors and the centralized statistic $T_n^{(c)}$: we can always write, $\forall\, n$:

$$T_n = T_n^{(c)}\, \mathbf{1} + e_n, \tag{7.7}$$

or, by zooming on a single sensor:

$$T_{n,s} = T_n^{(c)} + e_{n,s}, \tag{7.8}$$

where $e_{n,s}$ is the error term at instant n for the sth sensor.

The main contribution of this work is in showing that the decentralized statistic $T_{n,s}$, *for any node* s, is asymptotically equivalent to $T_n^{(c)}$, as claimed by the theorems in the next section.

7.3 Asymptotic performance

A meaningful setup for the asymptotic design and characterization of the detector is obtained by studying test (7.4) as the parameter θ of the alternative hypothesis approaches θ_0, and we investigate the issue by considering two different kinds of detection strategies: the Neyman–Pearson setup, or Fixed Sample Size (FSS) test, and the sequential paradigm (see, e.g., [45,53]).

In the former case, we refer to the standard framework of Pitman's ARE [50,51], and asymptotic optimality is proved by exploiting the boundedness of the second moment of $e_{n,s}$. Addressing the sequential case is considerably more involved. Therefore, we first introduce the proper asymptotic framework proposed by Lai [54] and then consider the rigorous running consensus asymptotic optimality theorem.

7.3.1 Fixed sample size test

Let us consider the hypothesis test of a simple alternative $\theta = \theta_0$ against the one-sided alternative $\theta > \theta_0$, with θ being a real, unknown parameter. Formally, we use model (7.4), with [49]:

$$\theta_n = \theta_0 + \frac{\gamma}{\sqrt{n}}, \quad \gamma > 0, \tag{7.9}$$

and the performance is studied in the limit of $n \to \infty$.

For a given n, the ideal centralized detector is identified by the pair $(T_n^{(c)}, \delta_n)$, in the sense that $\mathscr{H}_1$ is declared whenever $T_n^{(c)} \geq \delta_n$ and $\mathscr{H}_0$ is declared otherwise. The detection and false alarm probabilities are accordingly defined as $p_{dn} = \Pr[T_n^{(c)} \geq \delta_n | \mathscr{H}_1]$, $p_{fn} = \Pr[T_n^{(c)} \geq \delta_n | \mathscr{H}_0]$, and the detector is said to be of (asymptotic) size p_f if $\lim_{n\to\infty} p_{fn} = p_f$ [49]. Suppose now that the centralized detection statistic is asymptotically normal under both hypotheses, that is,

$$\frac{T_n^{(c)} - n\,S\mu(\theta_0)}{\sqrt{n\,S}\,\sigma(\theta_0)} \xrightarrow{f_{\theta_0}} \mathscr{N}(0,1), \quad \frac{T_n^{(c)} - n\,S\mu(\theta_n)}{\sqrt{n\,S}\,\sigma(\theta_n)} \xrightarrow{f_{\theta_n}} \mathscr{N}(0,1), \tag{7.10}$$

where $\xrightarrow{f_{\theta_n}}$ (resp. $\xrightarrow{f_{\theta_0}}$) denotes convergence in distribution under $f_{\theta_n}(x)$ (resp. under $f_{\theta_0}(x)$) as n diverges. Under this assumption, it can be shown that, for any detector of size p_f, $p_d = \lim_{n\to\infty} p_{dn} = Q\big(Q^{-1}(p_f) - \gamma d\big)$, where $Q(\cdot)$ is the area under the right tail of a standard Gaussian function, $Q^{-1}(\cdot)$ is its inverse function, and

$d = \sqrt{S}\,\mu'(\theta_0)/\sigma(\theta_0)$ is called efficacy [49,50]. We note explicitly that the asymptotic formula depends on the specific relationship between θ_n and θ_0 through the factor γ.

Comparison of two detectors in the asymptotic regime is usually accomplished by their ARE. The ARE of detector 2 with respect to detector 1 is defined as the asymptotic ratio of the sample size of detector 1, say n_1, divided by that of detector 2, n_2, for the same asymptotic probabilities p_f and p_d. We have ARE $= (n_1/n_2) = (d_2/d_1)^2$, where d_1 and d_2 are the efficacies of detectors 1 and 2, respectively [51,55]. As a consequence, we shall say that two detectors are asymptotically equivalent if they share the same efficacy d, and that a detection statistic is asymptotically optimal if it reaches the best attainable efficacy, which, under suitable regularity conditions, is $d_{\max} = \sqrt{S\,I(\theta_0)}$, where $I(\theta_0)$ is the Fisher information at θ_0 [50,56].

Up to now, we have summarized some known results referred to the ideal centralized statistic $T_n^{(c)}$. Let us consider the detector $(T_{n,s}, \delta_n)$, that is the running consensus test using the same threshold δ_n which is used for the ideal centralized detector. Conditions for asymptotic optimality of the running consensus statistic $T_{n,s}$ are provided by the following theorem.

Theorem 7.1

Suppose that:

- *the maximum eigenvalue* λ *of* $\mathrm{E}[\widetilde{\mathbf{W}}^T\widetilde{\mathbf{W}}]$ *is* < 1*;*
- $\mu(\theta)$ *is differentiable with* $\mu'(\theta_0) \neq 0$*, and* $\sigma(\theta)$ *is a continuous function;*
- *the centralized test statistic* $T_n^{(c)}$ *fulfills (7.10).*

Then, the decentralized detection statistic $T_{n,s}$ *in (7.8) is asymptotically equivalent to* $T_n^{(c)}$*, for any sensor* $s = 1, 2, \ldots, S$. •

The proof of Theorem 7.1 is given in [26].

Before ending this section, we would like to make a brief comment on the main hypotheses of the theorem. The first condition can be translated into a basic requirement of having a connected network [3]. The remaining ones are technical regularity conditions, usually adopted in the context of asymptotic detection [49], and focused on the asymptotic normality of the detection statistic.

7.3.2 Sequential detection

Let us recall some known results about sequential testing between continuous-time processes. Let $\mathscr{W}(t)$ be a standard Wiener process, where t ranges over the reals, and suppose that one wants to test the hypothesis that the Wiener process has a negative drift $-d/2$ against that of a positive drift $+d/2$. In formulas:

$$\mathscr{H}_0 : \mathscr{W}(t) = \mathscr{W}_{-d/2}(t), \qquad \mathscr{H}_1 : \mathscr{W}(t) = \mathscr{W}_{+d/2}(t),$$

Adopting a sequential approach, it is known that the optimum test, in the sense of minimizing the expected sample size for a prescribed pair of error probabilities p_f and p_d, is:

$$\begin{cases} \mathscr{W}(t) \geq \beta, & \text{declare } \mathscr{H}_1, \\ \mathscr{W}(t) \leq \alpha, & \text{declare } \mathscr{H}_0, \\ \mathscr{W}(t) \in (\alpha, \beta), & \text{take another sample,} \end{cases}$$

where the thresholds exactly take the form [43]:

$$\alpha = \frac{1}{d} \log \frac{1 - p_d}{1 - p_f}, \qquad \beta = \frac{1}{d} \log \frac{p_d}{p_f}, \tag{7.11}$$

The above test implicitly defines a (continuous) stopping time, viz.,

$$\tau = \inf\{t : \mathscr{W}(t) \notin (\alpha, \beta)\}$$

and the average times for making a decision are:

$$\mathrm{E}_{-d/2}[\tau] = 2\frac{\mathscr{D}_b(p_f, p_d)}{d^2}, \quad \mathrm{E}_{d/2}[\tau] = 2\frac{\mathscr{D}_b(p_d, p_f)}{d^2}, \tag{7.12}$$

where we used $\mathrm{E}_{\pm d/2}$ to denote expectation under the two hypotheses, and where $\mathscr{D}_b(p, q)$ stands for the Kullback–Leibler divergence between the binary probability mass functions $[p, 1-p]$ and $[q, 1-q]$ [57].

Let us now come back to our discrete-time problem, involving the statistic $T_n^{(c)}$, see (7.5). In parallel to the reasoning used in the FSS case, we study the asymptotic properties of a sequential decision rule as the hypotheses come close to each other. However, in sequential tests, the number of samples is a random quantity, and a natural modification of (7.9), see [54], is:

$$\theta_r = \theta_0 + \frac{1}{\sqrt{r}}, \tag{7.13}$$

where r is a positive number, and the asymptotic performance will be studied in the limit $r \to \infty$.

The sequential test is then implemented as follows. We fix a value of the parameter r in (7.13), and the decision rule is designed for such r. To this aim, it is expedient to shift the detection statistic such that its expectations under the two hypotheses are opposite in sign. This is easily accomplished by defining:

$$\frac{\mathrm{E}_{\theta_r}[T_n^{(c)}] + \mathrm{E}_{\theta_0}[T_n^{(c)}]}{2} = n\,S\,\frac{\mu(\theta_r) + \mu(\theta_0)}{2} \overset{def}{=} n\,S\,\eta_r,$$

so that a sequential decision rule can be formulated as:

$$\begin{cases} T_n^{(c)} - n\,S\,\eta_r \geq b_r, & \text{declare } \mathscr{H}_1, \\ T_n^{(c)} - n\,S\,\eta_r \leq a_r, & \text{declare } \mathscr{H}_0, \\ T_n^{(c)} - n\,S\,\eta_r \in (a_r, b_r), & \text{take another sample.} \end{cases} \tag{7.14}$$

Such a test clearly defines implicitly a random variable N_r representing the number of samples needed to terminate the testing procedure:

$$N_r = \inf\{n : T_n^{(c)} - n\,S\,\eta_r \notin (a_r, b_r)\}\,. \tag{7.15}$$

The false alarm and detection probabilities are $p_{dr} = \Pr[\text{declare } \mathcal{H}_1 | \mathcal{H}_1]$, $p_{fr} = \Pr[\text{declare } \mathcal{H}_1 | \mathcal{H}_0]$.

A suitable mathematical tool for dealing with the asymptotic performance of sequential tests, in the limit of $r \to \infty$, is provided in [54], which essentially formulates a LOD (locally optimum detection) theory in the sequential framework. The rationale behind the development used in [54] is as follows. For a prescribed pair of error probabilities, as the two hypotheses come close to each other (i.e., as r diverges), the (average) number of samples needed by the detection statistic for exceeding one of the thresholds is expected to increase. Therefore, the time evolution of the statistic can be regarded as a random walk which moves inside two barriers, with single steps that become smaller and smaller with respect to the distance between the barriers, as the hypotheses approach each other. Otherwise stated, as r grows, the random walk approaches a continuous-time process.

In [54], the above simple intuition is made precise by considering the following detection statistic:

$$T_{[rt]}^{(c)} - [rt]\,S\,\eta_r = T_n^{(c)} - n\,S\,\eta_r, \tag{7.16}$$

where $[x]$ stands for the integer part of x, and t defines a continuous-time axis. This means that we are looking at a piecewise constant random process which changes values at the integer time instants n/r (for integer n), where the elementary step $1/r$ goes to zero as r diverges. As to the limiting distribution of the above statistic for $r \to \infty$, we can invoke the functional central limit theorem, namely, the convergence to Wiener processes. In paralleling (7.10), suppose that the following convergences hold (d is the efficacy):

$$\frac{T_{[rt]}^{(c)} - [rt]\,S\,\eta_r}{\sqrt{r}\,S\,\sigma(\theta_0)} \xrightarrow{f_{\theta_0}} \mathcal{W}_{-d/2}(t), \qquad \frac{T_{[rt]}^{(c)} - [rt]\,S\,\eta_r}{\sqrt{r}\,S\,\sigma(\theta_0)} \xrightarrow{f_{\theta_r}} \mathcal{W}_{+d/2}(t), \tag{7.17}$$

where $\xrightarrow{f_{\theta_r}}$ (resp. $\xrightarrow{f_{\theta_0}}$) denotes weak convergence to a random process [58], under $f_{\theta_r}(x)$ (resp. $f_{\theta_0}(x)$) as r diverges.

In light of (7.17), the results on testing two continuous Wiener processes $\mathcal{W}_{\pm d/2}(t)$, summarized at the beginning of this section, can be exploited in our discrete-time set up. Indeed, a threshold comparison of the form $((T_{[rt]}^{(c)} - [rt]\,S\,\eta_r)/(\sqrt{r}\,S\,\sigma(\theta_0))) \notin (\alpha, \beta)$ yields explicit expressions for the thresholds to be used in (7.14):

$$a_r = \sqrt{r}\,S\,\sigma(\theta_0)\,\alpha, \qquad b_r = \sqrt{r}\,S\,\sigma(\theta_0)\,\beta. \tag{7.18}$$

It is possible to show that the detection and false alarm probabilities, used for setting the thresholds α and β, are asymptotically attained by the designed detector [54]. Namely:

$$\lim_{r\to\infty} p_{dr} = p_d, \qquad \lim_{r\to\infty} p_{fr} = p_f. \tag{7.19}$$

Furthermore, in view of the relationship $n = [rt]$, one would expect that $N_r \approx [r\tau]$. Indeed, it turns out that, in the light of (7.12), the expected sample size scales as [54]:

$$\lim_{r\to\infty} \frac{\mathrm{E}_{\theta_0}[N_r]}{r} = 2\frac{\mathscr{D}_b(p_f, p_d)}{d^2}, \qquad \lim_{r\to\infty} \frac{\mathrm{E}_{\theta_r}[N_r]}{r} = 2\frac{\mathscr{D}_b(p_d, p_f)}{d^2}. \tag{7.20}$$

By defining the ARE as the ratio of the expected sample sizes, and accordingly taking the limit for large r, it is immediate to see that these two detectors are simply compared in terms of their efficacies, just as happens in the FSS case. It is now natural to use, for the sequential framework, the same definitions of asymptotic equivalence and asymptotic optimality adopted in the FSS case.

In addition, the unbeatable efficacy in the sequential case is the same of d_{max} defined in the FSS context. This result can be obtained by using the expressions of the average stopping time for the optimal sequential probability ratio test ([45,46]), in the limit of large r, see [43,54].

The above reasoning has been focused on the ideal centralized statistic $T_n^{(c)}$. As done before, we switch now on the decentralized statistic available at the sth node $T_{n,s}$ and, as for the FSS test, we consider a sequential detector using the same thresholds a_r and b_r used for the ideal system. The main results about the sequential scenario is that the sequential test:

$$\begin{cases} T_{n,s} - n\,S\,\eta_r \geq b_r, & \text{declare } \mathscr{H}_1, \\ T_{n,s} - n\,S\,\eta_r \leq a_r, & \text{declare } \mathscr{H}_0, \\ T_{n,s} - n\,S\,\eta_r \in (a_r, b_r)\,, & \text{take another sample,} \end{cases} \tag{7.21}$$

behaves asymptotically as one in which the statistic $T_{n,s}$ is replaced by the ideal centralized statistic $T_n^{(c)}$. This statement is made precise in the following claim.

Theorem 7.2

Suppose that:

- *the maximum eigenvalue* λ *of* $\mathrm{E}[\widetilde{\mathbf{W}}^T\widetilde{\mathbf{W}}]$ *is* < 1;
- $\mu(\theta)$ *is differentiable with* $\mu'(\theta_0) \neq 0$, *and* $\sigma(\theta)$ *is a continuous function;*
- *the centralized test statistic* $T_n^{(c)}$ *fulfills (7.17);*
- *for any* $\varepsilon > 0$, *there exist* r_ε *and a function* $g_\varepsilon(t) > 0$ *such that, for all* $r \geq r_\varepsilon$ *and* $t \geq 1$:

$$\begin{aligned} &\mathrm{P}_{\theta_0}\left[\frac{T_{[rt]}^{(c)} - [rt]\,S\,\mu(\theta_0)}{\sqrt{r}\,S\,\sigma(\theta_0)} > \varepsilon t\right] \leq g_\varepsilon(t), \\ &\mathrm{P}_{\theta_r}\left[\frac{T_{[rt]}^{(c)} - [rt]\,S\,\mu(\theta_r)}{\sqrt{r}\,S\,\sigma(\theta_r)} \leq -\varepsilon t\right] \leq g_\varepsilon(t), \end{aligned} \tag{7.22}$$

with $\int_1^\infty g_\varepsilon(t) < \infty$, *and by* P_θ *we mean the probability corresponding to distribution* $f_\theta(x)$.

Then, the decentralized detection statistic $T_{n,s}$ *in (7.7) is asymptotically equivalent to* $T_n^{(c)}$, *for any sensor* $s = 1, 2, \ldots, S$. •

The proof of Theorem 7.2 is given in [26].

Please note that the first three conditions parallel the ones discussed just after Theorem 7.1. The additional technical assumption is needed to ensure convergence of the expected stopping times, see, e.g., [54].

7.4 Numerical examples

The network used for our experiments consists of ten sensors, arranged so as to form the topology in panel (a) of Figure 7.2, with constant consensus matrices following the Laplacian rule [59].

Let us consider the following hypothesis test involving local decision made at sensors: for $i = 1, 2, \dots, n$, and $s = 1, 2, \dots, S$:

$$\begin{aligned} \mathscr{H}_0 &: x_{i,s} \sim \mathscr{B}(\theta_0), \\ \mathscr{H}_1 &: x_{i,s} \sim \mathscr{B}(\theta), \end{aligned} \tag{7.23}$$

where $\mathscr{B}(p)$ is a Bernoulli distribution with parameter $p \in [0, 1]$.

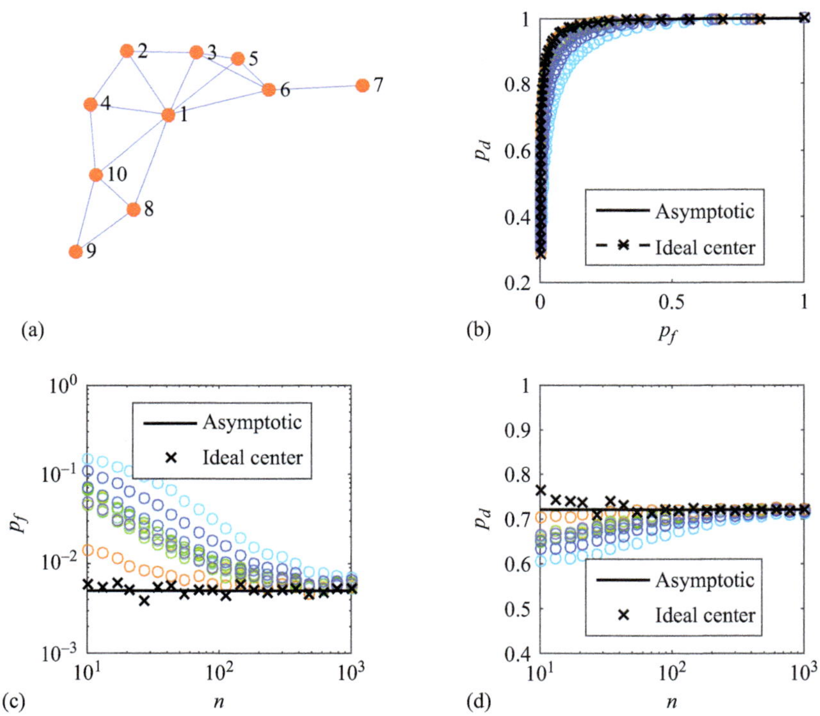

Figure 7.2 Parameters: $S = 10$, $\gamma = 0.5$ and $\theta_0 = 0.5$. Panel (a): network topology. Panel (b): ROC curves of running consensus ("o," colors identify sensors) and ideal centralized detector vs asymptotic ROC curves. Panels (c) and (d): Running consensus ("o," colors identify sensors) and ideal centralized system vs n

We shall use, as nonlinearity characterizing the statistic in (7.5), the well-known score function [49,50]:

$$t(x) = \frac{\partial}{\partial\theta} \log f_\theta(x)\Big|_{\theta=\theta_0} = \frac{x}{\theta_0} - \frac{1-x}{1-\theta_0}.$$

The first two moments of the score function and the related efficacy are the following:

$$\mu(\theta) = \frac{\theta}{\theta_0} - \frac{1-\theta}{1-\theta_0}, \quad \sigma^2(\theta) = I(\theta), \quad d = \sqrt{S\,I(\theta_0)}. \tag{7.24}$$

7.4.1 Fixed sample size test

In this subsection, we focus on the FSS test. From theory, we know that, in the asymptotic setting specified in the previous sections, with the hypotheses coming closer and closer as $n \to \infty$, the decentralized detector must approach the same performance as that of the ideal centralized one. We set the parameter under the alternative hypothesis $\theta = \theta_n = \theta_0 + \gamma/\sqrt{n}$, as prescribed by the asymptotic theory; the asymptotic ROC is as follows:

$$p_d = Q\left(Q^{-1}(p_f) - \gamma d\right). \tag{7.25}$$

In Figure 7.2, the detection and false alarm probabilities of a test based on the running consensus statistic $T_{n,s}$ for $s = 1, \dots, S$ as function of n are reported. We have assumed $\theta_n = \gamma/\sqrt{n}$ with $\gamma = 0.5$ and $\theta_0 = 0.5$. Specifically, panel (b) of Figure 7.2 reports the ROC curves related to running consensus and the ideal centralized system for $n = 90$ and the asymptotic ROC (7.25). Panels (c) and (d) of Figure 7.2 provide a sanity check for the asymptotic results and is useful to show the rate of convergence toward the ideal centralized detector and the asymptotic regime. The ideal detector seems to converge quickly than running consensus sensors. Differences in terms of convergence rate among the sensors are also observed. Such differences are related to the number of sensor edges in the network topology.

7.4.2 Sequential test

We analyze the sequential detection strategy with running consensus in the case of Bernoulli observations. Assuming $\theta_r = \theta_0 + 1/\sqrt{r}$, and using the detection statistic in (7.5), with decision rule and thresholds given by (7.14) and (7.18), the centralized system will compare the statistic:

$$T_n^{(c)} = \sum_{i=1}^{n}\sum_{s=1}^{S} x_{i,s} - n\,S\,\eta_r, \qquad \eta_r = \frac{\mu(\theta_r) + \mu(\theta_0)}{2}$$

with thresholds:

$$a_r = \sqrt{r\,S}\,\sigma_{\theta_0}\,\alpha, \qquad b_r = \sqrt{r\,S}\,\sigma_{\theta_0}\,\beta,$$

where σ_{θ_0} and $\mu(\theta)$ are given in (7.24).

In the computer simulations, and according to our theoretical findings, we use the above thresholds also for the running consensus scheme. Furthermore, we set

$p_f = 1 - p_d = p_e = 0.01$. As discussed in Section 7.3.2, the asymptotic performance of the ideal centralized detector (and, in view of Theorem 7.2, also of the running consensus) is just ruled by $d = \sqrt{S/\sigma^2}$, such that, following (7.20),

$$\mathrm{E}[N_r] \sim 2\,r\,\frac{\mathscr{D}_b(1-p_e,p_e)}{d^2}.$$

In panels (a) and (b) of Figure 7.3, the (scaled) expected stopping times are displayed, respectively, under $\mathscr{H}_0$ and $\mathscr{H}_1$, as functions of r, for the ideal centralized statistic and for each sensors. It can be seen that, as $r \to \infty$, the ratio $\mathrm{E}_\theta[N_r]/r$ of the sequential test with running consensus approaches that of the ideal centralized system, and both converge toward the asymptotic constant value $2\,\mathscr{D}_b(1-p_e,p_e)/d^2$.

Figure 7.3(a) and (b) also reveal that the expected sample number of the running consensus scheme is smaller than that of the ideal centralized scheme. This actually makes sense, given that $p_{fa,s}$ and $1 - p_{d,s}$ enforced by the decentralized scheme are

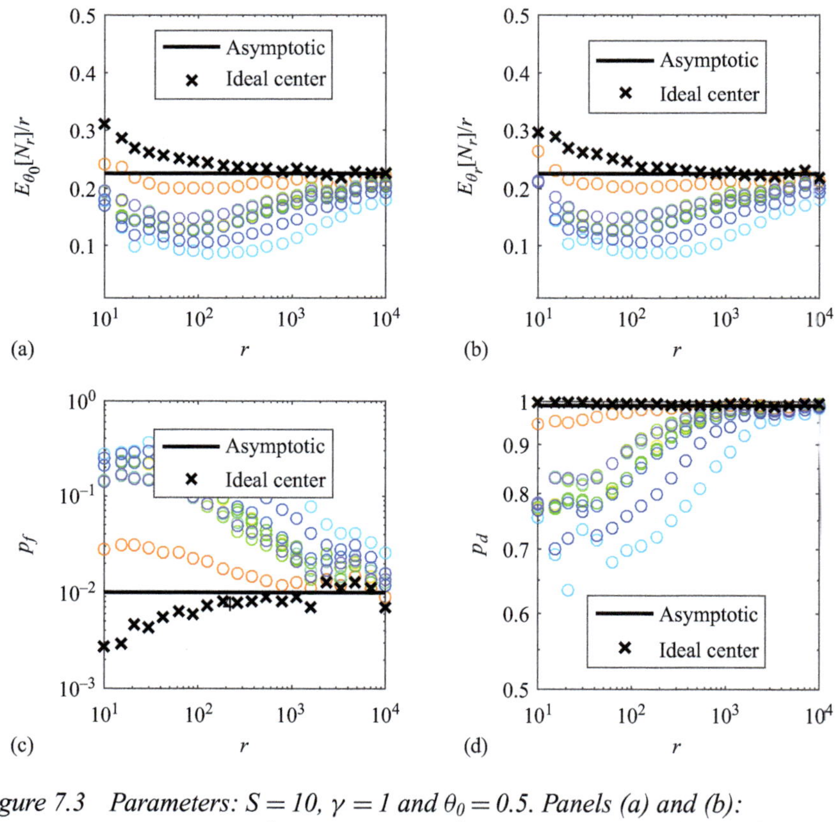

Figure 7.3 *Parameters: $S = 10$, $\gamma = 1$ and $\theta_0 = 0.5$. Panels (a) and (b): convergence of stopping times of running consensus ("o," colors identify sensors) and ideal centralized detector. Panels (c) and (d): false alarm probability and detection probability of running consensus ("o", colors identify sensors) and ideal centralized detector vs r, respectively*

always larger than that of the ideal system, thus explaining the decrease of $E_\theta[N_r]$ for the running scheme. The simulation results also show that both the running consensus scheme and the ideal centralized detector exhibit error probabilities that approach their nominal asymptotic values, as predicted by (7.19).

As already verified in the FSS case, each sensor exhibits different convergence rate to the asymptotic values of the false alarm and detection probabilities as well as of the expected decision stopping times.

7.5 Conclusions

In the recent years, a new paradigm, often referred to as *running consensus* or *consensus + innovation*, has been proposed for distributed inference in sensor networks. The main ingredient of such paradigm is that sensors collect sequentially new measurements and, at the same time, implement the consensus algorithm.

In this book chapter, we review the recent literature of the distributed inference via running consensus. Then we focus on the running consensus facing a binary hypothesis testing problem. Specifically, the state of each sensors acts as a detection statistic to infer an unknown binary state of the nature. When compared with an ideal, centrally computed, detection statistic, these local states are affected by an error term that, we show, is asymptotically negligible when the number of consensus steps grows.

We analyze such convergence exploiting mathematical tools from the locally optimum statistics. We show the asymptotic equivalence of the running consensus with the ideal system, in terms of detection capabilities. Our results are proved in two theorems that refer to fixed sample tests and to the sequential case. In the former case, the theory based upon Pitman ARE is used, relying upon asymptotic normality of the detection statistics, while in the latter, we resort to more advanced tools concerning the asymptotic behavior of sequential decision rules, which require convergence toward Wiener processes. The main message is that decentralized detection algorithms based on running consensus reach the same performance of the centralized schemes, provided that the system evolves for a sufficiently large amount of time.

References

[1] DeGroot MH. Reaching a Consensus. J Am Stat Assoc. 1974;69(345): 118–121.

[2] Dimakis AG, Kar S, Moura JMF, *et al.* Gossip Algorithms for Distributed Signal Processing. Proc IEEE. 2010;98(11):1847–1864.

[3] Boyd S, Ghosh A, Prabhakar B, *et al.* Randomized Gossip Algorithms. IEEE Trans Inf Theory. 2006;52(6):2508–2530.

[4] Olfati-Saber R, Fax A, and Murray RM. Consensus and Cooperation in Networked Multi-Agent Systems. Proc IEEE. 2007;95(1):215–233.

[5] Kashyap A, Basar T, and Srikant R. Quantized Consensus. Automatica. 2007;43(7): 1192–1203.

[6] Benezit F, Thiran P, and Vetterli M. The Distributed Multiple Voting Problem. IEEE J Sel Top Signal Process. 2011;5(4):791–804.

[7] Tsianos KI, Lawlor S, and Rabbat MG. Consensus-based distributed optimization: Practical issues and applications in large-scale machine learning. In: 2012 50th Annual Allerton Conference on Communication, Control, and Computing. Allerton; 2012. p. 1543–1550.

[8] Schizas ID, Ribeiro A, and Giannakis GB. Consensus in Ad Hoc WSNs With Noisy Links—Part I: Distributed Estimation of Deterministic Signals. IEEE Trans Signal Process. 2008;56(1):350–364.

[9] Kar S, and Moura JMF. Distributed Consensus Algorithms in Sensor Networks With Imperfect Communication: Link Failures and Channel Noise. IEEE Trans Signal Process. 2009;57(1):355–369.

[10] Olfati-Saber R. Kalman-consensus filter: Optimality, stability, and performance. In: Proceedings of the 48h IEEE Conference on Decision and Control (CDC) Held Jointly with 2009 28th Chinese Control Conference; 2009. p. 7036–7042.

[11] Hlinka O, Sluciak O, Hlawatsch F, *et al.* Likelihood Consensus and Its Application to Distributed Particle Filtering. IEEE Trans Signal Process. 2012;60(8):4334–4349.

[12] Hlinka O, Hlawatsch F, and Djuric P. Distributed Particle Filtering in Agent Networks: A Survey, Classification, and Comparison. IEEE Signal Process Mag. 2013;30(1):61–81.

[13] Battistelli G, Chisci L, Fantacci C, *et al.* Consensus CPHD Filter for Distributed Multitarget Tracking. IEEE J Sel Top Signal Process. 2013;7(3):508–520.

[14] Lagendijk RL, Erkin Z, and Barni M. Encrypted Signal Processing for Privacy Protection: Conveying the Utility of Homomorphic Encryption and Multiparty Computation. IEEE Signal Process Mag. 2013;30(1):82–105.

[15] Braca P, Lazzeretti R, Marano S, *et al.* Learning With Privacy in Consensus + Obfuscation. IEEE Signal Process Lett. 2016;23(9):1174–1178.

[16] Mo Y, and Murray RM. Privacy Preserving Average Consensus. IEEE Trans Autom Control. 2017;62(2):753–765.

[17] Gubner JA. Distributed Estimation and Quantization. IEEE Trans Inf Theory. 1993;39(4):1456–1459.

[18] Varshney PK. Distributed Detection and Data Fusion. New York, NY: Springer; 1997.

[19] Marano S, Matta V, Tong L, *et al.* A Likelihood Based Multiple Access for Estimation in Sensor Networks. IEEE Trans Signal Process. 2007;55(11): 5155–5166.

[20] Braca P, Marano S, and Matta V. Single-Transmission Distributed Detection via Order Statistics. IEEE Trans Signal Process. 2012;60(4):2042–2048.

[21] Braca P, Marano S, Matta V, *et al.* Asymptotic Efficiency of the PHD in Multitarget/Multisensor E estimation. IEEE J Sel Top Signal Process. 2013;7(3):553–564.

[22] Meyer F, Braca P, Willett P, *et al.* A Scalable Algorithm for Tracking an Unknown Number of Targets Using Multiple Sensors. IEEE Trans Signal Process. 2017;65(13):3478–3493.

[23] Barbarossa S, and Scutari G. Bio-inspired Sensor Network Design. IEEE Signal Process Mag. 2007;24(3):26–35.
[24] Braca P, Marano S, and Matta V. Running Consensus in Wireless Sensor Networks. In: Proc. of the Intern. Conf. on Information Fusion (FUSION 2008). Cologne, Germany; 2008. p. 152–157.
[25] Braca P, Marano S, and Matta V. Enforcing Consensus While Monitoring the Environment in Wireless Sensor Networks. IEEE Trans Signal Process. 2008;56(7):3375–3380.
[26] Braca P, Marano S, Matta V, *et al.* Asymptotic Optimality of Running Consensus in Testing Statistical Hypotheses. IEEE Trans Signal Process. 2010;58(2):814–825.
[27] Braca P, Marano S, Matta V, *et al.* Consensus-Based Page's Test in Sensor Networks. Signal Process. 2011;91(4):919–930.
[28] Bajovic D, Jakovetic D, Xavier J, *et al.* Distributed Detection via Gaussian Running Consensus: Large Deviations Asymptotic Analysis. IEEE Trans Signal Process. 2011;59(9):4381–4396.
[29] Cattivelli F, and Sayed A. Distributed Detection Over Adaptive Networks Using Diffusion Adaptation. IEEE Trans Signal Process. 2011;59(5):1917–1932.
[30] Jakovetic D, Moura JMF, and Xavier J. Distributed Detection Over Noisy Networks: Large Deviations Analysis. IEEE Trans Signal Process. 2012;60(8): 4306–4320.
[31] Sahu AK, Kar S, Moura JMF, *et al.* Distributed Constrained Recursive Nonlinear Least-Squares Estimation: Algorithms and Asymptotics. IEEE Trans Signal Inf Process Netw. 2016;2(4):426–441.
[32] Sayed AH. Adaptive Networks. Proc IEEE. 2014;102(4):460–497.
[33] Kar S, Moura JMF, and Ramanan K. Distributed Parameter Estimation in Sensor Networks: Nonlinear Observation Models and Imperfect Communication. IEEE Trans Inf Theory. 2012;58(6):3575–3605.
[34] Matta V, Braca P, Marano S, *et al.* Diffusion-Based Adaptive Distributed Detection: Steady-State Performance in the Slow Adaptation Regime. IEEE Trans Inf Theory. 2016;62(8):4710–4732.
[35] Dembo A, and Zeitouni O. Large Deviations Techniques and Applications. New York: Springer-Verlag; 1998.
[36] Matta V, Braca P, Marano S, *et al.* Distributed Detection over Adaptive Networks: Refined Asymptotics and the Role of Connectivity. IEEE Trans Signal Inf Process Netw. 2016;2(4):442–460.
[37] Kar S, and Moura JMF. Consensus + Innovations Distributed Inference over Networks: Cooperation and Sensing in Networked Systems. IEEE Signal Process Mag. 2013;30(3):99–109.
[38] Javadi SH. Detection over Sensor Networks: A Tutorial. IEEE Aerosp Electron Syst Mag. 2016;31(3):2–18.
[39] Longo M, Lookabaugh T, and Gray R. Quantization for Decentralized Hypothesis Testing under Communication Constraints. IEEE Trans Inf Theory. 1990; 36(2):241–255.
[40] Blum RS, Kassam A, and Poor HV. Distributed Detection With Multiple Sensors: Part II – Advanced Topics. Proc IEEE. 1997;85(1):64–79.

[41] Chamberland JF, and Veeravalli VV. Decentralized Detection in Sensor Networks. IEEE Trans Signal Process. 2003;51(2):407–416.
[42] Chen B, Tong L, and Varshney P. Channel-Aware Distributed Detection in Wireless Sensor Networks. IEEE Signal Process Mag. 2006;23(4):16–26.
[43] Basseville M, and Nikiforov IV. Detection of Abrupt Changes: Theory and Application. Englewood Cliffs, NJ: Prentice-Hall; 1993.
[44] Sahu AK, and Kar S. Recursive Distributed Detection for Composite Hypothesis Testing: Nonlinear Observation Models in Additive Gaussian Noise. IEEE Trans Inf Theory. 2017;63(8):4797–4828.
[45] Wald A. Sequential Analysis. New York, NY: Dover; 1947.
[46] Wald A, and Wolfowitz J. Optimum Character of the Sequential Probability Ratio Test. Ann Stat. 1948;19(3):326–339.
[47] Stankovic S, Ilic N, Stankovic M, *et al.* Distributed Change Detection Based on a Consensus Algorithm. IEEE Trans Signal Process. 2011;59(12):5686–5697.
[48] Ilic N, Stankovic S, Stankovic M, *et al.* Consensus Based Distributed Change Detection Using Generalized Likelihood Ratio Methodology. Signal Process. 2012;92(7):1715–1728.
[49] Kassam SA. Signal Detection in Non-Gaussian Noise. New York: Springer-Verlag; 1987.
[50] Lehmann EL, and Romano JP. Testing Statistical Hypotheses. New York: Springer; 2005.
[51] Pitman EJG. Lecture Notes on Nonparametric Inference. New York: Columbia Univ.; 1949 (unpublished).
[52] Bajovic D, Jakovetic D, Moura JMF, *et al.* Large Deviations Performance of Consensus + Innovations Distributed Detection With Non-Gaussian Observations. IEEE Trans Signal Process. 2012;60(11):5987–6002.
[53] Poor HV. An Introduction to Signal Detection and Estimation. New York, NY: Springer-Verlag; 1988.
[54] Lai TL. Pitman Efficiencies of Sequential Tests and Uniform Limit Theorems in Nonparametric Statistics. Ann Stat. 1978;6(5):1027–1047.
[55] Noether GE. On a Theorem of Pitman. Ann Math Stat. 1955;26:64–68.
[56] Van Trees H. Detection, Estimation, and Modulation Theory. Part I. New York, NY: John Wiley & Sons, Inc.; 1968 (reprinted, 2001).
[57] Cover T, and Thomas J. Elements of Information Theory. New York, NY: John Wiley & Sons; 1991.
[58] Billingsley P. Convergence of Probability Measures. 2nd ed. New York, NY: Wiley-Interscience; 1999.
[59] Xiao L, and Boyd S. Fast Linear Iterations for Distributed Averaging. Syst Control Lett. 2004;53:65–78.

Chapter 8

Distributed recursive testing of composite hypothesis in multi-agent networks

Anit Kumar Sahu[1] *and Soummya Kar*[1]

This chapter considers the problem of recursive composite hypothesis testing in a network of sparsely connected agents. In classical centralized composite hypothesis testing, procedures such as the generalized likelihood ratio test (GLRT), i.e., the detection procedure which uses the underlying parameter estimate based on all the collected samples as a plug-in estimate, may exhibit poor performance until a reasonably accurate parameter estimate (typically the maximum likelihood estimate of the underlying parameter/state) is obtained. Usually in setups that employ the classical (centralized) GLRTs, the data-collection phase precedes the parameter estimation and detection statistic update phase, thus rendering the testing an essentially *offline* batch procedure. The motivation behind studying distributed recursive *online* detection algorithms in contrast to offline batch processing based detection algorithms is that in most multi-agent networked scenarios, which are typically energy constrained, the priority is to obtain reasonable inference performance by expending fewer amount of resources.

8.1 Introduction

Wireless sensor networks has been one of the most pervasive architectures for inference and monitoring tasks in areas where a single sensor or a single data collection unit is ill-equipped to accomplish the task. Inference tasks in the context of wireless sensor networks need to be built around keeping several limitations in mind. To be specific, the data collection in a sensor network is inherently distributed; thus a sensor (also referred to as an agent in the sequel) in the network is individually in the network incapable of accomplishing an inference task as it has limited *observability*, i.e., potentially partial information of the field where the network is deployed. In addition to the inherently distributed nature of data, the ad hoc nature of the entire setup makes any inference task highly nontrivial in the context of wireless sensor networks. In this chapter, we specifically focus on composite hypothesis testing in the context of wireless sensor networks.

[1]Department of Electrical and Computer Engineering, Carnegie Mellon University, Pittsburgh, PA, USA

For illustration, consider a scenario in which a network of sensors is tasked to monitor an environment. The phenomena of interest can be in one of many states (hypotheses) and the goal of the network is to collaboratively determine the state of the environment based on the distributed measurements obtained by the sensors. Typically simple or finite multiple simple hypothesis testing involve hypotheses which are uniquely parameterized, i.e., each hypothesis corresponds to a unique probability distribution on the observed data and thus makes the hypotheses to be finitely parametrized. In contrast with finite simple hypotheses, a composite hypothesis testing involves at least one of the hypothesis to be infinitely parametrized, i.e., it corresponds to an infinite parametric family of probability distributions on data. For example, in a typical target tracking scenario, the hypotheses of interest could be the absence (the null hypothesis) or presence (the alternative hypothesis) of the target. Further, in the presence of the target (the alternative hypothesis), each possible location or configuration of the target would induce a possibly different probability distribution on the data observed by the sensors. Due to a continuum of possibilities on the location *parameter* of the target (if present), it follows that the alternative hypothesis in this case admits an infinite parameterization reflecting the different possibilities. In general, owing to the infinite parameterization under one of the hypothesis, the goal in a composite hypothesis testing problem is not only to detect as to which hypothesis is in force but also infer the parameter of interest, also referred to as the state of environment (e.g., the location of the target), whenever applicable, based on the sensed information across all the agents at all times. In this chapter, we are specifically interested in the design of distributed and *recursive* composite hypothesis testing procedures to decide between a null hypothesis and a composite alternative hypothesis parameterized by a continuous vector parameter. By *recursive*, we mean algorithms which continuously generate a decision at each agent as to which hypothesis is in force, whenever a new datum is collected. Technically speaking, our focus is on algorithms which process data samples as and when they are collected rather than waiting for all the data to be collected. The aforementioned class of detection algorithms takes into account the finite on board battery power and thus limited sensing and computation power in the sensors which are the basic building blocks of a wireless sensor network. Recursive composite hypothesis testing is particularly relevant to applications such as cooperative spectrum sensing [1–3] and multi-input multi-output (MIMO) radars [4], where it is of particular importance to achieve reasonable detection performance by frugal use of resources such as data samples, sensing and communication.

In the context of composite hypothesis testing, one of the classical approaches is the GLRT [5] which has an estimation scheme built into it so as to provide the decision statistic a plug-in estimate. However, in a setting which involves communicating the observations at each time to a fusion center from each sensor, so to estimate the underlying parameter for the composite hypothesis, which in turn is a maximization problem (for instance, if the maximum likelihood estimate is being used as a plug-in estimate) involves considerable communication overhead and might be impractical to implement in a wireless sensor network setting. In contrast to setups involving a fusion center, in this chapter, we consider a fully distributed setup where inter-sensor communication is dictated by a preassigned possibly sparse communication graph.

We focus on a setup, in which the agents in the network at each time step obtain conditionally Gaussian and identically and independently distributed observations and aim to determine as to which hypothesis among the null and the composite alternative is in force while estimating the underlying parameter which parameterizes the hypotheses. (The precise technical formulation is relegated to Section 8.2.) This coupled estimation and detection is achieved by the agents' iterative updating of their parameter estimates and decision statistics by simultaneous assimilation of the information obtained from the neighboring agents (*consensus*) and the latest locally sensed information (*innovation*). Going with the components instrumental to our algorithm, i.e., the obtained information from the neighborhood and the latest locally sensed information, namely consensus and innovations, respectively, we name our algorithm **CIGLRT**, a *consensus* + *innovations* type algorithm (see, e.g., [6]) of the form of the GLRT for implementing distributed composite hypothesis testing. Furthermore, in order to replicate real-world sensing scenarios and keeping the limited sensing resources of sensors in mind, we adopt an observation model where agents' measurements are noisy linear transformations of only a few components of the parameter. While such an observation model results in a frugal utilization of sensing energy of sensors, it renders the sensors observing only noise incapable of conducting the hypothesis test just based on their own data as the parameter of interest is *locally unobservable*. In this chapter, we propose a distributed recursive composite hypothesis testing algorithm, where the composite alternative abstracting out the state of the field is parameterized as a continuous unknown vector parameter. The interagent collaboration in our setup is achieved by iterative low-complexity message exchanges among *neighboring* sensors, the neighborhoods conforming to a possibly sparse preassigned communication graph. The algorithm **CIGLRT**, in spite of being a recursive and distributed algorithm, is shown to achieve provably exponentially decaying probabilities of miss and false alarm. We specifically analyze and characterize the feasible choice of thresholds and other algorithm design parameters for which an exponential decay of probabilities of errors in the large sample (time) limit can be guaranteed.

8.1.1 Related work

Distributed detection has garnered a lot of attention in the literature of late. The literature on distributed detection schemes in the context of networked setups can be categorized into three classes. The first class consists of architectures with a fusion center, and where all the agents send their local observations or decision statistics or the decision in their quantized or non-quantized form to the fusion center (see, e.g., [7,8]) and the fusion center assimilates all the received information and conducts the associated inference scheme. The second class consists of schemes with no fusion center utilizing consensus or gossip-based algorithms (see, e.g., [9–12]) which divide the algorithm timeline into a distinct data collection phase by the agents over a long period of time followed by the information exchange in the agents' neighborhoods; the neighborhoods in turn being specified by a preassigned communication graph or a sequence of possibly sparse time-varying communication graphs satisfying appropriate connectivity conditions. Finally, distributed inference schemes which perform simultaneous assimilation of information obtained from sensing and neighborhood communication (see, e.g., [13–15]) form the third class of aforementioned distributed detection schemes. Among various distributed inference schemes that have been employed for inference tasks in networked environments, the diffusion and *consensus+innovations* schemes have been versatile in terms of applications and have been utilized for tasks including distributed parameter estimation (see, e.g., [14,16,17]), distributed detection (see, e.g., [6,14,15,18–20]), distributed reinforcement learning (see, e.g., [21]), distributed information processing in presence of faulty agents or imperfect model information (see, e.g., [22]), and multitask learning (see, e.g., [23]) to name a few. The third class of distributed detection schemes which encompass subclasses such as the running consensus approach [24,25], the diffusion approach [3,15,26–30] and the *consensus+innovations* approach [13,14,18] are more relevant to the algorithm described in this chapter. The aforementioned works primarily address questions related to simple hypothesis testing and analyze the performance limits of detection schemes through large deviations analysis. Other relevant works include [31–34], where the focus is to determine as to which hypothesis is in force among multiple simple hypothesis involving finite parametric alternatives. Differently from the aforementioned references, the objective of the detection scheme in this chapter is to test between a simple null hypothesis and an alternative hypothesis which is parameterized by a vector parameter which takes values in a continuous space. In terms of characterization of the performance of recursive distributed detection schemes in a networked environment through decay exponents for probability of errors, the works most related to the algorithm **CIGLRT** are [13,18]. In the aforementioned works, the hypotheses under consideration for hypothesis testing are binary simple hypothesis and hence the incorporation of new sensed information (*innovation*) in the decision statistic update is independent and identically distributed (i.i.d.). However, the algorithm **CIGLRT** involves deciding between a simple null hypothesis and a composite alternative for which the underlying parameter is unknown a priori as the composite alternative is parameterized by a parameter that takes values in a continuous space. Hence, the algorithm involves estimation and detection schemes running in parallel which leads to incorporation of latest sensed information in the decision statistic update through an innovation sequence which is neither independent nor identically distributed. In the class of centralized detection schemes in the context of composite hypothesis testing, the weak convergence of the decision statistic sequence under the null and alternate hypothesis (see, e.g., [35]) is usually insufficient to infer about the decay exponents of the probability of errors. Technically speaking, we not only extend Wilks' Theorem to a stronger result so as to quantify and characterize the decay exponents of the probability of errors but also extend it to a distributed recursive networked setup which also involves characterizing the asymptotic normality of the decision statistic sequence. The coupling of the estimation and detection schemes in addition to the non-i.i.d innovation sequence for the decision statistic sequence warrants the development of technical tools in terms of concentration of measure for sums of non i.i.d. random variables that are instrumental in deriving the decay exponents of the error probabilities, which we study in this chapter.

8.1.2 Organization

We briefly discuss the organization of this chapter. In Section 8.1.3, we introduce the notation that we use throughout this chapter, while Section 8.1.4 introduces preliminaries about spectral graph theory. We introduce and setup the problem in Section 8.2, in which Sections 8.2.1 and 8.2.2 discuss the sensing models and preliminaries concerning classical centralized GLRT, respectively. In Section 8.3, distributed GLRT is introduced where in Section 8.3.1, we delve into the details of the algorithm **CIGLRT** and the different assumptions pertaining to the development of the algorithm **CIGLRT**. Section 8.3.2 presents a simplified setup for the algorithm **CIGLRT** and presents the main results pertaining to the algorithm which includes the characterization of the decay of the probability of errors as a function of the algorithm parameters of **CIGLRT**. Finally, Section 8.6 concludes the chapter.

8.1.3 Notation

We briefly summarize the notation we use in this chapter. We denote by $\mathbb{R}$ the set of reals, $\mathbb{R}_+$ the set of nonnegative reals, and by $\mathbb{R}^k$ the k-dimensional Euclidean space. The set of $k \times k$ real matrices is denoted by $\mathbb{R}^{k\times k}$. The set of integers is denoted by $\mathbb{Z}$, whereas, $\mathbb{Z}_+$ denotes the subset of nonnegative integers. We denote vectors and matrices by bold-faced characters. We denote by A_{ij} or $[A]_{ij}$ the (i,j)th entry of a matrix $\mathbf{A}$; a_i or $[a]_i$ the ith entry of a vector $\mathbf{a}$. The symbols $\mathbf{I}$ and $\mathbf{0}$ are used to denote the $k \times k$ identity matrix and the $k \times k$ zero matrix, respectively, the dimensions being clear from the context. We denote by $\mathbf{e_i}$ the ith column of $\mathbf{I}$. The symbol $\top$ denotes matrix transpose. We denote the determinant and trace of a matrix by $\det(\cdot)$ and $\mathrm{tr}(\cdot)$, respectively. The $k \times k$ matrix $\mathbf{J} = (1/k)\mathbf{1}\mathbf{1}^\top$ where $\mathbf{1}$ denotes the $k \times 1$ vector of ones. The operator $||\cdot||$ applied to a vector denotes the standard Euclidean $\mathscr{L}_2$ norm, while applied to matrices, it denotes the induced $\mathscr{L}_2$ norm, which is equivalent to the spectral radius for symmetric matrices. For a matrix A with real eigenvalues, the notation $\lambda_{\min}(A)$ and $\lambda_{\max}(A)$ will be used to denote its smallest and largest eigenvalues, respectively. Throughout the chapter, the true (but unknown) value of the parameter is denoted by θ^*. The estimate of θ^* at time t at agent n is denoted by $\theta_n(t) \in \mathbb{R}^{M\times 1}$. All the logarithms in the chapter are with respect to base e and represented as $\log(\cdot)$. The (simple) null hypothesis is denoted by $\mathscr{H}_0$, whereas, the (composite) alternative by $\mathscr{H}_\theta$ with the subscript θ indicating the parameter θ (a priori unknown) modeling or parametrizing the alternative. The operators $\mathbb{E}_0[\cdot]$ and $\mathbb{E}_\theta[\cdot]$ denote expectation conditioned on hypothesis $\mathscr{H}_0$ and $\mathscr{H}_\theta$. Similarly, $\mathbb{P}(\cdot)$ denotes the probability of an event, and $\mathbb{P}_0(\cdot)$ and $\mathbb{P}_\theta(\cdot)$ denote the probability of the event conditioned on the null hypothesis $\mathscr{H}_0$ and $\mathscr{H}_\theta$. For deterministic $\mathbb{R}_+$-valued sequences $\{a_t\}$ and $\{b_t\}$, the notation $a_t = O(b_t)$ denotes the existence of a constant $c > 0$ such that $a_t \leq cb_t$ for all t sufficiently large; the notation $a_t = o(b_t)$ denotes $a_t/b_t \to 0$ as $t \to \infty$. The order notations $O(\cdot)$ and $o(\cdot)$ will be used in the context of stochastic processes as well in which case they are to be interpreted almost surely or sample path-wise.

8.1.4 Spectral graph theory

For an undirected graph $G = (V, E)$, V denotes the set of agents or vertices with cardinality $|V| = N$, and E the set of edges with $|E| = M$. The unordered pair $(i,j) \in E$ if there exists an edge between agents i and j. We only consider simple graphs, i.e., graphs devoid of self-loops and multiple edges. A path between agents i and j of length m is a sequence $(i = p_0, p_1, \ldots, p_m = j)$ of vertices, such that $(p_t, p_{t+1}) \in E$, $0 \leq t \leq m-1$. A graph is connected if there exists a path between all the possible agent pairings. The neighborhood of an agent n is given by $\Omega_n = \{j \in V | (n,j) \in E\}$. The degree of agent n is given by $d_n = |\Omega_n|$. The structure of the graph may be equivalently represented by the symmetric $N \times N$ adjacency matrix $\mathbf{A} = [A_{ij}]$, where $A_{ij} = 1$ if $(i,j) \in E$, and 0 otherwise. The degree matrix is represented by the diagonal matrix $\mathbf{D} = diag(d_1 \cdots d_N)$. The graph Laplacian matrix is represented by:

$$\mathbf{L} = \mathbf{D} - \mathbf{A}. \tag{8.1}$$

The Laplacian is a positive semidefinite matrix; hence, its eigenvalues can be sorted and represented in the following manner:

$$0 = \lambda_1(\mathbf{L}) \leq \lambda_2(\mathbf{L}) \leq \cdots \leq \lambda_N(\mathbf{L}). \tag{8.2}$$

Furthermore, a graph is connected if and only if $\lambda_2(\mathbf{L}) > 0$ (see [36] for instance).

8.2 Problem formulation

8.2.1 System model and preliminaries

The networked environment consists of N sensing entities with onboard computation capabilities deployed in the network. At each discrete or slotted time step t, each agent makes a noisy measurement $\mathbf{y}_n(t)$ of the underlying parameter θ^* which is a vector parameter, i.e., $\theta^* \in \mathbb{R}^M$. To be specific, the measurement $\mathbf{y}_n(t) \in \mathbb{R}^{M_n}$ is a possibly low-dimensional noisy linear transformation of θ^*. The measurement is a sample from the probability distribution $\mathbb{P}_0$ under the hypothesis $\mathscr{H}_0$, whereas under the composite alternative $\mathscr{H}_1$ comes from a probability distribution which is a member of a parametric family $\{\mathbb{P}_{\theta^*}\}$. It is to be noted that the parameter θ^* is deterministic but unknown which points to the unavailability of any information a priori about the parameterization under the composite alternative. Formally,

Alternative and null hypothesis

$$\begin{aligned} \mathscr{H}_1 &: \mathbf{y}_n(t) = \mathbf{H}_n\theta^* + \gamma_n(t) \\ \mathscr{H}_0 &: \mathbf{y}_n(t) = \gamma_n(t), \end{aligned} \tag{8.3}$$

where $\mathbf{H}_n \in \mathbb{R}^{M_n \times M}$ denotes the sensing matrix (possibly a fat matrix), $\{\mathbf{y}_n(t)\}$ is a $\mathbb{R}^{M_n}$-valued observation sequence for the nth agent and $\{\gamma_n(t)\}$ is a zero-mean temporally i.i.d Gaussian noise sequence at the nth agent with nonsingular covariance matrix $\mathbf{R}_n$, where $\mathbf{R}_n \in \mathbb{R}^{M_n \times M_n}$. Moreover, the noise sequences at two agents n, l with $n \neq l$ are assumed to be independent.

The above formulation reduces to the case where the null hypothesis $\mathscr{H}_0$ corresponds to $\theta^* = 0$ and the alternative hypothesis $\mathscr{H}_1$ corresponds to the case $\theta^* \neq \mathbf{0}$.

In the observation model under consideration for our networked setup, the sources of randomness are the measurement processes $\{y_n(s)\}_{n=1}^N$ made by the sensing entities in the network. We define the natural filtration $\{\mathscr{F}_t\}$ generated by the random observations, i.e.,

$$\mathscr{F}_t = \sigma\left(\{\{y_n(s)\}_{n=1}^N\}_{s=0}^{t-1}\right), \tag{8.4}$$

which is the sequence of σ-algebras induced by the observation processes, in order to model the overall available network information at all times. Finally, a stochastic process $\{\mathbf{x}(t)\}$ is said to be $\{\mathscr{F}_t\}$-adapted if the $\mathbf{x}(t)$ is measurable with respect to $\mathscr{F}_t$ at each t (see, e.g., [37]).

8.2.2 Preliminaries: generalized likelihood ratio tests

In this subsection, we review concepts pertaining to the classical theory of composite hypothesis testing. In particular, we focus on GLRT. For the purpose of illustration, we consider a target-detection problem, in which the absence of target is modeled by a simple hypothesis $\mathscr{H}_0$. The presence of the target corresponds to the composite alternative $\mathscr{H}_1$, as it is parameterized by a continuous vector parameter including attributes such as location coordinates, velocity and acceleration coordinates of the target to name a few which is unknown a priori. Given a target detection setup, it is infeasible to monitor a geographical field by a single sensor or agent as it limits sensing abilities of the entire infrastructure and hence is a very natural use case for multiple sensing entities deployed in a network in the geographical area under consideration. Let the vectorized version of all the measurements collected by all agents at time t be denoted by $\mathbf{y}(t)$, i.e., $\mathbf{y}(t) = \left[\mathbf{y}_1^\top(t) \cdots \mathbf{y}_N^\top(t)\right]^\top$. However, owing to the location of a particular sensing entity in the network, certain attributes of the parameter pertaining to the target might not to be observable at an agent. Also, it is of particular interest to employ heterogeneous agents in such a networked setup so as to monitor different attributes of the target being monitored for in the geographical area. Hence, so as to replicate practical observation models, we assume that the observation at agent n is M_n dimensional, which makes the vectorized measurement $\mathbf{y}(t)$ a $\sum_{n=1}^N M_n$-dimensional vector. In such a setting, where there is a central information hub which is responsible for assimilating sensing information from all the sensing entities in the network and subsequently deciding on the absence or presence of a target, a classical approach is the GLRT (see, e.g., [5]). Specifically, the GLRT detection scheme decides on the absence or presence of the target[1] in the following manner:

Generalized likelihood ratio test:

$$\mathscr{H} = \begin{cases} \mathscr{H}_1, & \text{if } \max_\theta \sum_{s=0}^{t} \log \frac{f_\theta(\mathbf{y}(s))}{f_0(\mathbf{y}(s))} > \eta, \\ \mathscr{H}_0, & \text{otherwise,} \end{cases} \tag{8.5}$$

where η is a predefined threshold, t denotes time or equivalently the number of sensed observations obtained at each agent up to time t. Furthermore, exploiting conditional independence of the observation sequence at the sensing entities, we have:

$$f_{\theta^*}(\mathbf{y}(t)) = f_{\theta^*}^1(\mathbf{y}_1(t)) \cdots f_{\theta^*}^N(\mathbf{y}_N(t))$$

where $f_{\theta^*}(\mathbf{y}(t))$ denotes the likelihood of sampling the measurement $\mathbf{y}(t)$ under $\mathscr{H}_1$ parameterized by $\boldsymbol{\theta}^*$, and $f_{\theta^*}^n(y_n(t))$ denotes the likelihood of observing $y_n(t)$ at the nth agent under $\mathscr{H}_1$ the latter being parameterized by $\boldsymbol{\theta}^*$; similarly, under the null hypothesis exploiting conditional independence of the measurement sequence across agents, we have:

$$f_0(\mathbf{y}(t)) = f_0^1(\mathbf{y}_1(t)) \cdots f_0^N(\mathbf{y}_N(t)),$$

which denotes the likelihood of observing $\mathbf{y}(t)$ under $\mathscr{H}_0$, and $f_0^N(\mathbf{y}_N(t))$ denotes the likelihood of observing $y_n(t)$ at the nth agent under $\mathscr{H}_0$. One of the crucial steps involved in calculating the decision statistic in classical GLRT as formulated in (8.5) is the maximization:

$$\max_\theta \sum_{t=0}^{T} \log \frac{f_\theta(\mathbf{y}(t))}{f_0(\mathbf{y}(t))} = \max_\theta \sum_{t=0}^{T} \sum_{n=1}^{N} \log \frac{f_\theta^n(y_n(t))}{f_0^n(y_n(t))}. \tag{8.6}$$

The aforementioned maximization admits a maximizer which is unknown a priori as it depends on the instance of the measurement sequences collected at all times across all the agents. Such a maximization incurs a heavy communication overhead, requiring exchange of raw observation data between the sensing entities and the fusion center to obtain a decision regarding the absence or presence of a target at prespecified time instances. Moreover, owing to the lack of observability of the parameter at all the sensing entities, it might be undesirable for the fusion center to discard raw-sensed data from any particular sensing entity so as to achieve provably good detection performance. Technically speaking, a direct implementation of the maximization (8.6) requires access to the entire raw data $\mathbf{y}(t)$ at all times t at the fusion center, which thus makes the maximization the major implementation bottleneck of such an approach.

[1]It is important to note that the considered setup does not admit uniquely most powerful tests [38].

8.3 Distributed generalized likelihood ratio testing

The tremendous communication cost of a centralized GLRT procedure makes it highly infeasible to be applied to a networked setup. In order to substantially reduce the communication cost, we resort to a distributed peer-to-peer information exchange scheme, where the sensing entities instead of sending raw-sensed data to the fusion center exchange information only in their neighborhood. By such a distributed message passing scheme, the agents participate in a collaborative recursive mechanism to obtain a maximizing θ^*. In conjunction with the iterative maximization scheme, the agents maintain a local decision statistic so as to be able to make local decisions as to which hypothesis is in force at all times. The agents update their decision statistics by simultaneous assimilation of neighborhood information in the form of decision statistics of other agents in their neighborhood and their latest sensed information. In order to obtain provably reasonable decision performance with the communication and collaboration among agents constrained to a preassigned possibly sparse communication graph, we propose a distributed detector of the *consensus* + *innovations* form. To this end, we propose an algorithm, namely **CIGLRT**, in which the individual sensing entities measure low dimensional linear transformations of the underlying parameter corrupted by additive Gaussian noise. We not only show that the probabilities of errors go to zero asymptotically but also we characterize the large deviations exponent upper bounds for the probabilities of errors arising from the decision scheme under minimal assumptions of *global observability* and connectedness of the communication graph (to be made precise below).

The algorithms **CIGLRT** is motivated from the bottlenecks that one would encounter from centralized batch processing. Technically speaking, if a hypothetical fusion center which had access to all the agents' observations at all times were to conduct the parameter estimation in a recursive way, a natural recursive approach would be the following:

$$\theta_c(t+1) = \theta_c(t) + \underbrace{\alpha_t \sum_{n=1}^{N} \mathbf{H}_n^\top \mathbf{R}_n^{-1}(y_n(t) - \mathbf{H}_n\theta_c(t))}_{\text{global innovation}}, \tag{8.7}$$

where $\{\theta_c(t)\}$ represents the centralized sequence of estimates of the parameter θ^*. It is worth noting that because of the procedure being recursive, the fusion center generates an estimate, though suboptimal, at each time as opposed to a centralized batch processing scheme as in the maximization in (8.6) where an estimate is not generated until the data collection phase is over. Similarly, the fusion center at each slotted time index t generates its decision statistic in the following manner:

$$z_c(t+1) = \frac{t}{t+1} z_c(t) + \underbrace{\frac{1}{t+1} \sum_{n=1}^{N} \log \frac{f_{\theta_c(t)}(y_n(t))}{f_0(y_n(t))}}_{\text{global innovation}},$$

where $\{z_c(t)\}$ represents the centralized decision statistic sequence. In spite of being recursive, thereby ensuring that the fusion center generates an estimate of the underlying parameter and the decision statistic at all times, so as to be able to decide between the hypotheses at each time based on the available information, the centralized procedure needs access to all the agents' raw-sensed information at all times for the computation of the global innovation term. The requirement that raw-sensed data from the sensing entities be forwarded to the fusion center makes such a centralized procedure infeasible owing to the significant communication overhead. So as to ensure a long operating lifetime of such a sensing infrastructure, it is particularly desirable to restrict the communication and collaboration among agents to agents' neighborhoods and thus conforming to a possibly sparse interagent communication topology.

With the interagent collaboration and communication adhering to a possibly sparse communication topology, if an agent n in such a networked environment were to replicate the centralized estimate and local decision statistic update by replacing the global innovation in accordance with its local innovation, the updates for the parameter estimate and the decision statistic would be as follows:

$$\widehat{\theta}_n(t+1) = \widehat{\theta}_n(t) + \underbrace{\alpha_t \mathbf{H}_n^\top \mathbf{R}_n^{-1}\left(y_n(t) - \mathbf{H}_n\widehat{\theta}_n(t)\right)}_{\text{local innovation}},$$

where $\left\{\widehat{\theta}_n(t)\right\}$ represents the estimate sequence at agent n. Similarly, the decision statistic update at agent n would have been:

$$\widehat{z}_n(t+1) = \frac{t}{t+1}\widehat{z}_n(t) + \underbrace{\frac{1}{t+1}\log\frac{f_{\widehat{\theta}_n(t)}(y_n(t))}{f_0(y_n(t))}}_{\text{local innovation}},$$

where $\left\{\hat{z}_n(t)\right\}$ represents the decision statistic sequence at agent n. However, it is to be noted that in both the above updates, an agent does not collaborate at all and thus such an update corresponds to purely decentralized local processing. Owing to the local unobservability of the parameter at the individual agents, the estimate sequence and the decision statistic sequence would be rendered erroneous, and hence the detection performance would be highly suboptimal. Here, as a surrogate to the global innovation in the centralized recursions, the local estimators compute a local innovation based on the locally sensed data as an agent has access to the information in its neighborhood. However, there is an information loss due to local unobservability and localized innovation at the agents as compared to the centralized update. An agent makes up for the aforementioned information loss by incorporating an agreement or consensus potential into its updates. For the parameter estimate update and the decision statistic update, the incorporation of the consensus term is achieved by assimilating parameter estimates and decision statistics from the neighborhood respectively. We now present the algorithm **CIGLRT**.

8.3.1 Algorithm **CIGLRT***:* consensus+innovations *GLRT*

In this section, we will describe the **CIGLRT** algorithm. Consider the linear observation model described in (8.3). Our formulation does not assume a zone of indifference. In that, our setup considers the case in which $\mathscr{H}_1 \cup \mathscr{H}_0 = \mathbb{R}^M$. However, as we will see later, a zone of indifference falls off from our analysis, which can then be characterized in terms of the algorithm parameters and the specific instance of the parameter θ^* in force. Before mentioning the specifics of our algorithm, we formalize some identifiability assumptions on our sensing model.

Assumption A1. We require the following global observability condition. The matrix **G**:

$$\mathbf{G} = \sum_{n=1}^{N} \mathbf{H}_n^{\top} \mathbf{R}_n^{-1} \mathbf{H}_n \tag{8.8}$$

is full rank.

Intuitively speaking, such an assumption is needed for the network as a whole to be able to distinguish between any two instances of the parameter, say θ and θ^*. Owing to the local unobservability at each agent, it is unlikely for an individual agent to be able to distinguish between all possible pairs of parameters. Broadly speaking, assumption A1 ensures the parameter θ^* is observable to the network as a whole, in spite of being locally unobservable at individual agents. Ensuring global observability is particularly important in the case of inference procedures in a networked setup so as to be able to have reasonable performance guarantees. Even a centralized fusion-center-based architecture needs global observability.

We formalize an assumption on the interagent communication topology before proceeding further.

Assumption A2. The interagent communication graph is connected, i.e., $\lambda_2(\mathbf{L}) > 0$, where **L** denotes the associated graph Laplacian matrix.

Assumption A2 ensures consistent information flow among the agents. Technically speaking, the communication graph here is modeled as an undirected graph. The connectedness basically ensures that over time, the information from each agent in the graph reaches other agents over time in a symmetric fashion and thus ensuring information flow. Especially with the lack of observability of the parameter at individual agents, the connectivity is crucial to ensure the availability of sufficient information about the parameter at each agent which would then ensure the good performance of the detection scheme.

We now present the distributed recursive composite hypothesis testing, **CIGLRT** algorithm, which in turn is inspired from the **CIGLRT − L** algorithm in [6]. The **CIGLRT** algorithm is a sequential decision-making framework which consists of three interacting recursive processes running in parallel. To be specific, the sequential decision-making in **CIGLRT** is enabled by a distributed parameter estimation mechanism, a distributed decision statistic update mechanism and finally a local decision rule at each individual agent, as described below.

Parameter estimate update. The first building block of the sequential decision mechanism in **CIGLRT** is the parameter estimate update which aims to estimate the parameter which parameterizes the hypothesis, i.e., tries to find the specific $\mathbb{P}_{\boldsymbol{\theta}^*}$ from which the measurements of the sensing entities are sampled from. Technically speaking, the algorithm **CIGLRT** generates a sequence $\{\boldsymbol{\theta}_n(t)\} \in \mathbb{R}^M$ of estimates of the parameter θ^* at the nth agent according to the distributed recursive scheme:

Parameter estimate update:

$$\boldsymbol{\theta}_n(t+1) = \boldsymbol{\theta}_n(t) - \underbrace{\beta_t \sum_{l \in \Omega_n} (\boldsymbol{\theta}_n(t) - \boldsymbol{\theta}_l(t))}_{\text{neighborhood consensus}} + \underbrace{\alpha_t \nabla_\theta \log \frac{f_{\theta_n(t)}(\mathbf{y}_n(t))}{f_0(\mathbf{y}_n(t))}}_{\text{local innovation}}, \tag{8.9}$$

where Ω_n denotes the communication neighborhood of agent n, $\nabla(\cdot)$ denotes the gradient and $\{\beta_t\}$ and $\{\alpha_t\}$ are consensus and innovation weight sequences, respectively (to be specified shortly), and

$$\log \frac{f_{\theta_n(t)}(\mathbf{y}_n(t))}{f_0(\mathbf{y}_n(t))} = \theta_n(t)^\top \mathbf{H}_n^\top \mathbf{R}_n^{-1} \mathbf{y}_n(t) - \frac{\theta_n(t)^\top \mathbf{H}_n^\top \mathbf{R}_n^{-1} \mathbf{H}_n \theta_n(t)}{2}. \tag{8.10}$$

In the update in (8.9), the communication is strictly local, i.e., an agent only exchanges estimates in its neighborhood and does not exchange raw-sensed data at any time. At the same time, it is also to be noted that as compared to the estimate sequence $\{\theta_c(t)\}$ generated by the centralized update in (8.7), the estimate sequence $\{\boldsymbol{\theta}_n(t)\}$ is inherently suboptimal as the sensed information available to an agent in the update above is strictly a subset of the sensed information available to the fusion center. However, as we will see later, in spite of the suboptimality of the estimate sequence, the sequential decision-making scheme **CIGLRT** has probabilities of errors decaying exponentially in the large sample limit.

The update in (8.9) can be written in a compact manner as follows:

$$\theta(t+1) = \theta(t) - \beta_t(\mathbf{L} \otimes \mathbf{I}_M)\theta(t) + \alpha_t \mathbf{G}_H \mathbf{R}^{-1}\left(\mathbf{y}(t) - \mathbf{G}_H^\top \theta(t)\right), \tag{8.11}$$

where $\theta(t) = [\theta_1^\top(t) \ \ \theta_2^\top(t) \cdots \theta_N^\top(t)]^\top$, $\mathbf{G}_H = diag[\mathbf{H}_1^\top, \mathbf{H}_2^\top, \ldots, \mathbf{H}_N^\top]$, $\mathbf{y}(t) = [\mathbf{y}_1^\top(t) \, \mathbf{y}_2^\top(t) \cdots \mathbf{y}_N^\top(t)]^\top$ and $\mathbf{R} = diag\,[\mathbf{R}_1, \ldots, \mathbf{R}_N]$.

The update in (8.9)–(8.11) conforms to a mixed-time scale stochastic approximation procedure (see, e.g., [39]). By mixed time-scale, we mean that the interagent communication (consensus) occurs at the same rate as that of observation sampling or incorporation of the latest sensed information (innovation); however, the consensus and innovation terms in the iterative update are weighed by two different weight sequences, which decay to zero in a controlled manner at different rates. (It is to be noted that this mixed time-scale is different from stochastic approximation algorithms with coupling (see [40]), where a quickly switching parameter influences the relatively slower dynamics of another state, leading to averaged dynamics.) We also note that in [41], in the context of optimization, methods pertaining to mixed time-scale stochastic approximation algorithms are developed, albeit in the centralized context.

The corresponding innovation term in the algorithm proposed in [41] is a martingale difference term. However, in the *consensus + innovations* algorithms, see [42,43], and this work, the innovation term is not a martingale difference sequence, and hence it is of particular interest to characterize the rate of convergence of the innovation sequence to a martingale difference sequence, so as to establish convergence and consistency of the parameter estimate sequence.

We make the following assumption on the weight sequences $\{\alpha_t\}$ and $\{\beta_t\}$:

Assumption A3. The weight sequences $\{\alpha_t\}_{t\geq 0}$ and $\{\beta_t\}_{t\geq 0}$ are given by:

$$\alpha_t = \frac{a}{(t+1)} \; \beta_t = \frac{b}{(t+1)^{\tau_2}}, \tag{8.12}$$

where $a \geq 1$ and $0 < \tau_2 \leq 1$.

Decision statistic update. As the algorithm **CIGLRT** is a sequential decision-making scheme, it is crucial for the algorithm to have a decision statistic update which then allows it to generate a decision statistic so as to be able to decide as to which hypothesis is in force. Formally, the algorithm generates a decision statistic sequence $\{z_n(t)\}$ at the nth agent according to the distributed recursive scheme:

Decision statistic update:

$$\hat{z}_n(kt-k+1) = \theta_n(k(t-1))^\top \mathbf{H}_n^\top \mathbf{R}_n^{-1}\left(\mathbf{s}_n(k(t-1)) - \frac{\mathbf{H}_n\theta_n(k(t-1))}{2}\right), \tag{8.13}$$

where $\mathbf{s}_n(k(t-1)) = \sum_{i=0}^{k(t-1)} (y_n(i)/(k(t-1)+1))$, i.e., the time averaged sum of local measurements at the sensing entity n, and the underlying parameter estimate used in the decision statistic is the estimate at time $k(t-1)$. The incorporation of such a time averaged observation sequence in the decision statistic aids in decreasing the noise variance as compared to a stand-alone latest sensed observation. As noted above, the estimate sequence $\{\theta_n(t)\}$ is suboptimal and in particular for the first few time instants, the network-sensed data (possibly delayed version) is yet to percolate at individual agents. The incorporation of initial erroneous parameter estimates into the decision statistic is remedied to some extent by incorporation of the time averaged observation sequence. In other words, at every time instant $k(t-1)+1$ (times which are one modulo k), where k is a predetermined positive integer (k to be specified shortly), an agent n incorporates its latest sensed information. To be particular, an agent incorporates the latest sensed information into its decision statistic in the form of its local observations made in the past k time instants, in the manner as described in (8.13). The time-averaged observation sequence is maintained locally at an agent, independent of the decision statistic as and when a new measurement is made at agent n. After incorporating the local-sensed information, every agent n undergoes

$k-1$ rounds of consensus (see, e.g., [44,45] for a detailed treatment), which can be expressed in a compact form as follows:

$$\hat{\mathbf{z}}(kt) = \mathbf{W}^{k-1}\mathbf{G}_\theta(k(t-1))\mathbf{R}^{-1}\left(\mathbf{s}(k(t-1)) - \frac{\mathbf{G}_H^\top\theta(k(t-1))}{2}\right), \tag{8.14}$$

$\mathbf{z}(t) = [z_1(t)\cdots z_N(t)]$, $\mathbf{G}_\theta(t) = diag\left[\theta_1^\top(t)\mathbf{H}_1^\top, \theta_2^\top(t)\mathbf{H}_2^\top, \ldots, \theta_N^\top(t)\mathbf{H}_N^\top\right]$ and $\mathbf{s}(t) = \left[\mathbf{s}_1^\top(t)\,\mathbf{s}_2^\top(t)\cdots\mathbf{s}_n^\top(t)\right]^\top$, where $\mathbf{W}$ is a $N\times N$ weight matrix, where we assign $w_{ij}=0$, if $(i,j)\notin E$. The sequence $\{\hat{z}_n(t)\}$ is an auxiliary decision statistic sequence and the decision statistic sequence $\{z_n(t)\}$ is generated from $\{\hat{z}_n(t)\}$ in the following manner:

$$z_n(kt) = \hat{z}_n(kt), \quad \forall t, \tag{8.15}$$

where, in the interval $[k(t-1), kt-1]$, the value of the decision statistic stays constant corresponding to its value at $z_n(kt-k)$, $\forall t$. Broadly speaking, the decision statistic update happens in two phases as depicted in (8.13)–(8.15). In the first phase, as depicted in (8.13), an agent incorporates latest sensed information into its decision statistic in the form of a time-averaged measurement sequence. It is however to be noted that the incorporation of the latest sensed information occurs in intervals of k time steps. The second phase, in which an agent exchanges its decision statistics in its neighborhood, is sandwiched between two successive incorporation of the latest sensed information into the decision statistic. Intuitively speaking, the performance of the algorithm in terms of decay of the probability of miss is intricately related to the existence of a mechanism in the decision statistic update so as to get rid of the initial erroneous parameter estimates, which are weighed equally as the more accurate parameter estimates obtained later. The decision statistics update for the **CIGLRT** algorithm as depicted in (8.13)–(8.15) ensures that at any time after kt, the previous parameter estimates i.e., $\theta(ks), s = 1, \ldots, t-1$ do not contribute to the decision statistic. To be specific, the decision statistic update is carefully designed so as to ensure that the sequential decision-making scheme attains exponentially decaying probabilities of errors.

Moreover, the decision statistic update requires the agents to maintain a local copy of the running time-average of their local observations. In order to maintain a local copy of the running time-average of the observations, an agent's memory requirement stays constant over time as the running time-average can be updated off the fly as and when a new observation is made without storing the latest sensed information. Owing to the k rounds of consensus, where an agent just exchanges decision statistics in its neighborhood, the decision statistic update in (8.14) has time-delayed parameter estimates and observations. To be specific, in the ideal setting, the decision statistic update at a particular time instant t would be using the latest parameter estimate at time t and the latest sensed measurement at time t, but due to the k rounds of consensus, the algorithm uses parameter estimates which are delayed by k time steps. The k rounds of consensus between successive incorporation of latest sensed information into the decision statistic promote more interagent collaboration, thus leading to higher accuracy levels for the decision statistic. However, it is to be noted that there is an inherent trade-off between the performance as a function of the number of rounds of consensus as an algorithm parameter and the time delay. With more rounds of consensus and thus more interagent collaboration, the algorithm attains better detection performance asymptotically but the considerable lag in incorporating latest sensed information affects the transient characteristics of the probability of errors.

We now state some design assumptions on $\mathbf{W}$ which models the undirected interagent communication topology.

Assumption A4. The entries in the weight matrix $\mathbf{W}$ are designed in such a way that $\mathbf{W}$ is nonnegative, symmetric, irreducible and stochastic, i.e., each row of $\mathbf{W}$ sums to one.

We remark that, the enforcement of Assumption A4 automatically ensures that the second largest eigenvalue in magnitude of $\mathbf{W}$, denoted by r, turns out to be strictly less than one, see, for example, [45]. Owing to the stochasticity of $\mathbf{W}$, the quantity r satisfies:

$$r = ||\mathbf{W} - \mathbf{J}||, \tag{8.16}$$

where $\mathbf{J} = (1/N)\mathbf{1}_N\mathbf{1}_N^\top$. Intuitively speaking, the quantity r reflects the degree of connectivity of the interagent communication topology. In the case, when $r = 0$, it basically corresponds to the case when $\mathbf{W} = \mathbf{J}$, i.e., the interagent communication graph is completely connected and hence reduces to the case of the centralized setup. At the other end of the spectrum, when $r = 1$, it corresponds to the case $\mathbf{W} = \mathbf{I}$, i.e., no collaboration and communication among the agents and each agent relies on its local observations for inference. While the first case $r = 1$ results in optimal performance as the entire sensed data are available to each agent, it also results in the highest implementation complexity as agents need to exchange their parameter estimates and decision statistics with all the agents of the network. The case $r = 0$ leads to no communication between the agents but results in erroneous estimates. Hence, there is an intrinsic trade-off between communication and performance in such a networked setting.

We make an assumption on k which concerns the number of rounds of consensus in the decision statistic update of **CIGLRT**.

Assumption A5. Recall r as defined in (8.16). The number of rounds k of consensus between two updates of agent decision statistics satisfies:

$$k \geq 1 + \left\lfloor \frac{-3\log N}{2\log r} \right\rfloor. \tag{8.17}$$

Intuitively speaking, in (8.14), k rounds of consensus corresponds to one round of consensus with the weight matrix $\mathbf{W}^k$. It follows that $\left\|\mathbf{W}^k - \mathbf{J}\right\| = r^k < r$ and thus corresponds to an effective interagent communication graph which enables much more interagent collaboration as compared to $\mathbf{W}$. A simple yet intuitive way to design the entries of $\mathbf{W}$ is to assign equal combination weights, in which case we have,

$$\mathbf{W} = \mathbf{I}_N - \delta\mathbf{L}, \tag{8.18}$$

where $\delta \in (0, (2/\lambda_N(\mathbf{L})))$. For subsequent analysis, we specifically choose $\delta = (2/(\lambda_2(\mathbf{L}) + \lambda_N(\mathbf{L})))$. Such a design results in $r = ((\lambda_N(\mathbf{L}) - \lambda_2(\mathbf{L}))/(\lambda_2(\mathbf{L}) + \lambda_N(\mathbf{L})))$.

Decision rule. The final phase of the sequential decision procedure at hand is the decision rule at each agent, which lets every entity in the network to come up with its own decision at each time t. The following decision rule is adopted at all times t at an agent n:

Decision rule:

$$\mathcal{H}_n(t) = \begin{cases} \mathcal{H}_0 & z_n(t) \leq \eta \\ \mathcal{H}_1 & z_n(t) > \eta, \end{cases} \tag{8.19}$$

where $\mathcal{H}_n(t)$ is the local decision at time t at agent n.

The natural question which arises with such a localized decision rule is regarding the agreement of local decisions. It is to be noted in the present context that the parameter estimates and the decision statistics across all agents reach asymptotic agreement. Hence, in the initial few time steps, there might be disagreement among the agents in terms of local decisions. But, with subsequent cooperation among the agents, each agent gets to the same local decision eventually with probability one. Hence, the overall decision then is the decision of any local agent. With the decision rule in (8.19), we study error probabilities at each agent n:

$$\begin{aligned} \mathbb{P}^n_{M,\theta*}(t) &= \mathbb{P}_{1,\theta*}\left(z_n(t) \leq \eta\right) \\ \mathbb{P}^n_{FA}(t) &= \mathbb{P}_0\left(z_n(t) > \eta\right), \end{aligned} \tag{8.20}$$

where $\mathbb{P}^n_{M,\theta*}$ and $\mathbb{P}^n_{FA}$ refer to probability of miss and probability of false alarm, respectively, at agent n respectively.

We make an assumption on a, which is introduced in (8.12).

Assumption A6. Recall a as defined for the innovation weight sequence α_t. We assume that a satisfies:

$$a \geq \frac{1}{2c_1} + 2, \tag{8.21}$$

where c_1 is defined as:

$$\begin{aligned} c_1 &= \min_{\|\mathbf{x}\|=1} \mathbf{x}^\top\left(\mathbf{L} \otimes \mathbf{I}_M + \mathbf{G}_H\mathbf{R}^{-1}\mathbf{G}_H^\top\right)\mathbf{x} \\ &= \lambda_{\min}\left(\mathbf{L} \otimes \mathbf{I}_M + \mathbf{G}_H\mathbf{R}^{-1}\mathbf{G}_H^\top\right). \end{aligned} \tag{8.22}$$

8.3.2 *Simplified model for* CIGLRT

Before proceeding to the main results, we consider a special case of our observation model for the rest of this chapter so as to make the main results more intuitive (we refer the reader to [6] for the results on the general model (8.3)). In particular, we assume that the parameter of interest θ^* is a scalar, i.e., $\theta^* \in \mathbb{R}$. We divide the set of agents into two sets $\mathbf{S}_1$ and $\mathbf{S}_2$ with cardinalities N_1 and N_2, respectively. The agents in the set $\mathbf{S}_1$ observe a noisy linear-scaled version of the scalar parameter θ^*, while the agents in the set $\mathbf{S}_2$ observe just noise irrespective of which hypothesis in force. The noise power is taken to be σ^2 across all agents. On evaluating the matrix $\mathbf{G}$ in (8.8), with the aforementioned observation model, we have that the condition reduces to the condition that N_1 is strictly positive. Formally, for agents in set $\mathbf{S}_1$:

$$y_n(t) = h_n\theta^* + \gamma_n(t),$$

whereas for agents in set $\mathbf{S}_2$,

$$y_n(t) = \gamma_n(t).$$

In the above, we assume that $h_n > 0$, $\forall n$. The above setup corresponds to a setting where the agents in set $\mathbf{S}_1$ can conduct their recursive hypothesis testing scheme based on their own observation sequence. However, the network objective is for every agent to come up with local decisions which is not possible for agents in set $\mathbf{S}_2$ without collaboration as their own measurements do not provide any information as to which hypothesis in force. Moreover, the agents in set $\mathbf{S}_1$ have an incentive to collaborate with other network agents, in that, by assimilating the network information they can improve the performance of their local detectors. Hence, the setup described above is an apt example where interagent collaboration is necessary for the network objective of facilitating local decision-making. For the rest of the chapter, i.e., for the main results, we consider the observation model and the networked setup described above.

8.4 Main results: CIGLRT

In this section, we state the main results for the algorithm **CIGLRT**. The first main results concerns with the convergence of the parameter estimate sequences across all agents. We reiterate that we consider the observation model described in Section 8.3.2 for the main results.

Consistency of the estimate sequence

Theorem 8.1 ([42]). *Let the Assumptions A1–A6 hold. Consider the parameter estimation part of the* **CIGLRT** *algorithm in* (8.9). *The algorithm generates a consistent estimate sequence* $\{\theta_n(t)\}$ *for each agent n, i.e.,*

$$\mathbb{P}_{\theta^*}\left(\lim_{t\to\infty} \theta_n(t) = \theta^*\right) = 1, \quad \forall n. \tag{8.23}$$

The result in Theorem 8.1 ensures that even the agents in the set **S2** in spite of no local sensed information are able to obtain consistent parameter estimates due to interagent collaboration. However, it is to be noted that the above result is an asymptotic characterization, and the transients in the convergence of the estimate sequences across different agents might follow different trajectories owing to network connectivity. We provide a proof sketch of Theorem 8.1 before proceeding further.

The proof of almost sure convergence of the estimate sequence to $\boldsymbol{\theta}^*$ involves establishing the boundedness of the estimate sequence. However, it turns out that the estimate sequence is only sample path-wise bounded and not uniformly bounded across sample paths. Finally, the result follows by constructing a Lyapunov function and then studying the convergence of it. For a complete proof of Theorem 8.1, we direct the reader to [42]. The next result concerns with the large deviations exponent upper bounds for the associated probability of errors of the decision-making scheme. Broadly speaking, the large deviations exponent upper bounds quantify the exponential rate of decay of the probabilities of errors. Let $h_{\max} = \max_{n=1,\ldots,N} h_n$. We define the following quantities which play a key role in the statement of the next theorem: let c_4 and c_4^* be constants given by,

$$c_4 = \frac{\sigma^2}{h_{\max}^2\left(c_3((t_1+1)^{2c_1\alpha_0}/kt_1^{2c_1\alpha_0-1}) + (\alpha_0^2/kt_1) + (\alpha_0^2/(2c_1\alpha_0-1))\right)}, \tag{8.24}$$

and

$$c_4^* = \frac{\sigma^2(2c_1\alpha_0-1)}{\alpha_0^2 h_{\max}^2} - \frac{N}{2\eta_2} \tag{8.25}$$

respectively, where η_2 is given by

$$\eta_2 = \frac{-2N\sigma^2\eta + N_1 h_{\max}^2(\theta^*)^2\left(1 - N\sqrt{N}r^{k-1}\right)}{4h_{\max}^2\left(1 + N\sqrt{N}r^{k-1}\right)}, \tag{8.26}$$

c_3 is defined as

$$c_3 = \sum_{v=0}^{t_1-1}\alpha_v^2\prod_{u=v+1}^{t_1-1}\left\|\mathbf{I}_N - \beta_u\mathbf{L} - \alpha_u\mathbf{G}_H\mathbf{R}^{-1}\mathbf{G}_H^\top\right\|, \tag{8.27}$$

and t_1 defined as

$$t_1 = \max\{t_2, t_3\}, \tag{8.28}$$

with t_3 such that, $\forall t \geq t_3$,

$$\lambda_{\min}\left(\mathbf{L} + \mathbf{G}_H\mathbf{R}^{-1}\mathbf{G}_H^\top\right)\alpha_t < 1, \tag{8.29}$$

and t_2 such that[2], $\forall t \geq t_2$,

$$\beta_t\lambda_N(\mathbf{L}) + \alpha_t\frac{h_{\max}^2}{\sigma^2} < 1. \tag{8.30}$$

[2]It is to be noted that t_2 and t_3 exist as $\alpha_t, \beta_t \to 0$.

Error exponents for probabilities of error

Theorem 8.2. *Let the hypothesis of Theorem 8.1 hold. Consider, the decision statistic update of the* **CIGLRT** *algorithm in* (8.13). *For all θ^*, which satisfy the following condition:*

$$|\theta^*|^2 \geq \frac{1 + N\sqrt{N}r^{k-1}}{1 - N\sqrt{N}r^{k-1}} \left(\frac{N\sigma^2}{N_1 h_{\max}^2} + \frac{2N\alpha_0^2 h_{\max}^2}{N_1\sigma^2 (2c_1\alpha_0 - 1)} \right), \tag{8.31}$$

we have the following range of feasible thresholds,

$$\frac{\left((1/N) + \sqrt{N}r^{k-1}\right) N_1}{2} < \eta < \frac{N_1 (h_{\max}\theta^*)^2 \left(1 - N\sqrt{N}r^{k-1}\right)}{2N\sigma^2} - \frac{\alpha_0^2 h_{\max}^4 \left(1 + N\sqrt{N}r^{k-1}\right)}{\sigma^4 (2c_1\alpha_0 - 1)}, \tag{8.32}$$

for which we have the following large deviations upper bound characterization for the probability of false alarm $\mathbb{P}_{FA}$:

$$\limsup_{t\to\infty} \frac{1}{t} \log(\mathbb{P}_0 (z_n(t) > \eta)) \leq -\frac{\eta}{(1/N) + \sqrt{N}r^{k-1}} - \frac{N_1}{2}\left(1 + \log \frac{2\eta}{\left((1/N) + \sqrt{N}r^{k-1}\right) N_1}\right) = LD_0(\eta), \tag{8.33}$$

and the following large deviations upper bound characterization for the probability of miss $\mathbb{P}_M$:

$$\limsup_{t\to\infty} \frac{1}{t} \log\left(\mathbb{P}_{1,\theta^*}(z_n(t) < \eta)\right) \leq \max\left\{ -\left(\frac{\left(-(\eta/4) + ((N_1 h_{\max}^2(\theta^*)^2 \left((1/N) - \sqrt{N}r^{k-1}\right))/8\sigma^2)\right)^2}{2N_1 h_{\max}^2(\theta^*)^2\left((1/N) + \sqrt{N}r^{k-1}\right)^2/\sigma^2} \right) - LD\left(\min\{c_4, c_4^*\}\right)\right\} = LD_1(\eta), \tag{8.34}$$

where, $LD(\lambda)$ is given by:

$$LD(\lambda) = \lambda\eta_2 + \frac{N}{2}\log\left(1 - \frac{\lambda\alpha_0^2 h_{\max}^2}{\sigma^2(2c_1\alpha_0 - 1)}\right). \tag{8.35}$$

Theorem 8.2 ensures that the probability of errors decay exponentially and quantifies the exponents through large deviations upper bounds. We now discuss one of the most critical aspects of the decision-making scheme which is to choose the threshold. While designing the threshold, it is to be kept in mind that the underlying parameter is unknown a priori and hence has to be designed in terms of algorithm parameters, i.e., in terms of the sensing model, the number of agents in the network and the network connectivity. The threshold in the current context can be designed as $((((1/N) + \sqrt{N}r^{k-1})N_1)/2) + \varepsilon$, where $\varepsilon > 0$ can be chosen to be arbitrarily small. Such a choice of threshold would guarantee exponentially decaying probability of false alarm. Moreover, the aforementioned choice of thresholds is uniform across agents. Agent specific thresholds may be designed for such a scheme which would then be chosen based on an agent's connectivity with other agents in the network. However, given the ad hoc nature of networked wireless-sensing architectures, the connectivity of an individual agent is hard to track, thus making agent specific thresholds computationally intractable. In order to ensure decaying probability of miss, the underlying parameter needs to conform to a range as specified in (8.31). The above result also identifies a range of values of θ^*s for which exponentially decaying probability of miss can be guaranteed. In particular, as intuitively expected, signals with low $\|\theta^*\|$, i.e., setups corresponding to low signal-to-noise ratios exponential decay, may not be guaranteed for both the probabilities of errors. The range of θ^*s, for which Theorem 8.2 turns out to be inconclusive, can be specified as follows:

$$|\theta^*|^2 \leq \frac{1 + N\sqrt{N}r^{k-1}}{1 - N\sqrt{N}r^{k-1}} \left(\frac{N\sigma^2}{N_1 h_{\max}^2} + \frac{2N\alpha_0^2 h_{\max}^2}{N_1\sigma^2\,(2c_1\alpha_0 - 1)} \right)$$

Through the dependence of the exponents in terms of r, it can be seen that networks with lower r have faster exponential decay. At this juncture, we provide a proof sketch for Theorem 8.2.

It is to be noted that, in the context of classical centralized composite hypothesis testing, Wilks' Theorem establishes asymptotic distributions of the log-likelihood ratios under the null and alternate hypothesis (see [35]). However, convergence in distribution is not sufficient to characterize the exponents for the probabilities of errors. The crucial step in establishing the exponential decay of probabilities of errors involves studying the quantity $t\|\theta(t) - \theta\|^2$ and consequently showing the boundedness of it. Finally, the result follows by rewriting $\|\theta(t) - \theta\|^2$ as a Wishart matrix with one degree of freedom and perform analysis using the moment generating function of the Wishart distribution. We point the reader to [6] for a complete proof of Theorem 8.2.

8.5 Simulation experiments

To demonstrate the performance of the algorithm **CIGLRT**, we consider a network of ten agents. In particular, we consider a ring network, where each agent has two agents in its neighborhood. We consider the underlying parameter to be such that $\theta \in \mathbb{R}^5$ and $\theta^* = [1\ 0.9\ 1.2\ 1.1\ 1.5]$. The observation matrices for the agents are of the dimension

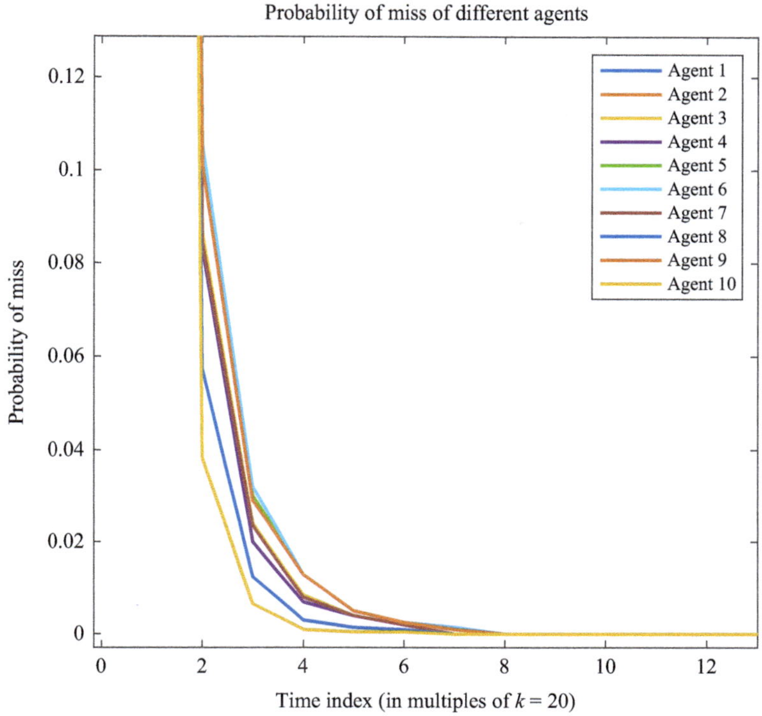

Figure 8.1 Probability of miss at each agent

5×1, i.e., $M_n = 1$, $\forall n$. Specifically the $\mathbf{H}_n$s are given by $\mathbf{H}_1 = [1\ 1\ 0\ 0\ 0]$, $\mathbf{H}_2 = [0\ 1\ 1\ 0\ 0]$, $\mathbf{H}_3 = [0\ 0\ 1\ 1\ 0]$, $\mathbf{H}_4 = [0\ 0\ 0\ 1\ 1]$, $\mathbf{H}_5 = [1\ 0\ 0\ 0\ 1]$, $\mathbf{H}_6 = [1\ 0\ 1\ 0\ 0]$, $\mathbf{H}_7 = [0\ 1\ 0\ 1\ 0]$, $\mathbf{H}_8 = [0\ 0\ 1\ 0\ 1]$, $\mathbf{H}_9 = [1\ 0\ 0\ 1\ 0]$, $\mathbf{H}_{10} = [0\ 1\ 0\ 0\ 1]$. Note, that each agent observes a scalar at each time, thus rendering the parameter locally unobservable at each agent. Furthermore, the scalar observation made by each agent is corrupted by noise of variance 3. However, note that $\sum_{n=1}^{10} \mathbf{H}_n^\top \mathbf{H}_n$ is invertible, thus ensuring that the parameter is globally observable. The network is poorly connected which in turn is reflected by the quantity $r = \|\mathbf{W} - \mathbf{J}\|$ and is given by 0.8404. In particular, for the parameter estimation algorithm, $a = 9.1$ and $\tau_2 = 0.4$, where a, τ_2 are as defined in Assumption A3. The time-lag k is taken to be $k = 20$.

For the analysis of the probability of miss, we run the algorithm for 2,000 sample paths. We choose the threshold as $\eta = ((((1/N) + \sqrt{N} r^{k-1}) \sum_{n=1}^{N} M_n)/2) + 0.01 = 0.8280$. Figure 8.1 demonstrates the exponential decay of the probability of miss. The exponent of the exponential decay of the probability of miss across different agents is plotted in Figure 8.2. The theoretical upper bound in (8.33) is found to be 0.045. The simulation results verify that the empirically estimated large deviations exponent upper bound for different agents is upper bounded by the theoretically derived upper bound.

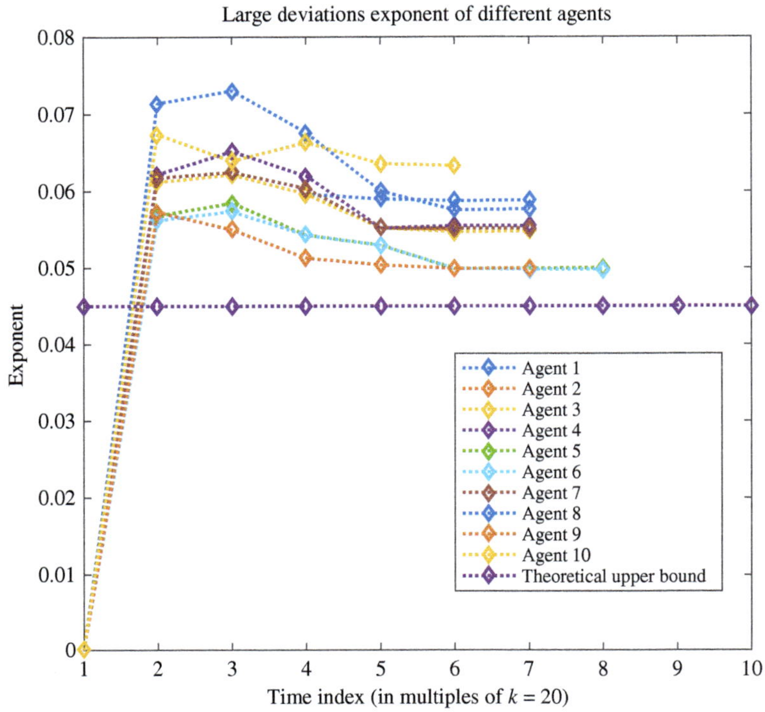

Figure 8.2 Large deviations exponent upper bounds

8.6 Discussions and future directions

In this chapter, we have considered the problem of recursive composite hypothesis testing in a networked setting, where the network objective is to test between a simple null hypothesis and a composite alternative parameterized by a parameter taking values in a continuous space. For the algorithm **CIGLRT**, which involves a parameter estimation scheme and a decision statistic update running in parallel and thus making it a recursive online sequential decision-making algorithm, we have established the consistency of the parameter estimate sequence and characterized the algorithm parameters for which the exponential decay of the probabilities of errors can be ensured. The algorithm discussed in this chapter is centered around the availability of a closed form expression of the log-likelihood ratio. An extension to noise distributions from the exponential family is thus amenable to be extended from the framework discussed in this chapter. However, the approach would need a major change in terms of the technical tools so as to extend to scenarios involving noise distributions in which log-likelihood ratios do not admit closed-form solutions. Especially, in the lookout for recursive sequential schemes in the scenario described above, approaches be inspired from sequential convex optimization as in [46] and data-driven approaches as in [47] might be helpful. It is worth noting that the algorithm **CIGLRT** is not truly *sequential*; in that, the algorithm is a fixed sample size testing scheme, and the sample size is fixed a priori and not decided on the fly subject to a stopping criteria. It is well known that sequential schemes like the sequential probability ratio test proposed by [48] lose their sample efficiency property when the hypotheses are not exactly specified. The distributed sequential composite hypothesis testing problem is largely unexplored and an initial step towards exploring the problem in the distributed setup would be following in the lines of [49].

References

[1] Zarrin S, and Lim TJ. Composite hypothesis testing for cooperative spectrum sensing in cognitive radio. In: IEEE International Conference on Communications, 2009. ICC'09. IEEE; 2009. p. 1–5.
[2] Font-Segura J, and Wang X. GLRT-based spectrum sensing for cognitive radio with prior information. IEEE Transactions on Communications. 2010;58(7):2137–2146.
[3] Zou Q, Zheng S, and Sayed AH. Cooperative sensing via sequential detection. IEEE Transactions on Signal Processing. 2010;58(12):6266–6283.
[4] Tajer A, Jajamovich GH, Wang X, and Moustakides GV. Optimal joint target detection and parameter estimation by MIMO radar. IEEE Journal of Selected Topics in Signal Processing. 2010;4(1):127–145.
[5] Zeitouni O, Ziv J, and Merhav N. When is the generalized likelihood ratio test optimal? IEEE Transactions on Information Theory. 1992;38(5):1597–1602.
[6] Sahu AK, and Kar S. Recursive distributed detection for composite hypothesis testing: Nonlinear observation models in additive Gaussian noise. IEEE Transactions on Information Theory. 2017;63(8):4797–4828.
[7] Blum RS, Kassam SA, and Poor HV. Distributed detection with multiple sensors I. Advanced topics. Proceedings of the IEEE. 1997;85(1):64–79.
[8] Tsitsiklis JN. Decentralized detection. Advances in Statistical Signal Processing. 1993;2(2):297–344.
[9] Kar S, and Moura JMF. Consensus based detection in sensor networks: Topology optimization under practical constraints. In: Proc 1st Intl Workshop Inform Theory Sensor Networks; 2007.
[10] Olfati-Saber R, Franco E, Frazzoli E, and Shamma JS. Belief consensus and distributed hypothesis testing in sensor networks. In: Networked Embedded Sensing and Control. Springer-Verlag, Berlin, Heidelberg, 2006. p. 169–182.
[11] Olfati-Saber R, Fax JA, and Murray RM. Consensus and cooperation in networked multi-agent systems. Proceedings of the IEEE. 2007;95(1):215–233.
[12] Jadbabaie A, Lin J, and Morse AS. Coordination of groups of mobile autonomous agents using nearest neighbor rules. IEEE Transactions on Automatic Control. 2003;48(6):988–1001.
[13] Bajovic D, Jakovetic D, Xavier J, Sinopoli B, and Moura JMF. Distributed detection via Gaussian running consensus: Large deviations asymptotic analysis. IEEE Transactions on Signal Processing. 2011;59(9):4381–4396.

[14] Kar S, Tandon R, Poor HV, and Cui S. Distributed detection in noisy sensor networks. In: IEEE International Symposium on Information Theory Proceedings (ISIT), 2011. IEEE; 2011. p. 2856–2860.
[15] Cattivelli FS, and Sayed AH. Distributed detection over adaptive networks using diffusion adaptation. IEEE Transactions on Signal Processing. 2011;59(5):1917–1932.
[16] Cattivelli FS, and Sayed AH. Diffusion LMS strategies for distributed estimation. IEEE Transactions on Signal Processing. 2010;58(3):1035–1048.
[17] Bajović D, Moura JMF, Xavier J, and Sinopoli B. Distributed inference over directed networks: Performance limits and optimal design. IEEE Transactions on Signal Processing. 2016;64(13):3308–3323.
[18] Jakovetic D, Moura JMF, and Xavier J. Distributed detection over noisy networks: Large deviations analysis. IEEE Transactions on Signal Processing. 2012;60(8):4306–4320.
[19] Sahu AK, and Kar S. Distributed sequential detection for Gaussian shift-in-mean hypothesis testing. IEEE Transactions on Signal Processing. 2016;64(1): 89–103.
[20] Sahu AK, and Kar S. Distributed generalized likelihood ratio tests: Fundamental limits and tradeoffs. In: IEEE International Conference on Acoustics, Speech and Signal Processing, 2016. ICASSP'16. IEEE; 2016.
[21] Kar S, Moura JMF, and Poor HV. QD-learning: A collaborative distributed strategy for multi-agent reinforcement learning through. IEEE Transactions on Signal Processing. 2013;61(7):1848–1862.
[22] Zhou Q, Kar S, Huie L, and Cui S. Distributed estimation in sensor networks with imperfect model information: An adaptive learning-based approach. In: IEEE International Conference on Acoustics, Speech and Signal Processing. Kyoto, Japan; 2012. p. 3109–3112.
[23] Chen J, Richard C, and Sayed AH. Multitask diffusion adaptation over networks. IEEE Transactions on Signal Processing. 2014;62(16):4129–4144.
[24] Braca P, Marano S, and Matta V. Enforcing consensus while monitoring the environment in wireless sensor networks. IEEE Transactions on Signal Processing. 2008;56(7):3375–3380.
[25] Braca P, Marano S, Matta V, and Willett P. Asymptotic optimality of running consensus in testing binary hypotheses. IEEE Transactions on Signal Processing. 2010;58(2):814–825.
[26] Cattivelli F, and Sayed AH. Diffusion LMS-based distributed detection over adaptive networks. In: Conference Record of the Forty-Third Asilomar Conference on Signals, Systems and Computers, 2009. IEEE; 2009. p. 171–175.
[27] Cattivelli FS, and Sayed AH. Distributed detection over adaptive networks based on diffusion estimation schemes. In: IEEE 10th Workshop on Signal Processing Advances in Wireless Communications, 2009. SPAWC'09. IEEE; 2009. p. 61–65.
[28] Matta V, Braca P, Marano S, and Sayed AH. Diffusion-based adaptive distributed detection: Steady-state performance in the slow adaptation regime. IEEE Transactions on Information Theory. 2016;62(8):4710–4732.

[29] Matta V, Braca P, Marano S, and Sayed AH. Exact asymptotics of distributed detection over adaptive networks. In: 2015 IEEE International Conference on Acoustics, Speech and Signal Processing (ICASSP). IEEE; 2015. p. 3377–3381.

[30] Braca P, Marano S, Matta V, and Sayed AH. Large deviations analysis of adaptive distributed detection. In: 2014 IEEE International Conference on Acoustics, Speech and Signal Processing (ICASSP). IEEE; 2014. p. 6112–6116.

[31] Lalitha A, Sarwate A, and Javidi T. Social learning and distributed hypothesis testing. In: IEEE International Symposium on Information Theory (ISIT), 2014. IEEE; 2014. p. 551–555.

[32] Lalitha A, Javidi T, and Sarwate A. Social learning and distributed hypothesis testing. IEEE Transactions on Information Theory. 2018;64(9): 6161–6179.

[33] Nedić A, Olshevsky A, and Uribe CA. Nonasymptotic convergence rates for cooperative learning over time-varying directed graphs. 2015 American Control Conference (ACC), Chicago, IL; 2015, p. 5884–5889.

[34] Jadbabaie A, Molavi P, Sandroni A, and Tahbaz-Salehi A. Non-Bayesian social learning. Games and Economic Behavior. 2012;76(1):210–225.

[35] Wilks SS. The large-sample distribution of the likelihood ratio for testing composite hypotheses. The Annals of Mathematical Statistics. 1938;9(1):60–62.

[36] Chung FR. Spectral graph theory. vol. 92. Providence, RI: American Mathematical Society; 1997.

[37] Jacod J, and Shiryaev AN. Limit theorems for stochastic processes. Berlin, Heidelberg: Springer-Verlag; 1987.

[38] Scharf LL. Statistical signal processing. vol. 98. Reading, MA: Addison-Wesley; 1991.

[39] Nevelson MB, and Khas'minskiĭ RZ. Stochastic approximation and recursive estimation. vol. 47. Providence, RI: American Mathematical Society; 1973.

[40] Borkar VS. Stochastic approximation: A dynamical systems viewpoint. Cambridge, UK: Cambridge University Press; 2008.

[41] Gelfand SB, and Mitter SK. Recursive Stochastic Algorithms for Global Optimization in $\mathbb{R}^d$. SIAM Journal on Control and Optimization. 1991;29(5): 999–1018.

[42] Kar S, and Moura JMF. Convergence rate analysis of distributed gossip (linear parameter) estimation: Fundamental limits and tradeoffs. IEEE Journal of Selected Topics in Signal Processing. 2011;5(4):674–690.

[43] Kar S, Moura JMF, and Ramanan K. Distributed parameter estimation in sensor networks: nonlinear observation models and imperfect communication. IEEE Transactions on Information Theory. 2012;58(6):3575–3605.

[44] Xiao L, Boyd S, and Kim SJ. Distributed average consensus with least-mean-square deviation. Journal of Parallel and Distributed Computing. 2007;67:33–46.

[45] Dimakis AG, Kar S, Moura JMF, Rabbat MG, and Scaglione A. Gossip algorithms for distributed signal processing. Proceedings of the IEEE. 2010;98(11):1847–1864.

[46] Goldenshluger A, Juditsky A, Nemirovski A, *et al.* Hypothesis testing by convex optimization. Electronic Journal of Statistics. 2015;9(2):1645–1712.

[47] Inglot T, Kallenberg WC, and Ledwina T. Data driven smooth tests for composite hypotheses. The Annals of Statistics. 1997;25(3):1222–1250.

[48] Wald A, Sequential tests of statistical hypotheses. Annals of Mathematical Statistics. 1945;16(2):117–186.

[49] Tartakovsky A, Nikiforov I, and Basseville M. Sequential analysis: Hypothesis testing and changepoint detection. Boca Raton, FL: Chapman and Hall/CRC; 2014.

Chapter 9
Expectation–maximisation based distributed estimation in sensor networks

Roberto López-Valcarce[1] and Alba Pagès-Zamora[2]

Estimating the unknown parameters of a statistical model based on the observations collected by a sensor network is an important problem with application in multiple fields. In this setting, distributed processing, by which computations are carried out within the network in order to avoid raw data transmission to a fusion centre, is a desirable feature resulting in improved robustness and energy savings.

In the presence of incomplete data, the expectation–maximisation (EM) algorithm is a popular means to iteratively compute the maximum likelihood (ML) estimate. It has found application in diverse fields such as computational biology, anomaly detection, speech segmentation, reinforcement learning, and motion estimation, among others. In this chapter we will review the formulation of the centralised EM estimation algorithm as a starting point and then discuss distributed versions well suited for implementation in sensor networks. The first class of these distributed versions requires specialised routing through the network in terms of a linear or circular path visiting all nodes, whereas the second class does away with this requirement by using the concept of network consensus to diffuse information through the network. Our focus will be on a relevant sensor network application, in which the parameter of a linear model is to be estimated in the presence of an unknown number of randomly malfunctioning sensors.

9.1 Introduction

Sensor networks typically consist of a potentially large number of cheap, battery-powered nodes with limited sensing, computation, and communication capabilities. These nodes can measure the environment over time, producing data from which relevant information is to be extracted. The range of applications is wide, including military, environmental, health, domestic, automotive, and industrial scenarios [1].

[1]Research Centre for Information and Communication Technologies (atlanTTic), Universidade de Vigo, Spain
[2]SPCOM Group, Universitat Politècnica de Catalunya, Spain

Since sensors can only monitor their local surroundings, their data must be jointly processed in order to obtain a global description of the environment.

In centralised approaches, nodes transmit their local data, possibly after some compression, to a fusion centre where joint processing is performed. Typically, this transmission is wireless and may be either direct or via multihop routing through the network. Direct transmission limits the network size to the node transmission range and will quickly drain the battery of sensors located farther from the fusion centre. Thus, multihop relaying schemes are essential to achieve certain degree of scalability in the network, although at the cost of increasingly complex routing protocols. Additionally, the centralised architecture presents drawbacks in terms of robustness since failure of some nodes can be fatal for network operation.

In decentralised sensor networks, rather than merely reporting their readings to a fusion centre, nodes communicate with their neighbours and cooperate with each other in processing their data, so that only the final result of the inference task (typically including parameter estimation, model fitting, and event detection) is transmitted to the end user. The distributed algorithms employed in this approach are usually iterative, and their final goal is for each node to eventually reach, as iterations proceed, the same result as that which would be obtained by a centralised processor. In this way, the outcome of a query to the network will be consistent, regardless of the sensor that happens to process the query. Distributed algorithms have a longstanding tradition with roots in the field of parallel computation [2].

When the task of the sensor network is to estimate some environmental parameters, the ML method is an appealing choice because of its good asymptotic performance properties [3,4]. Given a parameter-dependent probabilistic model for the data collected by the sensors, the ML estimate of the parameters is the value maximising the likelihood function that is the corresponding probability density function (pdf) evaluated at the data and regarded as a function of the parameters. Unfortunately, in many practical cases, the likelihood function is analytically intractable, and direct maximisation is not possible: one has to resort to numerical approaches such as e.g. gradient descent or Newton-type methods.

In some cases, however, the problem can be recast by assuming the existence of additional missing variables which cannot be directly observed, but which would result in a tractable expression of the likelihood function if their values were revealed. This scenario often lends itself to the application of the EM algorithm [5], which has become a standard device in the statistical signal processing toolbox, and whose formulation in its present generality is due to [6]. The main concept underlying the EM algorithm is intuitively appealing and consists of iteratively alternating between prediction of the missing variables based on a current estimate of the parameter, and reestimation of the parameter using such predicted values of the missing data. In principle, this general idea can be implemented in many different ways, among which the EM algorithm stands out due to its link to a more general class of optimisation techniques known as majorization–minimisation methods [7,8]: in contrast with other *ad hoc* procedures based on the previous iterative mechanism, the EM algorithm can be shown to produce a sequence of parameter estimates which monotonically increase the likelihood function.

In this chapter, we discuss a number of implementations of the EM algorithm as applied to parameter estimation problems in sensor networks. We start by presenting its general formulation in Section 9.2 as would be performed at the fusion centre of a centralised architecture. To illustrate these general concepts, we consider in Section 9.3 the particular problem of estimating the parameters of a linear model from the data gathered by the network, when an unknown subset of these data is corrupted due to malfunctioning of the corresponding sensors. Our discussion of distributed EM implementations begins in Section 9.4, where a class of decentralised methods requiring specialised network routing is presented. A second class avoiding this requirement and based on network consensus ideas is discussed in Sections 9.5 and 9.6. Some numerical examples are presented in Section 9.7.

We adopt the following notation throughout the chapter. Lower and uppercase boldface symbols respectively denote vectors and matrices. The vector of all ones is denoted by $\mathbf{1}$, whereas superscript ^{T} denotes transposition. The elements of matrix $\mathbf{A}$ and vector $\mathbf{x}$ are denoted by A_{ij} and x_i, respectively. The spectral radius of a square matrix $\mathbf{A}$, i.e. the largest absolute value of its eigenvalues, is denoted by $\rho(\mathbf{A})$. For a symmetric $n \times n$ matrix $\mathbf{A}$, its ordered eigenvalues are denoted $\lambda_1(\mathbf{A}) \geq \lambda_2(\mathbf{A}) \geq \cdots \geq \lambda_n(\mathbf{A})$. Statistical expectation is denoted by $\mathbb{E}\{\cdot\}$, and $\mathcal{N}(\boldsymbol{\mu}, \boldsymbol{\Sigma})$ denotes the Gaussian distribution with mean $\boldsymbol{\mu}$ and covariance matrix $\boldsymbol{\Sigma}$.

9.2 The EM algorithm

Consider a parameter estimation problem in which each sensor collects a (possibly vector-valued) measurement $\mathbf{y}_i$, $i = 1, \ldots, N$, where N is the number of sensors in the network. We let $\mathbf{y} = \begin{bmatrix} \mathbf{y}_1^T & \cdots & \mathbf{y}_N^T \end{bmatrix}^T$ denote the whole dataset of observations. A statistical model is assumed for the data, whose pdf, denoted by $p_Y(\mathbf{y}; \boldsymbol{\theta})$, is parameterised by the unknown parameter vector $\boldsymbol{\theta}$. When seen as a function of $\boldsymbol{\theta}$, $p_Y(\mathbf{y}; \boldsymbol{\theta})$ is termed the likelihood function. Since the logarithm is a monotonically increasing function, the ML estimate of $\boldsymbol{\theta}$ given $\mathbf{y}$ is:

$$\hat{\boldsymbol{\theta}}_{\text{ML}} = \arg\max_{\boldsymbol{\theta}} \ln p_Y(\mathbf{y}; \boldsymbol{\theta}). \tag{9.1}$$

Solving (9.1) may be easy or hard, depending on the form of $p_Y(\mathbf{y}; \boldsymbol{\theta})$. Among other cases, when the problem is such that the data present missing values, it is generally not possible to obtain $\hat{\boldsymbol{\theta}}_{\text{ML}}$ in closed form, and the EM algorithm provides an efficient means to numerically approach (9.1).

Suppose that a different dataset $\mathbf{z}$ exists, related to the available dataset $\mathbf{y}$ by a many-to-one transformation $\mathbf{y} = g(\mathbf{z})$ which, in general, is non-invertible. Typically, $\mathbf{z}$ includes $\mathbf{y}$ in addition to some additional unobserved variables, also known as *hidden* or *latent* variables; for this reason, $\mathbf{y}$ and $\mathbf{z}$ are termed the *incomplete* and *complete* datasets, respectively. If $\mathbf{z}$ were available, as an alternative to (9.1) we may consider instead the maximisation of $\ln p_Z(\mathbf{z}; \boldsymbol{\theta})$ with respect to (w.r.t.) $\boldsymbol{\theta}$, assuming that this results in an easier problem. Selection of the complete dataset $\mathbf{z}$ is done under such premise.

The main obstacle, of course, is that $\mathbf{z}$ is not available. A sensible approach is to replace the log-likelihood function (LLF) $\ln p_Z(\mathbf{z};\boldsymbol{\theta})$ by its *conditional expectation*:

$$\mathbb{E}_{Z|Y}\{\ln p_Z(\mathbf{z};\boldsymbol{\theta})\,|\,\mathbf{y}\} = \int \ln p_Z(\mathbf{z};\boldsymbol{\theta})\, p_{Z|Y}(\mathbf{z}|\mathbf{y};\boldsymbol{\theta})\, d\mathbf{z}, \tag{9.2}$$

where the conditioning is on the incomplete dataset $\mathbf{y}$. In that way, the dependence of $\ln p_Z(\mathbf{z};\boldsymbol{\theta})$ on unavailable data $\mathbf{z}$ is averaged out. However, in order to compute (9.2), one needs the conditional pdf $p_{Z|Y}(\mathbf{z}|\mathbf{y};\boldsymbol{\theta})$, which depends on the unknown parameter $\boldsymbol{\theta}$. At this point, we may use our current guess in lieu of $\boldsymbol{\theta}$ and then proceed iteratively. The final form of the EM algorithm is formulated as follows. Given the dataset $\mathbf{y}$ and an initial guess $\hat{\boldsymbol{\theta}}_0$, at iteration k, one performs the following two steps:

- **Expectation (E) step:** Compute the conditional expectation of the complete data LLF, based on the current guess $\hat{\boldsymbol{\theta}}_k$:

$$\begin{aligned} Q(\boldsymbol{\theta};\hat{\boldsymbol{\theta}}_k) &\triangleq \mathbb{E}_{Z|Y}\left\{\ln p_Z(\mathbf{z};\boldsymbol{\theta})\,|\,\mathbf{y};\hat{\boldsymbol{\theta}}_k\right\} \\ &= \int \ln p_Z(\mathbf{z};\boldsymbol{\theta})\, p_{Z|Y}(\mathbf{z}|\mathbf{y};\hat{\boldsymbol{\theta}}_k) d\mathbf{z}. \end{aligned} \tag{9.3}$$

- **Maximisation (M) step:** Obtain the next guess of the parameters by maximising the conditional expectation of the complete data LLF:

$$\hat{\boldsymbol{\theta}}_{k+1} = \arg\max_{\boldsymbol{\theta}} Q(\boldsymbol{\theta};\hat{\boldsymbol{\theta}}_k). \tag{9.4}$$

The above derivation of the EM algorithm may seem rather *ad hoc,* but it is possible to show that the resulting iteration is numerically stable in the sense that the LLF is increased at each step. The reason is that:

$$g_k(\boldsymbol{\theta}) \triangleq -\ln p_Y(\mathbf{y};\hat{\boldsymbol{\theta}}_k) + Q(\hat{\boldsymbol{\theta}}_k,\hat{\boldsymbol{\theta}}_k) - Q(\boldsymbol{\theta},\hat{\boldsymbol{\theta}}_k) \tag{9.5}$$

is a *majorizing function* for the negative LLF of the incomplete-data problem, that is, it satisfies $g_k(\hat{\boldsymbol{\theta}}_k) = -\ln p_Y(\mathbf{y};\hat{\boldsymbol{\theta}}_k)$ and $g_k(\boldsymbol{\theta}) \geq -\ln p_Y(\mathbf{y};\boldsymbol{\theta})$ for all $\boldsymbol{\theta}$, as can be shown using Jensen's inequality [6,7]. It follows that the sequence of estimates $\hat{\boldsymbol{\theta}}_{k+1} = \arg\min_{\boldsymbol{\theta}} g_k(\boldsymbol{\theta})$, which is the same as that produced by the EM algorithm, results in a monotonically increasing sequence of LLF values $\ln p_Y(\mathbf{y};\hat{\boldsymbol{\theta}}_k) = -g_k(\hat{\boldsymbol{\theta}}_k)$ [7,8]. Although this property does not ensure convergence to the global maximum of the likelihood function, the EM algorithm usually works well in practice, provided that the iteration is judiciously initialised. In some cases, one may even have the freedom to run the EM algorithm multiple times with different initialisations and then keep the result that achieves the largest LLF.

In the next section, the EM algorithm is applied to a specific estimation problem based on a sensor network with faulty nodes. This is the illustrative application that will be used throughout the chapter to present a number of distributed implementations of the EM algorithm. It must be emphasised that the main ideas behind these procedures can be applied to other problems typically dealt with by means of the EM algorithm.

To avoid cumbersome notation, and as long as there is no room for confusion, in the sequel the pdf $p_X(x)$ of a random variable X will be denoted simply as $p(x)$.

9.3 Linear parameter estimation with anomalous data

Consider a sensor network with N nodes where each node i senses and records a noisy measurement $\mathbf{y}_i \in \mathbb{R}^M$. Under normal operation conditions, $\mathbf{y}_i$ is a linear function of a deterministic but unknown parameter of interest $\boldsymbol{\phi} \in \mathbb{R}^L$. We further assume that the nodes are subject to random faults, so that the observations are not completely reliable. Let us introduce Bernoulli random variables $\{a_i : i = 1, \dots, N\}$ such that $a_i = 0$ indicates an anomalous reading, whereas $a_i = 1$ corresponds to a correct reading. Data faults are assumed statistically independent and with equal probability for all sensors; thus, the variables $\{a_i\}$ are assumed independent and identically distributed (i.i.d.) with $\Pr\{a_i = 0\} = q$.

In the absence of a data fault, the observation from the ith sensor is given by:

$$\mathbf{y}_i = \mathbf{H}_i\boldsymbol{\phi} + \mathbf{w}_i, \qquad \text{if } a_i = 1, \tag{9.6}$$

with $\mathbf{H}_i \in \mathbb{R}^{M\times L}$ a known matrix, and $\mathbf{w}_i$ a random vector modelling measurement noise. The noise is assumed zero-mean Gaussian with covariance matrix $\sigma^2\mathbf{I}_M$, that is, $\mathbf{w}_i \sim \mathcal{N}(\mathbf{0}, \sigma^2\mathbf{I}_M)$. Therefore, the sample $\mathbf{y}_i$ is conditionally Gaussian, i.e., $\mathbf{y}_i \mid a_i = 1 \sim \mathcal{N}(\mathbf{H}_i\boldsymbol{\phi}, \sigma^2\mathbf{I}_M)$. It is assumed that a_i, $\mathbf{w}_j$ are statistically independent for all i, j; and that $\mathbf{w}_i$, $\mathbf{w}_j$ are statistically independent for $i \neq j$.

The problem of estimating $\boldsymbol{\phi}$ from the data is hindered by the presence of random data faults. With completely reliable sensors, i.e., $q = 0$, the data are Gaussian distributed and the standard least squares (LS) estimate:

$$\hat{\boldsymbol{\phi}}_{\text{LS}} = \left(\sum_{i=1}^{N} \mathbf{H}_i^T\mathbf{H}_i\right)^{-1} \left(\sum_{i=1}^{N} \mathbf{H}_i^T\mathbf{y}_i\right) \tag{9.7}$$

is also the minimum variance unbiased (MVU) estimate [4]. However, with unreliable sensors, i.e., $q > 0$, it is not possible in general to obtain an MVU estimate for this problem, and the performance of (9.7) degrades as q increases.

We model anomalous data probabilistically, in terms of some pdf $p(\mathbf{y}_i \mid a_i = 0; \boldsymbol{\omega})$ possibly depending on some unknown parameter $\boldsymbol{\omega}$. For example, in [9–11], readings from faulty nodes were assumed to contain only noise with the same variance as in (9.6), i.e., $\mathbf{y}_i \mid a_i = 0 \sim \mathcal{N}(0, \sigma^2\mathbf{I}_M)$; in that case, $\boldsymbol{\omega} = \sigma^2$. Alternatively, data faults could be assimilated to a larger noise variance, so that $\mathbf{y}_i \mid a_i = 0 \sim \mathcal{N}(\mathbf{H}_i\boldsymbol{\phi}, \gamma^2\mathbf{I}_M)$, with $\gamma^2 \gg \sigma^2$, and thus $\boldsymbol{\omega} = \begin{bmatrix}\boldsymbol{\phi}^T & \gamma^2\end{bmatrix}^T$. In the sequel, we adopt a *non-informative model* under which readings $\mathbf{y}_i = \begin{bmatrix}y_{i,1} & \cdots & y_{i,M}\end{bmatrix}^T$ from defective sensors have i.i.d. entries $y_{i,m}$ which are uniformly distributed in some unknown interval $[A, B]$, and hence $\boldsymbol{\omega} = \begin{bmatrix}A & B\end{bmatrix}^T$. This probabilistic model accounts for anomalous data from a wide spectrum of sensor faults, as well as communication errors. Data fault types are discussed at length in [12].

At this point, we note that if we had knowledge about which readings are anomalous, that is, if the values of the a_i's were known, then we could discard them and construct the LS estimate of $\boldsymbol{\phi}$ using only non-defective data. Anomalous data do not carry information about $\boldsymbol{\phi}$, and therefore they can be safely discarded. Thus, the corresponding *clairvoyant* (CV) estimator is MVU, and can be written as:

$$\hat{\boldsymbol{\phi}}_{\mathrm{CV}} = \left(\sum_{i=1}^{N} a_i \mathbf{H}_i^T \mathbf{H}_i \right)^{-1} \left(\sum_{i=1}^{N} a_i \mathbf{H}_i^T \mathbf{y}_i \right). \tag{9.8}$$

In practice, however, knowledge of the a_i's is not available and (9.8) cannot be computed. An appealing alternative to improve on (9.7) is ML estimation under the adopted statistical model. Letting $\boldsymbol{\theta} = \left[\boldsymbol{\phi}^T \ \sigma^2 \ q \ A \ B\right]^T$ collect all unknown parameters, for convenience we shall denote:

$$p_1(\mathbf{y}_i; \boldsymbol{\theta}) \triangleq p(\mathbf{y}_i \mid a_i = 1; \boldsymbol{\theta}) = \frac{1}{(2\pi\sigma^2)^{M/2}} \mathrm{e}^{-(1/2\sigma^2)\|\mathbf{y}_i - \mathbf{H}_i \boldsymbol{\phi}\|^2}, \tag{9.9}$$

$$p_0(\mathbf{y}_i; \boldsymbol{\theta}) \triangleq p(\mathbf{y}_i \mid a_i = 0; \boldsymbol{\theta}) = \prod_{m=1}^{M} \mathscr{U}(y_{i,m}; A, B), \tag{9.10}$$

where $\mathscr{U}(y; A, B)$ is the pdf of the uniform distribution in $[A, B]$, i.e., $\mathscr{U}(y; A, B) = (1/(B-A))$ if $A \le y \le B$, and $\mathscr{U}(y; A, B) = 0$ otherwise. Thus, $\mathbf{y}_i$ has pdf $p(\mathbf{y}_i; \boldsymbol{\theta}) = q p_0(\mathbf{y}_i; \boldsymbol{\theta}) + (1-q) p_1(\mathbf{y}_i; \boldsymbol{\theta})$, which is a mixture of a Gaussian density and a uniform density, similarly to [13]. Since readings from different nodes are statistically independent, then given the available data $\mathbf{y}$, the LLF reads as:

$$\ln p(\mathbf{y}; \boldsymbol{\theta}) = \sum_{i=1}^{N} \ln\left[q p_0(\mathbf{y}_i; \boldsymbol{\theta}) + (1-q) p_1(\mathbf{y}_i; \boldsymbol{\theta})\right]. \tag{9.11}$$

Although it is generally not possible to maximise (9.11) in closed form, this problem naturally lends itself to application of the EM algorithm as follows. In this case, the complete dataset is taken as $\mathbf{z} = \{\mathbf{y}, \mathbf{a}\}$, where $\mathbf{a} = \left[a_1 \ \cdots \ a_N\right]^T$ collects the unobserved variables reflecting our lack of knowledge about the faulty or faultless state of each sensor. Due to independence, the corresponding pdf satisfies $p(\mathbf{z}; \boldsymbol{\theta}) = \prod_{i=1}^{N} p(\mathbf{y}_i, a_i; \boldsymbol{\theta})$, and since $a_i \in \{0, 1\}$, we can write:

$$\begin{aligned} p(\mathbf{y}_i, a_i; \boldsymbol{\theta}) &= p(a_i; \boldsymbol{\theta}) p(\mathbf{y}_i \mid a_i; \boldsymbol{\theta}) \\ &= \left[q p_0(\mathbf{y}_i; \boldsymbol{\theta})\right]^{1-a_i} \cdot \left[(1-q) p_1(\mathbf{y}_i; \boldsymbol{\theta})\right]^{a_i}. \end{aligned} \tag{9.12}$$

Therefore, the complete data LLF can be written as:

$$\begin{aligned} \ln p(\mathbf{z}; \boldsymbol{\theta}) &= \sum_{i=1}^{N} (1-a_i) \ln p_0(\mathbf{y}_i; \boldsymbol{\theta}) + \sum_{i=1}^{N} a_i \ln p_1(\mathbf{y}_i; \boldsymbol{\theta}) \\ &\quad + \left(\sum_{i=1}^{N} a_i \right) \ln \frac{1-q}{q} + N \ln q. \end{aligned} \tag{9.13}$$

E-step: The E-step of the EM algorithm requires the computation of the conditional expectation of (9.13), where the conditioning is on the available dataset $\mathbf{y}$, and based on the current estimate $\hat{\boldsymbol{\theta}}_k = \left[\hat{\boldsymbol{\phi}}_k^T \ \hat{\sigma}_k^2 \ \hat{q}_k \ \hat{A}_k \ \hat{B}_k\right]^T$. When $\mathbf{y}$ is given, the only random terms in (9.13) are the unobserved variables $\{a_i\}$; hence, upon defining:

$$\hat{a}_i(\hat{\boldsymbol{\theta}}_k) \triangleq \mathbb{E}\{a_i \,|\, \mathbf{y}\,;\,\hat{\boldsymbol{\theta}}_k\}, \qquad i = 1, \ldots, N, \tag{9.14}$$

it follows from (9.3) and (9.13) that:

$$\begin{aligned} Q(\boldsymbol{\theta};\hat{\boldsymbol{\theta}}_k) &= \sum_{i=1}^{N}\left(1 - \hat{a}_i(\hat{\boldsymbol{\theta}}_k)\right)\ln p_0(\mathbf{y}_i;\boldsymbol{\theta}) + \sum_{i=1}^{N}\hat{a}_i(\hat{\boldsymbol{\theta}}_k)\ln p_1(\mathbf{y}_i;\boldsymbol{\theta}) \\ &\quad + \left(\sum_{i=1}^{N}\hat{a}_i(\hat{\boldsymbol{\theta}}_k)\right)\ln\frac{1-q}{q} + N\ln q. \end{aligned} \tag{9.15}$$

In order to complete the E-step, it remains to compute (9.14). Note that $\hat{a}_i(\hat{\boldsymbol{\theta}}_k)$ can be interpreted as a *soft estimate* of the value taken by $a_i \in \{0, 1\}$, based on the observations $\mathbf{y}$ and the current guess $\hat{\boldsymbol{\theta}}_k$. In fact, this soft estimate is a conditional mean that closely resembles the minimum mean squared error estimate $\mathbb{E}\{a_i \,|\, \mathbf{y}\,;\,\boldsymbol{\theta}\}$ [4]; the only difference is that $\boldsymbol{\theta}$ is replaced by its current guess $\hat{\boldsymbol{\theta}}_k$. It is given by:

$$\begin{aligned} \hat{a}_i(\hat{\boldsymbol{\theta}}_k) &= 1 \cdot \Pr\{a_i = 1 \,|\, \mathbf{y}\,;\,\hat{\boldsymbol{\theta}}_k\} + 0 \cdot \Pr\{a_i = 0 \,|\, \mathbf{y}\,;\,\hat{\boldsymbol{\theta}}_k\} \\ &= \Pr\{a_i = 1 \,|\, \mathbf{y}\,;\,\hat{\boldsymbol{\theta}}_k\}, \end{aligned} \tag{9.16}$$

which is the *a posteriori* probability of the ith sensor being non-defective, given the data $\mathbf{y}$ and based on the current guess $\hat{\boldsymbol{\theta}}_k$. Since a_i, $\mathbf{y}_j$ are statistically independent for $i \neq j$:

$$\begin{aligned} \hat{a}_i(\hat{\boldsymbol{\theta}}_k) &= \Pr\{a_i = 1 \,|\, \mathbf{y}_i\,;\,\hat{\boldsymbol{\theta}}_k\} \\ &= \frac{p(\mathbf{y}_i \,|\, a_i = 1\,;\,\hat{\boldsymbol{\theta}}_k)\Pr\{a_i = 1\,;\,\hat{\boldsymbol{\theta}}_k\}}{p(\mathbf{y}_i \,|\, a_i = 0\,;\,\hat{\boldsymbol{\theta}}_k)\Pr\{a_i = 0\,;\,\hat{\boldsymbol{\theta}}_k\} + p(\mathbf{y}_i \,|\, a_i = 1\,;\,\hat{\boldsymbol{\theta}}_k)\Pr\{a_i = 1\,;\,\hat{\boldsymbol{\theta}}_k\}} \\ &= \frac{\left(1 - \hat{q}_k\right)p_1(\mathbf{y}_i\,;\,\hat{\boldsymbol{\theta}}_k)}{\hat{q}_k p_0(\mathbf{y}_i\,;\,\hat{\boldsymbol{\theta}}_k) + \left(1 - \hat{q}_k\right)p_1(\mathbf{y}_i\,;\,\hat{\boldsymbol{\theta}}_k)}. \end{aligned} \tag{9.17}$$

Thus, with (9.15) and (9.17), the E-step is completed. As it will be seen shortly, explicit computation of $Q(\boldsymbol{\theta};\hat{\boldsymbol{\theta}}_k)$ at each iteration is actually not required.

M-step: The M-step consists of maximising (9.15) w.r.t. $\boldsymbol{\theta}$. Since neither $p_0(\mathbf{y}_i;\boldsymbol{\theta})$ nor $p_1(\mathbf{y}_i;\boldsymbol{\theta})$ depend on the probability of sensor malfunctioning q (see (9.9) and (9.10)), maximisation w.r.t. q is easily accomplished:

$$\hat{q}_{k+1} = 1 - \frac{1}{N}\sum_{i=1}^{N}\hat{a}_i(\hat{\boldsymbol{\theta}}_k). \tag{9.18}$$

The dependence of $Q(\boldsymbol{\theta};\hat{\boldsymbol{\theta}}_k)$ on $\boldsymbol{\phi}$ and σ^2 is through the second term in the right-hand side of (9.15) only, given by:

$$\sum_{i=1}^{N} \hat{a}_i(\hat{\boldsymbol{\theta}}_k) \ln p_1(\mathbf{y}_i;\boldsymbol{\theta}) = -\frac{M}{2}\left(\sum_{i=1}^{N} \hat{a}_i(\hat{\boldsymbol{\theta}}_k)\right) \ln 2\pi\sigma^2 - \frac{1}{2\sigma^2}\sum_{i=1}^{N} \hat{a}_i(\hat{\boldsymbol{\theta}}_k)\|\mathbf{y}_i - \mathbf{H}_i\boldsymbol{\phi}\|^2. \quad (9.19)$$

The corresponding values maximising (9.19) are:

$$\hat{\boldsymbol{\phi}}_{k+1} = \left(\sum_{i=1}^{N} \hat{a}_i(\hat{\boldsymbol{\theta}}_k)\mathbf{H}_i^T\mathbf{H}_i\right)^{-1}\left(\sum_{i=1}^{N} \hat{a}_i(\hat{\boldsymbol{\theta}}_k)\mathbf{H}_i^T\mathbf{y}_i\right), \quad (9.20)$$

$$\hat{\sigma}_{k+1}^2 = \frac{\sum_{i=1}^{N} \hat{a}_i(\hat{\boldsymbol{\theta}}_k)\|\mathbf{y}_i - \mathbf{H}_i\hat{\boldsymbol{\phi}}_{k+1}\|^2}{M\sum_{i=1}^{N} \hat{a}_i(\hat{\boldsymbol{\theta}}_k)}. \quad (9.21)$$

Note the similarity between the structures of (9.8) and (9.20).

Finally, we note that only the first term in the right-hand side of (9.15) depends on A, B. This term reads as:

$$\sum_{i=1}^{N}\left(1-\hat{a}_i(\hat{\boldsymbol{\theta}}_k)\right) \ln p_0(\mathbf{y}_i;\boldsymbol{\theta}) = \sum_{i=1}^{N}\sum_{m=1}^{M}\left(1-\hat{a}_i(\hat{\boldsymbol{\theta}}_k)\right) \ln \mathscr{U}(y_{i,m};A,B). \quad (9.22)$$

The values of A, B maximising (9.22) do not depend on $\hat{\boldsymbol{\theta}}_k$ and, since $1-\hat{a}_i(\hat{\boldsymbol{\theta}}_k) \geq 0$ for all i, they are given by:

$$\hat{A}_{k+1} = \min_{\substack{1\leq i\leq N\\ 1\leq m\leq M}} \{y_{i,m}\}, \qquad \hat{B}_{k+1} = \max_{\substack{1\leq i\leq N\\ 1\leq m\leq M}} \{y_{i,m}\}. \quad (9.23)$$

This completes the M-step.

Initialisation: To start the EM algorithm, we need to specify $\hat{\boldsymbol{\theta}}_0$. Clearly, since (9.23) do not change along the iterations, we can directly take $\hat{A}_0 = \min\{y_{i,m}\}$, $\hat{B}_0 = \max\{y_{i,m}\}$. The LS estimate (9.7) can be used as starting point for $\hat{\boldsymbol{\phi}}_0$, with the sample variance of the resulting residual initialising the estimate of the noise variance σ^2, i.e., $\hat{\sigma}_0^2 = (1/MN)\sum_{i=1}^{N}\|\mathbf{y}_i - \mathbf{H}_i\hat{\boldsymbol{\phi}}_{\text{LS}}\|^2$. A reasonable choice in the absence of any *a priori* knowledge is $\hat{q}_0 = (1/2)$. Note from (9.17) that setting $\hat{q}_0 = 0$ would result in $\hat{a}_i^{(0)} = 1$ for all i, which in turn would yield $\hat{q}_1 = 0$ in view of (9.18), preventing the algorithm from evolving. A similar phenomenon takes place with $\hat{q}_0 = 1$, so these choices are to be avoided.

Example 9.1

As an example, consider a network with $N = 100$ nodes gathering observations under the model considered in this section. The dimension of the observations is $M = 2$, whereas that of the unknown parameter is $L = 3$. The matrices $\mathbf{H}_i$ and the parameter $\boldsymbol{\phi}$ are randomly generated with i.i.d. zero-mean unit-variance Gaussian entries and then fixed throughout. Anomalous data reported by malfunctioning

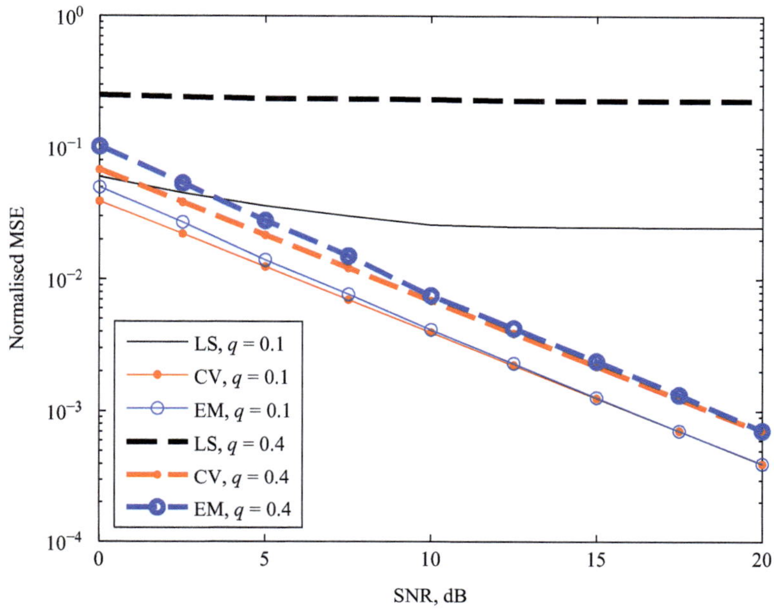

Figure 9.1 NMSE of the LS, clairvoyant, and EM estimates in a network of 100 sensors. $L=3$, $M=2$

sensors is uniformly distributed in the interval $[A, B] = [-10, 10]$. The signal-to-noise ratio (SNR) is defined as:

$$\text{SNR} = \frac{1}{NM\sigma^2} \sum_{i=1}^{N} \|\mathbf{H}_i \boldsymbol{\phi}\|^2. \tag{9.24}$$

Figure 9.1 shows the normalised mean squared error (NMSE)[1] of the LS estimate (9.7), the CV estimate (9.8), and the EM estimate, as a function of SNR, and for two different levels of sensor reliability: $q = 0.1$ and $q = 0.4$. The LS estimate, which is oblivious to the presence of anomalous data, exhibits a significant error floor. The NMSE of the CV estimate, in contrast, decays as the SNR improves. The performance of this estimate, which assumes perfect knowledge about which sensors are working properly and which are not, can be used as a benchmark for any practical estimate. In particular, the NMSE of the EM estimate is seen to asymptotically converge to this benchmark for sufficiently high SNR, which clearly motivates its use in practice.

[1]For an estimator $\hat{\boldsymbol{\phi}}$ of the parameter $\boldsymbol{\phi}$, the NMSE is obtained by Monte Carlo simulation as:

$$\text{NMSE} = \frac{(1/R)\sum_{r=1}^{R} \|\boldsymbol{\phi} - \hat{\boldsymbol{\phi}}_r\|^2}{\|\boldsymbol{\phi}\|^2},$$

with R the number of Monte Carlo runs (5, 000 in this example), and $\hat{\boldsymbol{\phi}}_r$ the estimate obtained in run r.

The EM algorithm described above is suited to a centralised implementation, that is, each node sends its local data $\{\mathbf{y}_i, \mathbf{H}_i\}$ to a fusion centre where the corresponding iterations are run. It is therefore of interest to explore *distributed* versions which avoid the robustness and scalability drawbacks associated to the centralised architecture.

9.4 Distributed routing-based approaches

The design of distributed versions of the EM algorithm is generally based on the following observations. On the one hand, if at iteration k all nodes have the current estimate $\hat{\boldsymbol{\theta}}_k$, then the E-step can be performed locally: each node simply computes $\hat{a}_i(\hat{\boldsymbol{\theta}}_k)$ via (9.17) using its local data $\{\mathbf{y}_i, \mathbf{H}_i\}$. On the other hand, the M-step requires global information to update the parameter estimate, but nevertheless, the required global variables can be obtained by suitable *aggregation* of local variables.

With these in mind, in this section, we present two distributed methods which hinge on the availability of special types of paths through the network. Information flows through the nodes making use of these paths, which must be established and maintained by specialised routing mechanisms. The first method assumes a *linear* path through the network in order $1, 2, \ldots, N$, and is referred to as the *distributed forward–backward implementation* of the EM algorithm. When a *cyclic* path through the network exists, say visiting nodes in order $1, 2, \ldots, N, 1$, a different *distributed cyclic implementation* is possible, whose basis is due to [14] and [15, Section IV], and which does away with the need for the forward and backward message-passing processes. These two approaches are discussed next.

9.4.1 Distributed forward–backward implementation

To see how the global information required by the M-step can be obtained by aggregating local variables, let us introduce the following global variables:

$$s_k = \sum_{i=1}^{N} \hat{a}_i(\hat{\boldsymbol{\theta}}_k), \qquad c_k = \sum_{i=1}^{N} \hat{a}_i(\hat{\boldsymbol{\theta}}_k)\|\mathbf{y}_i\|^2, \qquad \underline{\ell}_k = \min_{i,m}\{y_{i,m}\}, \tag{9.25}$$

$$\mathbf{v}_k = \sum_{i=1}^{N} \hat{a}_i(\hat{\boldsymbol{\theta}}_k)\mathbf{H}_i^T\mathbf{y}_i, \qquad \mathbf{R}_k = \sum_{i=1}^{N} \hat{a}_i(\hat{\boldsymbol{\theta}}_k)\mathbf{H}_i^T\mathbf{H}_i, \qquad \overline{\ell}_k = \max_{i,m}\{y_{i,m}\}. \tag{9.26}$$

Note that in fact $\underline{\ell}_k$ and $\overline{\ell}_k$ do not depend on k.

Then the parameter updates in the M-step can be rewritten in terms of (9.25) and (9.26) as:

$$\hat{q}_{k+1} = 1 - \frac{s_k}{N}, \qquad \hat{A}_{k+1} = \underline{\ell}_k, \qquad \hat{B}_{k+1} = \overline{\ell}_k, \tag{9.27}$$

$$\hat{\boldsymbol{\phi}}_{k+1} = \mathbf{R}_k^{-1}\mathbf{v}_k, \qquad \hat{\sigma}_{k+1}^2 = \frac{c_k - \mathbf{v}_k^T \mathbf{R}_k^{-1} \mathbf{v}_k}{M\, s_k}. \tag{9.28}$$

This suggests a distributed implementation similar in spirit to the one proposed in [15, Sec. III]. Assuming that all nodes have the current estimate $\hat{\boldsymbol{\theta}}_k$, they can compute $\hat{a}_i(\hat{\boldsymbol{\theta}}_k)$ via (9.17); then, to compute $\hat{\boldsymbol{\theta}}_{k+1}$, two message-passing stages are performed. It is assumed that a path through the network exists going through all nodes and visiting each node just once, so that messages can be passed from node to node in order $1, 2, \ldots, N$. The global variables $\{s_k^0, c_k^0, \mathbf{v}_k^0, \mathbf{R}_k^0, \underline{\ell}_k^0, \overline{\ell}_k^0\}$ are initialised to zero. In the forward pass, node i updates these global variables as follows:

$$s_k^i = s_k^{i-1} + \hat{a}_i(\hat{\boldsymbol{\theta}}_k), \qquad c_k^i = c_k^{i-1} + \hat{a}_i(\hat{\boldsymbol{\theta}}_k)\|\mathbf{y}_i\|^2, \tag{9.29}$$

$$\mathbf{v}_k^i = \mathbf{v}_k^{i-1} + \hat{a}_i(\hat{\boldsymbol{\theta}}_k)\mathbf{H}_i^T\mathbf{y}_i, \qquad \mathbf{R}_k^i = \mathbf{R}_k^{i-1} + \hat{a}_i(\hat{\boldsymbol{\theta}}_k)\mathbf{H}_i^T\mathbf{H}_i, \tag{9.30}$$

$$\underline{\ell}_k^i = \min\left\{\underline{\ell}_k^{i-1}, \min_m\{y_{i,m}\}\right\}, \qquad \overline{\ell}_k^i = \max\left\{\overline{\ell}_k^{i-1}, \max_m\{y_{i,m}\}\right\}, \tag{9.31}$$

and then passes the updated variables to node $i+1$. At the last node N, the global variables $\{s_k^N, c_k^N, \mathbf{v}_k^N, \mathbf{R}_k^N, \underline{\ell}_k^N, \overline{\ell}_k^N\}$ take the desired values (9.25) and (9.26). Then these are passed back in order $N, N-1, \ldots, 2, 1$. After this backward pass, the global variables are available to all nodes, which can complete the M-step according to (9.27) and (9.28). Therefore, each EM iteration requires a full round trip along the linear path. As a simple initialisation strategy, each node sets $\hat{a}_i(\hat{\boldsymbol{\theta}}_0) = (1/2)$ in the first pass of the first iteration.

Note that this approach assumes that all nodes know the network cardinality N in order to compute $\hat{s}_{k+1}$ in (9.27). The fact that a linear path of length N is assumed to have been established beforehand makes this a reasonable assumption. In any case, N can be easily determined by a similar incremental approach in which a global variable is initialised to zero and incremented in one unit before passing it to the next node in the path.

9.4.2 Distributed cyclic implementation

When a cycle through the network exists, say visiting nodes in order $1, 2, \ldots, N, 1$, a distributed cyclic implementation is possible. The corresponding communication load among nodes is lighter than for the forward–backward approach because the forward pass is avoided.

Recall the conditional expectation (9.3) computed in the general E-step. Suppose that the complete dataset $\mathbf{z}$ can be written as $\mathbf{z} = \{\mathbf{z}_1, \ldots, \mathbf{z}_N\}$ such that $\mathbf{z}_i$ is associated to node i only and $\mathbf{z}_i$, $\mathbf{z}_j$ are statistically independent for $i \neq j$; this is the case for the problem at hand, for which $\mathbf{z}_i = \{\mathbf{y}_i, a_i\}$. Then (9.3) can be written as:

$$Q(\boldsymbol{\theta}; \hat{\boldsymbol{\theta}}) = \sum_{i=1}^{N} Q_i(\boldsymbol{\theta}; \hat{\boldsymbol{\theta}}), \tag{9.32}$$

$$Q_i(\boldsymbol{\theta}; \hat{\boldsymbol{\theta}}) \triangleq \mathbb{E}_{Z_i|Y_i} \left\{ \ln p_{Z_i}(\mathbf{z}_i\,;\,\boldsymbol{\theta}) \,|\, \mathbf{y}_i\,;\,\hat{\boldsymbol{\theta}} \right\}. \tag{9.33}$$

Each term $Q_i(\boldsymbol{\theta}; \hat{\boldsymbol{\theta}})$ depends only on local data at node i, and they are all conditional expectations based on the same estimate $\hat{\boldsymbol{\theta}}$.

Suppose we allow each node i to maintain its own local estimate $\hat{\boldsymbol{\theta}}^i$, and then consider iteratively maximising the function:

$$Q\left(\boldsymbol{\theta}; \hat{\boldsymbol{\theta}}^1, \ldots, \hat{\boldsymbol{\theta}}^N\right) \triangleq \sum_{i=1}^{N} Q_i\left(\boldsymbol{\theta}; \hat{\boldsymbol{\theta}}^i\right). \tag{9.34}$$

As before, we denote the iteration number by k. We proceed in a cyclic fashion, letting $n = k \bmod N$ be the index of the "active" node at iteration k. Then, the proposed update is as follows. First, set:

$$\hat{\boldsymbol{\theta}}_k^i = \begin{cases} \hat{\boldsymbol{\theta}}_k, & \text{if} \quad i = n = k \bmod N, \\ \hat{\boldsymbol{\theta}}_{k-1}^i, & \text{if} \quad i \neq n. \end{cases} \tag{9.35}$$

Then find the next estimate as:

$$\hat{\boldsymbol{\theta}}_{k+1} = \arg\max_{\boldsymbol{\theta}} Q\left(\boldsymbol{\theta}; \hat{\boldsymbol{\theta}}_k^1, \ldots, \hat{\boldsymbol{\theta}}_k^N\right). \tag{9.36}$$

Note that if this iterative scheme converges, all nodes will arrive at the same value of the parameter estimate. In general, the sequence of estimates $\{\hat{\boldsymbol{\theta}}_k\}$ generated by (9.36) will be different from the corresponding sequence obtained by the centralised EM algorithm. Convergence properties are analysed at length in [15]; basically, (9.35) and (9.36) present the same set of fixed points as the standard EM algorithm, although their convergence rates may be different in general.

The solution to (9.36) is still given by (9.27) and (9.28), but with $\hat{a}_i(\hat{\boldsymbol{\theta}}_k)$ replaced by $\hat{a}_i(\hat{\boldsymbol{\theta}}_k^i)$ in the expressions of the global variables (9.25) and (9.26). For example, the global variable s_k is now given by:

$$\begin{aligned} s_k &= \sum_{i=1}^{N} \hat{a}_i(\hat{\boldsymbol{\theta}}_k^i) = \sum_{i \neq n}^{N} \hat{a}_i(\hat{\boldsymbol{\theta}}_{k-1}^i) + \hat{a}_n(\hat{\boldsymbol{\theta}}_k^n) \\ &= s_{k-1} + \hat{a}_n(\hat{\boldsymbol{\theta}}_k^n) - \hat{a}_n(\hat{\boldsymbol{\theta}}_{k-1}^n). \end{aligned} \tag{9.37}$$

Analogously, the global variables c_k, $\mathbf{v}_k$, $\mathbf{R}_k$ satisfy:

$$c_k = c_{k-1} + \left(\hat{a}_n(\hat{\boldsymbol{\theta}}^n_k) - \hat{a}_n(\hat{\boldsymbol{\theta}}^n_{k-1})\right) \|\mathbf{y}_n\|^2, \tag{9.38}$$

$$\mathbf{v}_k = \mathbf{v}_{k-1} + \left(\hat{a}_n(\hat{\boldsymbol{\theta}}^n_k) - \hat{a}_n(\hat{\boldsymbol{\theta}}^n_{k-1})\right) \mathbf{H}_n^T \mathbf{y}_n, \tag{9.39}$$

$$\mathbf{R}_k = \mathbf{R}_{k-1} + \left(\hat{a}_n(\hat{\boldsymbol{\theta}}^n_k) - \hat{a}_n(\hat{\boldsymbol{\theta}}^n_{k-1})\right) \mathbf{H}_n^T \mathbf{H}_n, \tag{9.40}$$

whereas the expressions of $\underline{\ell}_k$, $\overline{\ell}_k$ are as in (9.25) and (9.26), since they do not depend on the parameter estimate.

Based on these facts, the distributed cyclic implementation of the EM algorithm is as follows. At iteration k, node $n = k \bmod N$ receives the global variables s_{k-1}, c_{k-1}, $\mathbf{v}_{k-1}$, $\mathbf{R}_{k-1}$, $\underline{\ell}_{k-1}$, $\overline{\ell}_{k-1}$ from the previous node in the cycle. With these, it computes $\hat{\boldsymbol{\theta}}_k$ as per (9.27) and (9.28), sets $\hat{\boldsymbol{\theta}}^n_k = \hat{\boldsymbol{\theta}}_k$, and computes $\hat{a}_n(\hat{\boldsymbol{\theta}}^n_k)$ via (9.17). Finally, the global variables s_k, c_k, $\mathbf{v}_k$, $\mathbf{R}_k$ are updated as in (9.37)–(9.40), whereas the updates for $\underline{\ell}_k$ and $\overline{\ell}_k$ are $\underline{\ell}_k = \min\{\,\underline{\ell}_{k-1}, \min_m\{y_{n,m}\}\,\}$, $\overline{\ell}_k = \max\{\,\overline{\ell}_{k-1}, \max_m\{\,y_{n,m}\}\,\}$. These global variables are then passed along to the next node in the cycle, with index $(k+1) \bmod N$, and the process is repeated. To start the algorithm, one can simply initialise the global variables to zero and set $\hat{a}_i = (1/2)$ in the first cycle.

The two distributed implementations discussed in this section are scalable in the sense that each node communicates with two other nodes in the network, independently of the network size. However, they require that the global variables be propagated through the whole network, which may slow down convergence if the network is large. Furthermore, since they are based on routing through either a linear or a cyclic path, the establishment and maintenance of these routs becomes an issue. Additionally, these implementations may be compromised if one or more nodes in the network stop functioning. For these reasons, alternative distributed approaches not requiring specialised routing may be preferable, particularly in wireless sensor networks due to the unreliability of wireless links.

9.5 A distributed consensus-based approach

In this section, we present a distributed implementation of the EM algorithm based on the concept of *network consensus* [16–20]. Consensus algorithms are simple, decentralised iterative schemes which require only minimal computation and communication to obtain aggregated values of local quantities that reside in each node of the network, in such a way that all of the nodes eventually reach an agreement on such values. The aggregation function can be the average, the maximum, or the minimum, precisely what is needed in the computation of the M-step. In consensus schemes, each node maintains appropriate local variables and exchanges information only with a subset of nodes at close range, i.e., its *neighbours*. At each step, each node updates its local variables based on the data received from its neighbours, and in this way, information diffuses across the network without requiring specific routing mechanisms.

9.5.1 Distributed consensus basics

It is useful to describe the sensor network as a *graph* [21] $\mathcal{G} = \{V, E\}$, where $V = \{1, \ldots, N\}$ is the set of N nodes and E is the set of edges, that is, the set of pairs (i,j) for which nodes i and j can communicate directly. The set of neighbours of node i is denoted as $\mathcal{N}_i = \{j \in V \,:\, (i,j) \in E\}$. Communication is assumed bidirectional, so that if node i can send data to node j, then j can also send data to i; the graph $\mathcal{G}$ is then said to be *undirected.* The *degree* of a node, $d_i = |\mathcal{N}_i|$, is the number of its neighbours. A *path* is a sequence of nodes such that there is an edge from any one of them to the next. It is assumed that the network is *connected,* that is, there exists a path between any pair of nodes. The *distance* between two nodes is the number of edges in a shortest path connecting them, and the *diameter* of the graph is the largest distance between any two of its nodes. The Laplacian matrix of $\mathcal{G}$ is the $N \times N$ symmetric matrix $\mathbf{L}$ with entries given by:

$$L_{ij} = \begin{cases} -1, & \text{if } (i,j) \in E, \\ d_i, & \text{if } j = i, \\ 0, & \text{otherwise,} \end{cases} \tag{9.41}$$

which is positive semidefinite. If $\mathcal{G}$ is connected, $\mathbf{L}$ has a simple eigenvalue zero, with associated eigenvector $\mathbf{1}$.

Distributed maximum- and minimum-consensus algorithms are straightforward and intuitive [22,23]. In the problem at hand, the goal is for each node in the network to obtain the values:

$$\underline{\ell} \triangleq \min_{\substack{1\le i\le N \\ 1\le m\le M}} \{y_{i,m}\}, \qquad \overline{\ell} \triangleq \max_{\substack{1\le i\le N \\ 1\le m\le M}} \{y_{i,m}\}. \tag{9.42}$$

To this end, node i initialises local variables $\underline{\ell}_{i,0} = \min_m\{y_{i,m}\}$ and $\overline{\ell}_{i,0} = \max_m\{y_{i,m}\}$. Then, for $i = 1, \ldots, N$ and $t \ge 0$, node i broadcasts its current values $\{\underline{\ell}_{i,t}, \overline{\ell}_{i,t}\}$, receives the corresponding values from its neighbours, and aggregates this information to update its local variables as follows:

$$\underline{\ell}_{i,t+1} = \min_{j\in\{i\}\cup\mathcal{N}_i} \left\{\underline{\ell}_{j,t}\right\}, \qquad \overline{\ell}_{i,t+1} = \max_{j\in\{i\}\cup\mathcal{N}_i} \left\{\overline{\ell}_{j,t}\right\}. \tag{9.43}$$

Clearly, (9.43) only require communication between neighbouring nodes. Since the network is connected, it is easy to see that, for all i, the corresponding local variables $\underline{\ell}_{i,t}, \overline{\ell}_{i,t}$, respectively, converge to $\underline{\ell}$ and $\overline{\ell}$ in a number of iterations equal to the diameter of the graph, at most, so that the network reaches consensus [23].

The idea behind distributed *average*-consensus is similar in spirit to (9.43). To illustrate it, suppose that each node has a scalar observation y_i, and that it is desired that all nodes obtain the average value $\bar{y} = (1/N)\sum_{i=1}^{N} y_i$ through local communications only. A possible approach is to simply replace the min or max operations in (9.43) by some sort of (weighted) averaging operation. Thus, the local variables are initialised as $x_{0,i} = y_i$, and for $t \ge 0$ nodes exchange the values $x_{t,i}$ with their neighbours and then update them as $x_{t+1,i} = W_{ii}x_{t,i} + \sum_{j\in\mathcal{N}_i} W_{ij}x_{t,j}$, where $W_{ij} \ge 0$ are suitable weights such that $W_{ii} + \sum_{j \in \mathcal{N}_i} W_{ij} = 1$. Stacking the local variables at step t in the $N \times 1$ vector $x_t = [\, x_{t,1} \ \cdots \ x_{t,N} \,]^T$, this linear iteration can be written as:

$$\boldsymbol{x}_{t+1} = \mathbf{W}\boldsymbol{x}_t, \qquad \boldsymbol{x}_0 = [\, y_1 \ \cdots \ y_N \,]^T, \tag{9.44}$$

where the weight matrix $\mathbf{W} \in \mathbb{R}^{N \times N}$ has elements W_{ij}, and we set $W_{ij} = 0$ if $j \notin \{i\} \cup \mathcal{N}_i$. This last condition reflects that only communication between neighbours is allowed.

The convergence properties of (9.44) have been established under a broad range of conditions [18,24]. One would like to have $\lim_{t \to \infty} \boldsymbol{x}_t = \bar{y}\mathbf{1}$, meaning that the local variables $x_{i,t}$ of all nodes asymptotically become equal to the desired average value $\bar{y}$. In this regard, the following result from [24, Th. 1] is relevant. First, let us define the averaging matrix $\mathbf{J}$ as:

$$\mathbf{J} \triangleq \frac{1}{N}\mathbf{1}\mathbf{1}^T. \tag{9.45}$$

Note that $\mathbf{J}\boldsymbol{x} = ((1/N)\mathbf{1}^T\boldsymbol{x})\mathbf{1}$ is a vector with all entries equal to the average of the entries of $\boldsymbol{x}$.

Lemma 9.1. *Consider the linear iteration* (9.44). *Then* $\lim_{t \to \infty} \boldsymbol{x}_t = \bar{y}\mathbf{1}$ *if and only if the matrix* $\mathbf{W}$ *satisfies:*

$$\mathbf{1}^T\mathbf{W} = \mathbf{1}^T, \qquad \mathbf{W}\mathbf{1} = \mathbf{1}, \qquad \rho(\mathbf{W} - \mathbf{J}) < 1, \tag{9.46}$$

where $\rho(\mathbf{A})$ *is the spectral radius of* $\mathbf{A}$. *Moreover, the convergence rate is dictated by* $\rho(\mathbf{W} - \mathbf{J})$, *in the sense that the distance from* $\boldsymbol{x}_t$ *to the limiting point* $\bar{y}\mathbf{1}$ *obeys:*

$$\|\boldsymbol{x}_t - \bar{y}\mathbf{1}\| \approx [\rho(\mathbf{W} - \mathbf{J})]^t \, \|\boldsymbol{x}_0 - \bar{y}\mathbf{1}\| \qquad \text{as } t \to \infty. \tag{9.47}$$

The first two conditions in (9.46) say that $\lambda = 1$ must be an eigenvalue of both $\mathbf{W}$ and $\mathbf{W}^T$, with corresponding eigenvector $\mathbf{1}$ in both cases; equivalently, the sum of all elements in any row or any column of $\mathbf{W}$ must equal 1. A matrix with nonnegative elements meeting these conditions is said to be *doubly stochastic.* Two popular designs [17,24] for a doubly stochastic matrix with the desired sparsity pattern $W_{ij} = 0$ if $j \notin \{i\} \cup \mathcal{N}_i$, and satisfying $\rho(\mathbf{W} - \mathbf{J}) < 1$, are as follows:

1. *Constant weight assignment:*

$$\mathbf{W} = \mathbf{I} - \mu\mathbf{L}, \qquad \text{with } \mu \text{ such that} \quad 0 < \mu < \frac{2}{\lambda_1(\mathbf{L})}, \tag{9.48}$$

   where $\mathbf{L}$ is the Laplacian matrix. In terms of convergence speed, the optimal value of μ is the one minimising $\rho(\mathbf{W} - \mathbf{J})$, and it is given by $\mu = (2/(\lambda_1(\mathbf{L}) + \lambda_{N-1}(\mathbf{L})))$ [24].
2. *Metropolis design:*

$$W_{ij} = \begin{cases} \frac{1}{1+\max\{d_i, d_j\}}, & \text{if } j \in \mathcal{N}_i, \\ 1 - \sum_{k \in \mathcal{N}_i} W_{ik}, & \text{if } j = i, \\ 0, & \text{otherwise.} \end{cases} \tag{9.49}$$

Both (9.48) and (9.49) result in a symmetric weight matrix $\mathbf{W}$. The Metropolis design has the desirable property that, in order to determine its weights, each node only needs to know the degrees of its neighbours.

For a fully connected network where every node is connected to every other node, (9.49) results in $\mathbf{W} = \mathbf{J}$, and also (9.48) with $\mu = (1/N)$, so that $\rho(\mathbf{W} - \mathbf{J}) = 0$, yielding convergence in a *single* iteration: $\boldsymbol{x}_1 = \mathbf{W}\boldsymbol{x}_0 = \mathbf{J}\boldsymbol{x}_0 = \bar{y}\mathbf{1}$. In a more practical setting, the spectral radius $\rho(\mathbf{W} - \mathbf{J})$ becomes closer to 1 as the network becomes 'less connected'. This is true as well for other designs of $\mathbf{W}$ [24] and reflects the fact that it takes longer for the information to diffuse across the network as the network becomes sparser.

9.5.2 Distributed consensus-based EM algorithm

The application of the previous ideas to obtain a distributed version of the EM algorithm of Section 9.3 is simple. Let us first redefine the global variables in (9.25) and (9.26) slightly so that the corresponding sums become averages:

$$s_k = \frac{1}{N}\sum_{i=1}^{N} \hat{a}_i(\hat{\boldsymbol{\theta}}_k), \qquad c_k = \frac{1}{N}\sum_{i=1}^{N} \hat{a}_i(\hat{\boldsymbol{\theta}}_k)\|\mathbf{y}_i\|^2, \quad \underline{\ell}_k = \min_{i,m}\{y_{i,m}\}, \quad (9.50)$$

$$\mathbf{v}_k = \frac{1}{N}\sum_{i=1}^{N} \hat{a}_i(\hat{\boldsymbol{\theta}}_k)\mathbf{H}_i^T\mathbf{y}_i, \quad \mathbf{R}_k = \frac{1}{N}\sum_{i=1}^{N} \hat{a}_i(\hat{\boldsymbol{\theta}}_k)\mathbf{H}_i^T\mathbf{H}_i, \quad \overline{\ell}_k = \max_{i,m}\{y_{i,m}\}. \quad (9.51)$$

Then, at each EM step k, a total of T min-, max-, and average-consensus iterations is run, so that the nodes can obtain the aggregated values in (9.50) and (9.51). With these, the M-step as described in (9.27) and (9.28) remains valid, except for the estimate of the *a priori* probability of sensor malfunctioning q, which now becomes simply $\hat{q}_{k+1} = 1 - s_k$. This procedure incorporates a consensus *inner loop* of T iterations inside the EM *outer loop*.

An important difference between the min- or max-consensus schemes (9.43) and the average consensus iteration (9.44) is that, whereas (9.43) converge in a finite number of steps, in general the convergence of (9.44) is only asymptotic. Thus, if the number of inner iterations T is at least as large as the diameter of the network, at the end of this inner loop, all the nodes will have obtained the values $\min_{i,m}\{y_{i,m}\}$ and $\max_{i,m}\{y_{i,m}\}$. In fact, since these values do not change with the outer iteration index k, this process needs to be run just once. In contrast, after T inner iterations, the average-consensus scheme will only provide an approximation to the corresponding average value. This approximation becomes more accurate as T is increased at the cost of a larger number of inter-neighbour communication exchanges. Hence, there is a trade-off between communication cost and accuracy. In the next section, a related approach is presented, which allows to perform a single inner iteration by appropriately modifying the average-consensus mechanism.

9.6 Distributed implementation based on consensus filters

To avoid the nested structure for the distributed implementation of the EM algorithm in Section 9.5, we note that in principle it should not be necessary to enforce consensus on the nodes about the averages in (9.50) and (9.51) *at each iteration k,* as long as they do reach consensus *asymptotically.* Based on this observation, the basic average-consensus building block in (9.44) is replaced by a different mechanism, as described next.

Let $\boldsymbol{u}_k$, $k \geq 1$ be a convergent sequence of $N \times 1$ vectors, with $\lim_{k\to\infty} \boldsymbol{u}_k = \boldsymbol{u}$, and let $\bar{u} \triangleq (1/N)\mathbf{1}^T\boldsymbol{u}$ be the average of the elements of $\boldsymbol{u}$. Assume that $\boldsymbol{u}_k$ is distributed throughout the network, so that its ith entry is available at node i only. Our goal is to construct another sequence $\boldsymbol{x}_k = f(\boldsymbol{u}_k, \boldsymbol{u}_{k-1}, \ldots)$ in a distributed fashion, i.e., node i computes the ith entry of $\boldsymbol{x}_k$, and such that:

$$\lim_{k\to\infty} \boldsymbol{x}_k = \mathbf{J}\boldsymbol{u} = \bar{u}\mathbf{1}. \tag{9.52}$$

The system $\boldsymbol{x}_k = f(\boldsymbol{u}_k, \boldsymbol{u}_{k-1}, \ldots)$ having $\boldsymbol{u}_k$ as input and $\boldsymbol{x}_k$ as output is termed as *consensus filter.*

In the context of distributed EM implementation, consider for example the computation of the average of the *a posteriori* probabilities $\hat{a}_i(\hat{\boldsymbol{\theta}}_k)$, i.e., s_k in (9.50). This can be attempted by means of a consensus filter in which the ith entry of the input vector $\boldsymbol{u}_k$ is set to $u_{k,i} = \hat{a}_i(\hat{\boldsymbol{\theta}}_k)$, a variable local to node i. Node i also maintains a local variable $x_{k,i}$, the ith entry of the output vector $\boldsymbol{x}_k$. The consensus filter should be designed so that each of the entries of $\boldsymbol{x}_k$ *tracks* the average of the entries of $\boldsymbol{u}_k$. With such design, should the sequences $\hat{a}_i(\hat{\boldsymbol{\theta}}_k)$ converge for all $i = 1, \ldots, N$, it is clear that (9.52) will hold. The same scheme can be applied for the computation of each of the remaining averages in (9.50) and (9.51).

In this section, we discuss different approaches to the problem of constructing the sequence $\boldsymbol{x}_k$. Note that (9.52) trivially holds by just setting $\boldsymbol{x}_k = \mathbf{J}\boldsymbol{u}_k$; however, this would require each node to communicate its entry of $\boldsymbol{u}_k$ to every other node in the network. Thus, we add the additional constraint that only communication between neighbouring nodes is to be allowed.

9.6.1 Linear time-invariant consensus filter

A possible way to approach this problem is by linearly filtering the sequence $\boldsymbol{u}_k$:

$$\boldsymbol{x}_k = \mathbf{A}\boldsymbol{x}_{k-1} + \mathbf{B}\boldsymbol{u}_k, \qquad k \geq 1, \tag{9.53}$$

with given initial condition $\boldsymbol{x}_0$, and where the entries of the matrices $\mathbf{A}$, $\mathbf{B}$ satisfy $A_{ij} = B_{ij} = 0$ if $j \notin \{i\} \cup \mathcal{N}_i$. With this sparsity pattern, only local communications are required throughout, as desired.

Note that in order to achieve (9.52), the following two conditions must hold:

1. The average of the entries of $\boldsymbol{x}_k$, say $\bar{x}_k \triangleq (1/N)\mathbf{1}^T\boldsymbol{x}_k$, must converge to $\bar{u}$.
2. Consensus must be asymptotically reached, so that all of the entries of $\boldsymbol{x}_k$ become equal as $k \to \infty$.

Thus, $\mathbf{A}, \mathbf{B}$ must be selected to meet these conditions. The design of these matrices is discussed next, under the assumption that the input sequence $\boldsymbol{u}_k$ is convergent with limit $\boldsymbol{u}$.

We assume that $\mathbf{A}, \mathbf{B}$ are symmetric with the same eigenvectors; specifically, let:

$$\mathbf{A} = \gamma_1 \mathbf{J} + \mathbf{U}\boldsymbol{\Gamma}\mathbf{U}^T, \qquad \mathbf{B} = \delta_1 \mathbf{J} + \mathbf{U}\boldsymbol{\Delta}\mathbf{U}^T, \tag{9.54}$$

where $\mathbf{U}$ is an $N \times (N-1)$ matrix with orthonormal columns such that $\mathbf{U}^T\mathbf{1} = \mathbf{0}$, and $\boldsymbol{\Gamma}, \boldsymbol{\Delta}$ are $(N-1) \times (N-1)$ diagonal:

$$\boldsymbol{\Gamma} = \mathrm{diag}\{\gamma_2\ \gamma_3\ \cdots\ \gamma_N\}, \qquad \boldsymbol{\Delta} = \mathrm{diag}\{\delta_2\ \delta_3\ \cdots\ \delta_N\}. \tag{9.55}$$

In this way, the common eigenvectors of $\mathbf{A}$ and $\mathbf{B}$ are given by the columns of the orthonormal matrix $[(1/\sqrt{N})\mathbf{1}\ \ \mathbf{U}]$, whereas their eigenvalues are $\{\gamma_1, \ldots, \gamma_N\}$ and $\{\delta_1, \ldots, \delta_N\}$, respectively.

Regarding condition (1), note from (9.54) that $\mathbf{A}, \mathbf{B}$ satisfy:

$$\mathbf{1}^T\mathbf{A} = \gamma_1 \mathbf{1}^T, \qquad \mathbf{1}^T\mathbf{B} = \delta_1 \mathbf{1}^T. \tag{9.56}$$

Multiplying both sides of (9.53) from the left by $(1/N)\mathbf{1}^T$ and letting $\bar{u}_k \triangleq (1/N)\mathbf{1}^T\boldsymbol{u}_k$, yields:

$$\bar{x}_k = \gamma_1 \bar{x}_{k-1} + \delta_1 \bar{u}_k, \qquad k \geq 1. \tag{9.57}$$

The following result, which can be proved using [25, p. 45, Lemma 3], is now useful.

Lemma 9.2. *Let β_k be a convergent sequence with $\lim_{k\to\infty} \beta_k = \beta$, and let*

$$\alpha_k = a\alpha_{k-1} + b\beta_k, \qquad k \geq 1, \qquad \textit{with}\ \ |\alpha_0| < \infty, \quad |a| < 1. \tag{9.58}$$

Then α_k converges, and $\lim_{k\to\infty} \alpha_k = (b/(1-a))\beta$.

Since $\lim_{k\to\infty} \bar{u}_k = \bar{u} = (1/N)\mathbf{1}^T\boldsymbol{u}$, if $|\gamma_1| < 1$ then we can apply Lemma 9.2 to (9.57), and it follows that the average $\bar{x}_k$ converges. To ensure that the convergence is to $\bar{u}$ as desired, we must set $\delta_1 = 1 - \gamma_1$.

Let us focus now on condition (2), which can be checked through the deviation of the entries of $\boldsymbol{x}_k$ from the average of their values. That is, we would like the vector $\boldsymbol{x}_k - \mathbf{J}\boldsymbol{x}_k$ to converge to zero. Since $[(1/\sqrt{N})\mathbf{1}\ \ \mathbf{U}]$ is orthonormal, and since $\mathbf{1}^T\mathbf{J} = \mathbf{1}^T$, $\mathbf{U}^T\mathbf{J} = \mathbf{0}$, the squared norm of the deviation vector $\boldsymbol{x}_k - \mathbf{J}\boldsymbol{x}_k$ satisfies:

$$\begin{aligned} \|\boldsymbol{x}_k - \mathbf{J}\boldsymbol{x}_k\|^2 &= \left(\frac{1}{\sqrt{N}}\mathbf{1}^T(\boldsymbol{x}_k - \mathbf{J}\boldsymbol{x}_k)\right)^2 + \|\mathbf{U}^T(\boldsymbol{x}_k - \mathbf{J}\boldsymbol{x}_k)\|^2 \\ &= \|\mathbf{U}^T\boldsymbol{x}_k\|^2. \end{aligned} \tag{9.59}$$

Thus, let us define the $(N-1) \times 1$ vectors $\tilde{\boldsymbol{x}}_k \triangleq \mathbf{U}^T\boldsymbol{x}_k$ and $\tilde{\boldsymbol{u}}_k \triangleq \mathbf{U}^T\boldsymbol{u}_k$. Multiplying now both sides of (9.53) from the left by $\mathbf{U}^T$ yields:

$$\tilde{\boldsymbol{x}}_k = \boldsymbol{\Gamma}\tilde{\boldsymbol{x}}_{k-1} + \boldsymbol{\Delta}\tilde{\boldsymbol{u}}_k \ \Rightarrow\ \tilde{x}_{k,i} = \gamma_{i+1}\tilde{x}_{k-1,i} + \delta_{i+1}\tilde{u}_{k,i}, \quad i = 1, \ldots, N-1. \tag{9.60}$$

Since $\lim_{k\to\infty} \tilde{\boldsymbol{u}}_k = \tilde{\boldsymbol{u}} = \mathbf{U}^T\boldsymbol{u}$, we can apply Lemma 9.2 to these $N-1$ scalar recursions and conclude that they are convergent if $|\gamma_i| < 1$, $i = 2, \ldots, N$, with limiting points:

$$\lim_{k\to\infty} \tilde{x}_{i,k} = \frac{\delta_{i+1}}{1-\gamma_{i+1}}\tilde{u}_i, \qquad i = 1, \ldots, N-1. \tag{9.61}$$

Therefore, (9.59) satisfies:

$$\lim_{k\to\infty} \|\boldsymbol{x}_k - \mathbf{J}\boldsymbol{x}_k\|^2 = \sum_{i=1}^{N-1} \left(\frac{\delta_{i+1}}{1-\gamma_{i+1}}\right)^2 \tilde{u}_i^2, \tag{9.62}$$

showing that the deviation vector $\boldsymbol{x}_k - \mathbf{J}\boldsymbol{x}_k$ does not necessarily converge to zero, since $\tilde{\boldsymbol{u}} \neq \mathbf{0}$ in general. Note that we cannot set $\delta_2 = \cdots = \delta_N = 0$ because it would yield $\mathbf{B} = \delta_1\mathbf{J}$, which requires communication among non-neighbouring nodes. Therefore, condition (2) above can only be approximately satisfied.

In summary, the consensus filter matrices $\mathbf{A}$, $\mathbf{B}$ as in (9.54) should be such that:

1. They have the required sparsity pattern $A_{ij} = B_{ij} = 0$ for $j \notin \{i\} \cup \mathcal{N}_i$.
2. $\rho(\mathbf{A}) < 1$, to ensure asymptotic convergence.
3. $\gamma_1 + \delta_1 = 1$, for convergence of the average to the correct value.
4. $|\delta_i|/|1-\gamma_i|$ are small, $i = 2, \ldots, N$, for the deviation from consensus (9.62) to be small.

The design of matrices $\mathbf{A}$, $\mathbf{B}$ under these premises can be approached using as building block a symmetric weight matrix $\mathbf{W}$, e.g. the constant weight assignment (9.48), the Metropolis matrix (9.49), or any other satisfying the assumptions in Lemma 9.1 [9,11,26]. Four possible choices are given in Table 9.1, all of them parameterised by a constant η, and satisfying $\gamma_1 + \delta_1 = 1$. For designs 1 and 2, the condition $\rho(\mathbf{A}) < 1$ is satisfied if $0 < \eta < 1 + \lambda_N(\mathbf{W})$, whereas for designs 3 and 4, $\rho(\mathbf{A}) < 1$ holds if $0 < \eta < 2$.

For each of the four designs, the fourth column in Table 9.1 indicates the variables that node i must transmit at time k to its neighbours $j \in \mathcal{N}_i$. It is clear that in all cases, the consensus filter can be implemented using local communications only.

Table 9.1 Some consensus filter matrix designs

Design #	A	B	Data sent by node i to its neighbours	$\lvert(\delta_i/(1-\gamma_i))\rvert$, $i = 2, \ldots, N$
1	$\mathbf{W} - \eta\mathbf{I}$	$\eta\mathbf{I}$	$x_{i,k-1}$	$\eta/(\lvert 1+\eta-\lambda_i(\mathbf{W})\rvert)$
2	$\mathbf{W} - \eta\mathbf{I}$	$\eta\mathbf{W}$	$x_{i,k-1} + \eta u_{i,k}$	$(\eta\lvert\lambda_i(\mathbf{W})\rvert)/(\lvert 1+\eta-\lambda_i(\mathbf{W})\rvert)$
3	$(1-\eta)\mathbf{W}$	$\eta\mathbf{I}$	$x_{i,k-1}$	$\eta/(\lvert 1-(1-\eta)\lambda_i(\mathbf{W})\rvert)$
4	$(1-\eta)\mathbf{W}$	$\eta\mathbf{W}$	$(1-\eta)x_{i,k-1} + \eta u_{i,k}$	$(\eta\lvert\lambda_i(\mathbf{W})\rvert)/(\lvert 1-(1-\eta)\lambda_i(\mathbf{W})\rvert)$

Note that all the terms in the last column of Table 9.1 increase with η, so that η should be chosen small. However, as η is decreased, the matrix $\mathbf{A}$ approaches $\mathbf{W}$ for all four designs, and therefore $\rho(\mathbf{A})$ approaches $\rho(\mathbf{W}) = 1$, resulting in slower convergence. Thus, for a given weight matrix $\mathbf{W}$, a trade-off between convergence speed and deviation from consensus can be achieved by selecting η.

9.6.2 Linear time-varying consensus filter

The linear time-invariant consensus filter (9.53) cannot reach asymptotic consensus, as (9.62) shows. A possible means to overcome this drawback is to use instead a linear *time-varying* filter:

$$\boldsymbol{x}_k = \mathbf{A}_k \boldsymbol{x}_{k-1} + \mathbf{B}_k \boldsymbol{u}_k, \qquad k \geq 1, \tag{9.63}$$

in which the matrices $\mathbf{A}_k$, $\mathbf{B}_k$ still have the required sparsity pattern, but now they change with k. For each of the designs in Table 9.1, this can be readily done by replacing the constant $\eta > 0$ with a sequence of values $\eta_k \geq 0$. The matrices $\mathbf{A}_k$, $\mathbf{B}_k$ obtained in this way satisfy:

$$\mathbf{1}^T \mathbf{A}_k = \gamma_{1,k} \mathbf{1}^T, \quad \gamma_{1,k} = (1 - \eta_k), \qquad \mathbf{1}^T \mathbf{B}_k = \delta_{1,k} \mathbf{1}^T, \quad \delta_{1,k} = \eta_k. \tag{9.64}$$

In fact, these matrices all have the same set of eigenvectors, given by those of $\mathbf{W}$, which do not change with time. As before, we collect such eigenvectors in the orthonormal matrix $[(1/\sqrt{N})\mathbf{1} \;\; \mathbf{U}]$. The corresponding eigen decompositions are therefore:

$$\mathbf{A}_k = \gamma_{1,k} \mathbf{J} + \mathbf{U} \boldsymbol{\Gamma}_k \mathbf{U}^T, \qquad \mathbf{B}_k = \delta_{1,k} \mathbf{J} + \mathbf{U} \boldsymbol{\Delta}_k \mathbf{U}^T, \tag{9.65}$$

where the eigenvalues $\gamma_{i,k}$, $\delta_{i,k}$ depend on the particular design in Table 9.1.

Note that the squared norm of the deviation vector still satisfies (9.59), i.e. $\|\boldsymbol{x}_k - \mathbf{J}\boldsymbol{x}_k\|^2 = \|\mathbf{U}^T \boldsymbol{x}_k\|^2$. Following analogous steps to those in the time-invariant case leading to (9.57) and (9.60), the evolution of the average $\bar{x}_k = (1/N)\mathbf{1}^T \boldsymbol{x}_k$ and of $\tilde{\boldsymbol{x}}_k = \mathbf{U}^T \boldsymbol{x}_k$ is now seen to obey the time-varying recursions:

$$\bar{x}_k = (1 - \eta_k)\bar{x}_{k-1} + \eta_k \bar{u}_k, \qquad \tilde{\boldsymbol{x}}_k = \boldsymbol{\Gamma}_k \tilde{\boldsymbol{x}}_{k-1} + \boldsymbol{\Delta}_k \tilde{\boldsymbol{u}}_k, \qquad k \geq 1. \tag{9.66}$$

Therefore, assuming that $\bar{u}_k$ and $\tilde{\boldsymbol{u}}_k$ converge to some values $\bar{u}$ and $\tilde{\boldsymbol{u}}$, respectively, the goal is to design a sequence $\eta_k \geq 0$ such that $\bar{x}_k \to \bar{u}$ and $\tilde{\boldsymbol{x}}_k \to \mathbf{0}$ in (9.66). It can be anticipated that if the sequence η_k decays to zero asymptotically, then $\mathbf{A}_k \to \mathbf{W}$ and $\mathbf{B}_k \to \mathbf{0}$, so that for large k, the time-varying recursion (9.63) becomes $\boldsymbol{x}_k \approx \mathbf{W}\boldsymbol{x}_{k-1}$, i.e., a standard consensus iteration, which has the potential to drive the deviation from consensus to zero.

Consider for example the sequence $\eta_k = (1/k)$, $k \geq 1$. In that case, it is readily checked that $\bar{x}_k$ in (9.66) reduces to the average of the values of $\bar{u}_n$ up to time k:

$$\bar{x}_k = \frac{1}{k} \sum_{n=1}^{k} \bar{u}_n, \tag{9.67}$$

which are known as the *Cesàro means* of the sequence $\bar{u}_k$ [27]. It is well known that, if $\bar{u}_k$ converges, then the Cesàro means also converge to the same limit [27], so $\bar{x}_k \to \bar{u}$

as desired. Now, in order to analyse the convergence of $\tilde{\boldsymbol{x}}_k$ in (9.66), the following result is useful.

Lemma 9.3. *Let β_k be a bounded sequence, with $|\beta_k| \leq c$, $k \geq 1$, and let:*

$$\alpha_k = a\left(1 - \frac{1}{k}\right)\alpha_{k-1} + \frac{1}{k}\beta_k, \quad k \geq 1, \quad \textit{with } |\alpha_0| < \infty, \quad |a| < 1. \tag{9.68}$$

Then $\lim_{k\to\infty} \alpha_k = 0$.

Proof. From (9.68), one has $\alpha_k = (1/k)\sum_{n=1}^{k} a^{k-n}\beta_n$, $k \geq 1$. This can be bounded as:

$$|\alpha_k| \leq \frac{1}{k}\sum_{n=1}^{k} |a|^{k-n}c = \frac{c}{k}\frac{1-|a|^k}{1-|a|}, \tag{9.69}$$

and the right-hand side of (9.69) goes to zero as $k \to \infty$. □

Assume now that design 3 is adopted and consider the ith entry of $\tilde{\boldsymbol{x}}_k$ in (9.66). Then, for $\eta_k = (1/k)$, $k \geq 1$, one has:

$$\tilde{x}_{i,k} = \left(1 - \frac{1}{k}\right)\lambda_{i+1}(\mathbf{W})\tilde{x}_{i,k-1} + \frac{1}{k}\tilde{u}_{i,k}, \quad k \geq 1, \quad i = 1, \ldots, N-1. \tag{9.70}$$

If $\boldsymbol{u}_k$ is convergent, so is $\tilde{\boldsymbol{u}}_k$, and thus $\tilde{u}_{i,k}$ is bounded. Since $|\lambda_{i+1}(\mathbf{W})| < 1$ for $i = 1, \ldots, N-1$, Lemma 9.3 can be applied to (9.70) to conclude that $\tilde{x}_{i,k} \to 0$, and therefore $\|\boldsymbol{x}_k - \mathbf{J}\boldsymbol{x}_k\|^2 \to 0$ as desired. This can be proved along similar lines for the remaining designs in Table 9.1 with $\eta_k = (1/k)$.

It should be mentioned that other designs of the time-varying consensus filter are possible. For example, [11] considered a more general choice of sequence $\eta_k = (b/(k+b-1))$, where $b > 0$. Alternatively, the matrix sequences can be chosen so that $\mathbf{A}_k$ converges to $\mathbf{W}$ at a slower rate than $\mathbf{B}_k$ converges to zero [9,10,28].

To close this section, an important remark regarding the application to the distributed EM algorithm is in order. When designing the consensus filter, for both time-invariant and time-varying cases, our goal has been to asymptotically drive the filter output $\boldsymbol{x}_k$ as close as possible to a consensus, given by the average of the entries of the limiting value of the filter input sequence $\boldsymbol{u}_k$; this sequence was explicitly assumed to be convergent. When the consensus filter is used in the distributed implementation of the EM algorithm, however, the filter input $\boldsymbol{u}_k$ is computed based on the previous filter output $\boldsymbol{x}_{k-1}$. In other words, the consensus filter is operating in *closed loop,* so that convergence of $\boldsymbol{u}_k$ cannot be taken for granted. Observe that, in order to track the averages in (9.50) and (9.51), and since $\mathbf{R}_k$ is symmetric, a total of $(L^2/2) + (L/2) + 2$ consensus filters are required, whose inputs and outputs are coupled by an involved nonlinear function as specified in the E-step. This nonlinear feedback makes the analysis of the closed-loop system difficult; some results in this direction can be found in [11].

9.7 Numerical examples

To illustrate the behaviour of the distributed algorithms, consider a network with $N = 100$ nodes. As in Example 9.1, nodes collect observations of dimension $M = 2$, the parameter $\boldsymbol{\phi}$ to estimate has dimension $L = 3$, and anomalous data is uniformly distributed in $[-10, 10]$. The entries of $\boldsymbol{\phi}$ and $\mathbf{H}_i$ are randomly and independently generated following a $\mathcal{N}(0, 1)$ distribution, and then their values are fixed through the simulations. The SNR is defined as in (9.24).

The behaviour of the cyclic distributed implementation of Section 9.4.2 in this setting with an SNR of 5 dB is illustrated in Figures 9.2 and 9.3. Two cases are shown, corresponding to sensors of medium ($q = 0.1$) and low reliability ($q = 0.4$). Figure 9.2 shows the evolution of the *a posteriori* probabilities $\hat{a}_i(\hat{\boldsymbol{\theta}}_k)$ for a single run in each case, for a total of six cycles. Recall that this distributed scheme requires the establishment of a cyclic path through all nodes of the network, so that in this example, it takes 100 iterations to complete one cycle. For the same runs, Figure 9.3 shows the evolution of the parameter estimate $\hat{\boldsymbol{\phi}}_k = \mathbf{R}_k^{-1}\mathbf{v}_k$ with the iterations. In the two experiments shown, the cyclic distributed scheme correctly converges to the estimate produced by the centralised version of the EM algorithm operating on the same data (dashed lines in the figure). For $q = 0.1$, convergence takes about 2–3 cycles (200–300 iterations), whereas for $q = 0.4$, convergence is somewhat slower, requiring about five cycles. Thus, it is seen that the quality of the sensors has a direct impact on convergence speed.

In the following experiments, we consider the consensus- and consensus filter based distributed approaches of Sections 9.5 and 9.6, respectively, which do not require a cyclic path in the network. The topology of the sensor network, shown in

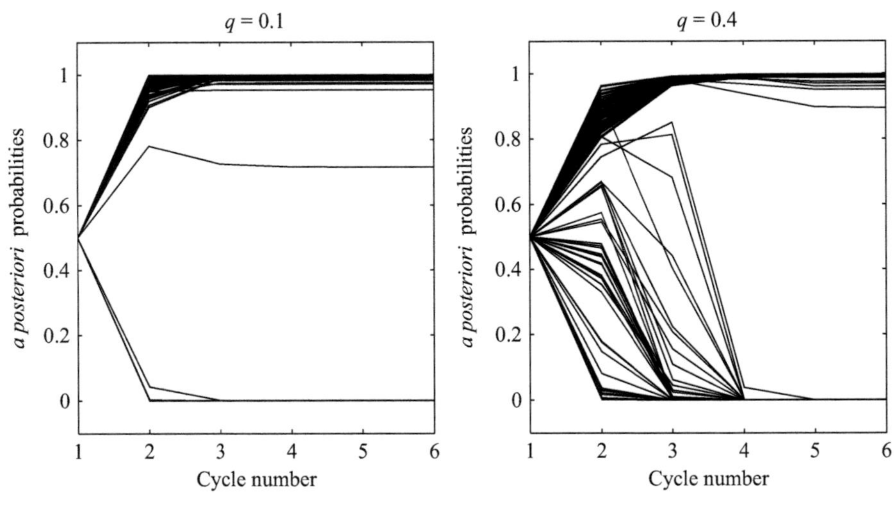

Figure 9.2 Evolution of the a posteriori *probabilities* $\hat{a}_i(\hat{\boldsymbol{\theta}}_k)$ *of the distributed cyclic implementation of the EM algorithm*

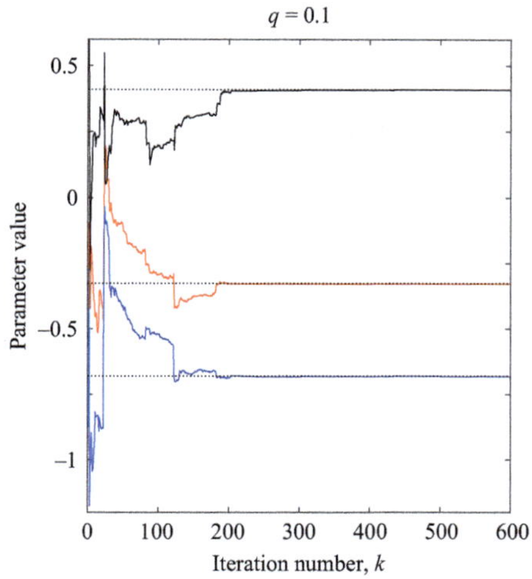

Figure 9.3 Evolution of the parameter estimate of the distributed cyclic implementation of the EM algorithm. Dashed lines correspond to the convergent point of the centralised implementation

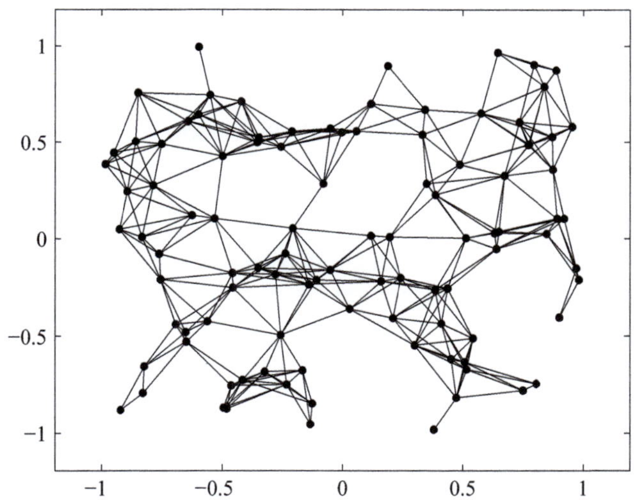

Figure 9.4 Graph of the sensor network used in the simulations

Figure 9.4 is generated by randomly choosing the locations of the $N = 100$ nodes in the $[-1, 1] \times [-1, 1]$ square, in such a way that any two nodes are able to communicate if their distance is less than 0.33. The resulting graph is connected, with minimum, maximum, and average node degrees of 1, 11, and 7.3, respectively, and a diameter of 10 edges.

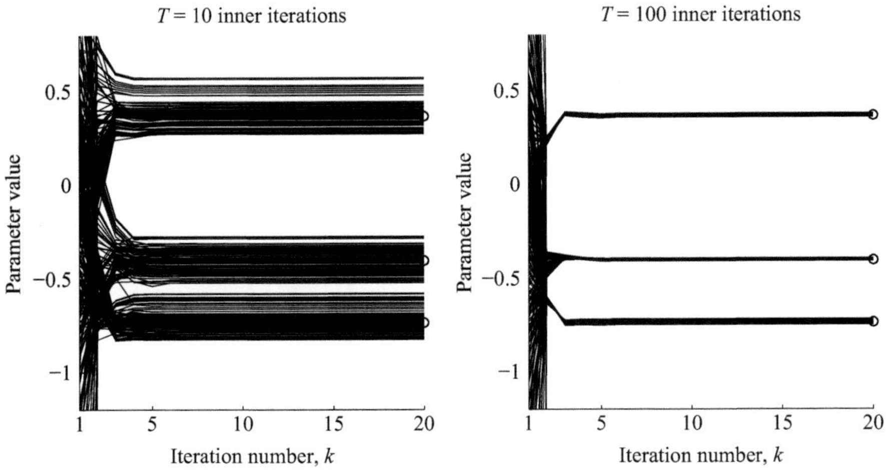

Figure 9.5 Evolution of the parameter estimates at all nodes for the distributed consensus-based implementation of the EM algorithm. Circles correspond to the convergent point of the centralised implementation

In this scenario, the consensus-based scheme of Section 9.5 was run, setting the SNR to 5 dB and $q = 0.1$. Local parameters were randomly initialised at each node, and a Metropolis weight matrix $\mathbf{W}$ was adopted as per (9.49). Figure 9.5 shows the evolution of the parameter estimate $\hat{\boldsymbol{\phi}}_k$ at all nodes in terms of the EM outer iterations, indexed by k, and for two different choices of the number of consensus inner iterations T performed for each outer iteration. With a small number of inner iterations ($T = 10$), the average-consensus mechanism is unable to reach its asymptotic regime, and it only provides a rough approximation to the desired average value. The result is a significant dispersion of the estimates across the different nodes about the convergent point of the centralised scheme. This dispersion can be reduced by increasing the number of inner iterations, as can be seen in Figure 9.5 for $T = 100$. Therefore, the choice of T should strike the correct balance between complexity, in terms of computation and communication costs, and performance, in terms of agreement on the estimate among the nodes.

The distributed schemes form Section 9.6 based on consensus filters are considered next, in the same network of Figure 9.4 with 5 dB SNR and $q = 0.1$. When linear time-invariant consensus filters as in Table 9.1 are employed with a Metropolis weight matrix $\mathbf{W}$, the behaviour is determined by the parameter η, as shown in Figure 9.6. As expected, smaller values of η result in smaller asymptotic deviation from consensus among the nodes, at the cost of slower convergence. Recall that, with linear time-invariant filters, the asymptotic deviation from consensus is given by (9.62), which depends on the asymptotic values of the input sequence, cf. $\tilde{u}_i$. If we assume that $\tilde{u}_i^2$ take similar values for $i = 1, \ldots, N-1$, then, for design $d \in \{1, 2, 3, 4\}$ in Table 9.1, (9.62) is approximately proportional to the sum, for $i = 2, \ldots, N$, of the squares of the terms in the last column of the dth row in that table. In the scenario

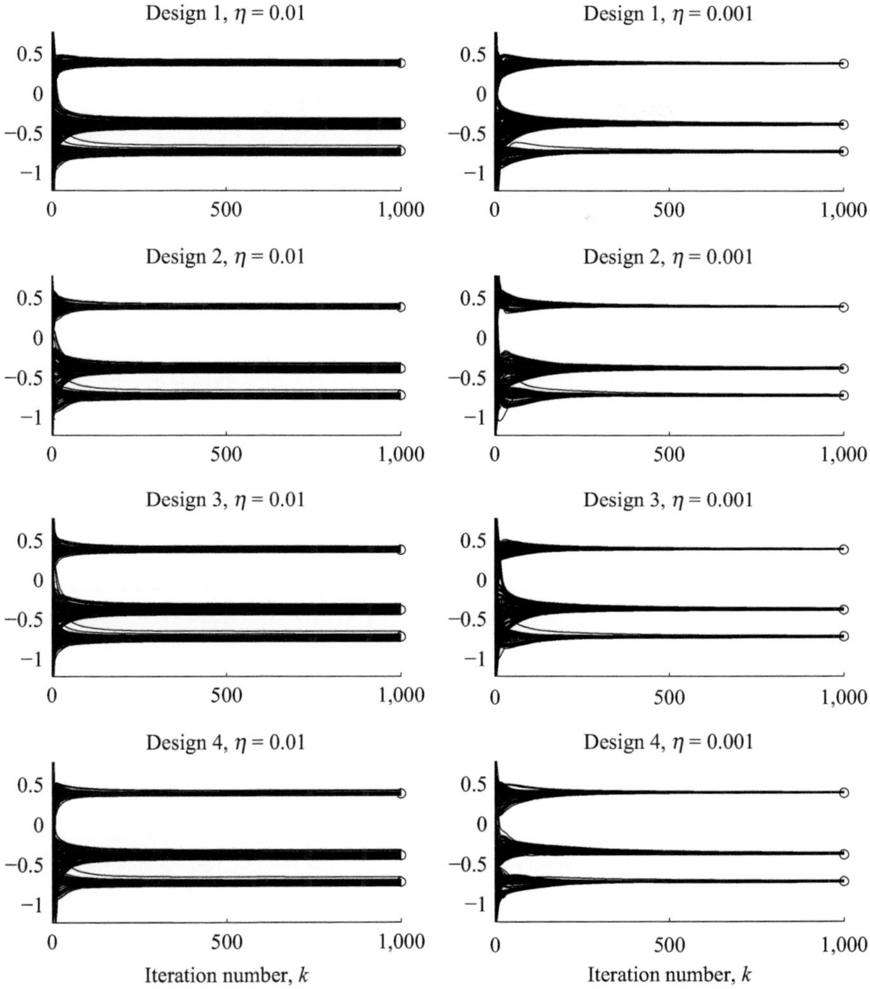

Figure 9.6 Evolution of the parameter estimates at all nodes for the distributed implementation of the EM algorithm based on linear time-invariant consensus filters. Circles correspond to the convergent point of the centralised implementation

of Figure 9.6, these factors are similar for all four designs: for $\eta = 0.01$, they are given by 0.3735, 0.3349, 0.3791, and 0.3398, respectively, for designs 1 through 4; and for $\eta = 0.001$, the corresponding values are 0.006833, 0.006301, 0.006846, and 0.006313. This suggests that the behaviour of the four designs for a given value of η should be similar; this is in good agreement with Figure 9.6. In Figure 9.7, the results obtained in the same setting, but employing linear time-varying consensus filters, are shown. The consensus filters were again taken from Table 9.1 with a Metropolis

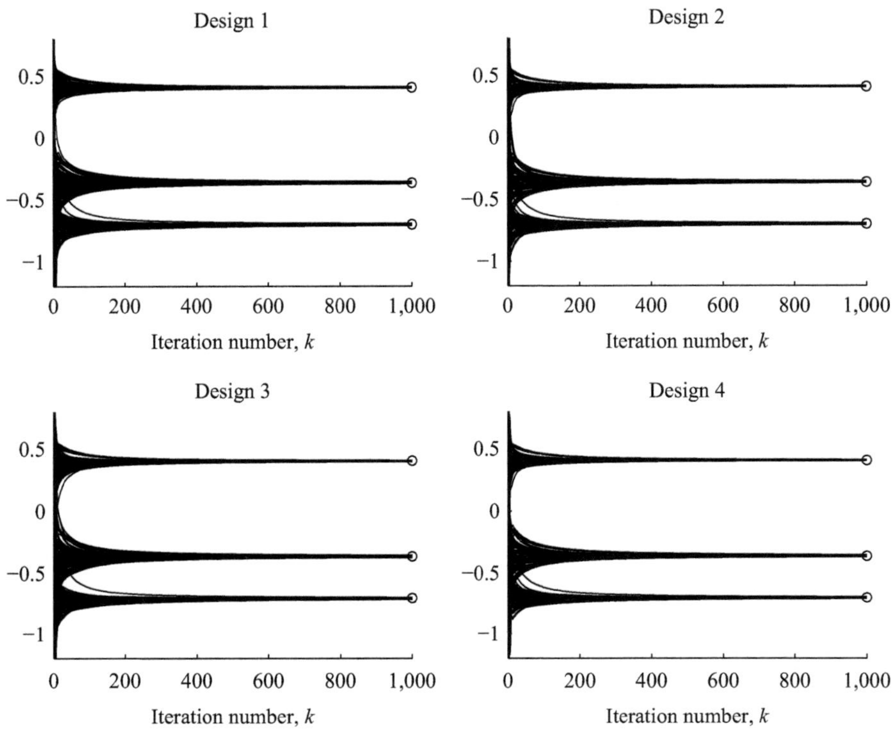

Figure 9.7 Evolution of the parameter estimates at all nodes for the distributed implementation of the EM algorithm based on linear time-varying consensus filters. Circles correspond to the convergent point of the centralised implementation

weight matrix $\mathbf{W}$, but with a decreasing constant $\eta_k = (1/k)$. In this case, the nodes asymptotically reach consensus, and the four designs behave similarly. To highlight the difference in asymptotic behaviour between time-invariant and time-varying consensus filters, Figure 9.8 shows the evolution of the total deviation from consensus in the parameter estimate, which is defined as:

$$\text{Deviation from consensus} = \sqrt{\sum_{j=1}^{L} \|\hat{\boldsymbol{\phi}}_{j,k} - \mathbf{J}\hat{\boldsymbol{\phi}}_{j,k}\|^2}, \tag{9.71}$$

where, for $j = 1, \ldots, L$, $\hat{\boldsymbol{\phi}}_{j,k} \in \mathbb{R}^N$ denotes the vector containing all local estimates at iteration k of ϕ_j, the jth component of $\boldsymbol{\phi}$. It is clear that for the linear time-invariant approaches, the deviation settles at some positive value, which can be reduced by decreasing η, whereas for the time-varying schemes with $\eta_k = (1/k)$, the deviation goes to zero at a rate $\sim 1/k$. No significant differences are perceived among the designs from Table 9.1.

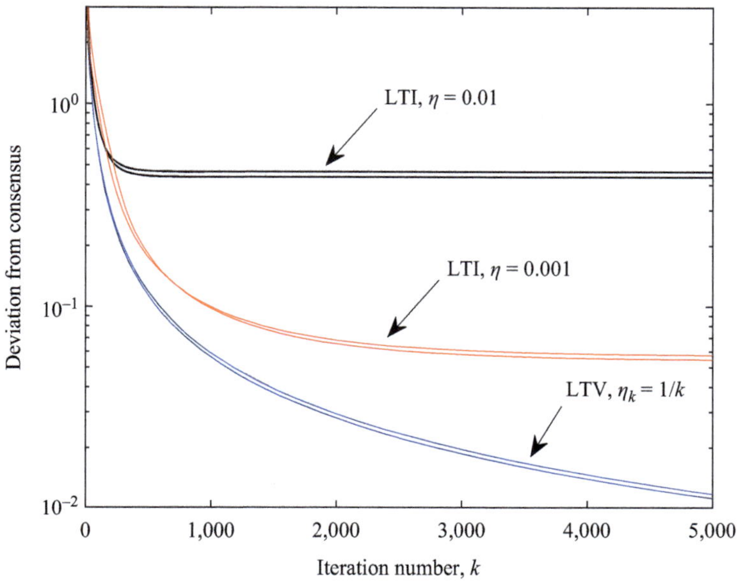

Figure 9.8 Evolution of the deviation from consensus for the distributed implementation of the EM algorithm based on consensus filters. For each choice of η, the trajectories for each of the four designs from Table 9.1 are shown

9.8 Concluding remarks

The EM algorithm is a powerful method for obtaining ML estimates in problems which may seem intractable at first sight. In addition, it may be amenable to a number of distributed implementations which are very appealing for sensor networks. It can be envisioned that whenever the network is sensing an environment which can be modelled as a mixture or superposition of some basic components, the problem of estimating the parameters of such components could be tackled by means of the EM algorithm. As an illustrative example, in this chapter we have focused on a particular application, namely, a linear estimation problem based on data collected by the network when an unknown subset of sensors delivers corrupted measurements, in which the basic components are the correct or incorrect functioning status of the sensors. Another typical application of the EM algorithm in sensor networks for which distributed versions have been proposed in the literature is the estimation of the parameters of a mixture of Gaussian densities, which is a standard tool for unsupervised clustering [15,26]. The same paradigm has also been applied to Total Least Squares estimation in sensor networks, again with randomly malfunctioning sensors [29].

In the aforementioned applications, the only steps in the EM algorithm which require global information from the whole network can be written in terms of global variables which are obtained by aggregation of local variables. Such aggregation functions, namely the average, the minimum, and the maximum, are special in the sense that their computation can be performed in a decentralised manner. This paved the way to a variety of distributed EM implementations, among which we have reviewed several routing-based and consensus-based approaches; other approaches not discussed here include gossip-based methods [30] as well as schemes based on the alternating-direction method of multipliers [31,32].

Throughout the presentation, it has been assumed that the sensors acquire a single snapshot of data on which estimation is to be based; an interesting open line is the extension to scenarios with streaming data [33]. We have also assumed ideal communication links among nodes, which need not be the case in practical sensor networks. Distributed methods robust to link failures, noisy exchanges, and quantisation can be found in the literature [34–36], and their application to the setting of this chapter is a research direction of definite interest.

References

[1] Akyildiz IF, and Vuran MC. Wireless sensor networks. Chichester, West Sussex, UK: John Wiley & Sons; 2010.
[2] Bertsekas DP, and Tsitsiklis JN. Parallel and distributed computation: numerical methods. Nashua, NH, USA: Athena Scientific; 1989.
[3] Scharf LL. Statistical signal processing: detection, estimation, and time series analysis. Reading, MA, USA: Addison-Wesley; 1990.
[4] Kay SM. Fundamentals of statistical signal processing, vol. I: estimation theory. Upper Saddle River, NJ, USA: Prentice-Hall; 1993.
[5] McLachlan GJ, and Krishnan T. The EM algorithm and extensions. New York, NY, USA: Wiley; 1997.
[6] Dempster AP, Laird NM, and Rubin DB. Maximum likelihood from incomplete data via the EM algorithm. J R Stat Soc Ser B. 1977;39(1):1–38.
[7] Stoica P, and Selén Y. Cyclic minimizers, majorization techniques, and the expectation–maximization algorithm: a refresher. IEEE Signal Process Mag. 2004;21(1):112–114.
[8] Sun Y, Babu P, and Palomar DP. Majorization–minimization algorithms in signal processing, communications, and machine learning. IEEE Trans Signal Process. 2017;65(3):794–816.
[9] Zhou Q, Kar S, Huie L, *et al.* Distributed estimation in sensor networks with imperfect model information: an adaptive learning-based approach. In: Proc. IEEE Int. Conf. Acoust. Speech Signal Process.; Brisbane, Australia; 2012. p. 3109–3112.
[10] Silva Pereira S, López-Valcarce R, and Pagès-Zamora A. A diffusion-based EM algorithm for distributed estimation in unreliable sensor networks. IEEE Signal Process Lett. 2013;20(6):595–598.
[11] Silva Pereira S, López-Valcarce R, and Pagès-Zamora A. Parameter estimation in wireless sensor networks with faulty transducers: a distributed EM approach. Signal Process. 2018;144:226–237.

[12] Ni K, Ramanathan N, Chehade MNH, *et al.* Sensor network data fault types. ACM Trans Sens Netw. 2009;5(3).
[13] Banfield JD, and Raftery AE. Model-based Gaussian and non-Gaussian clustering. Biometrics. 1993;49(3):803–821.
[14] Neal RM, and Hinton GE. A view of the EM algorithm that justifies incremental, sparse, and other variants. In: Jordan MI, editor. Learning in Graphical Models. Nato Science Series D, vol. 89. The Netherlands: Springer; 1998. p. 355–368.
[15] Nowak RD. Distributed EM algorithms for density estimation and clustering in sensor networks. IEEE Trans Signal Process. 2003;51(8):2245–2253.
[16] DeGroot MH. Reaching a consensus. J Am Stat Assoc. 1974;69(345):118–121.
[17] Xiao L, Boyd S, and Lall S. A scheme for robust distributed sensor fusion based on average consensus. In: Proc. Int. Symp. Inf. Process. Sensor Netw.; 2005. p. 63–70.
[18] Olfati-Saber R, Fax JA, and Murray RM. Consensus and cooperation in networked multi-agent systems. Proc IEEE. 2007;95(1):215–233.
[19] Wei R, Beard RW, and Atkins EM. Information consensus in multivehicle cooperative control. IEEE Control Syst Mag. 2007;27(2):71–82.
[20] Garin F, and Schenato L. A survey on distributed estimation and control applications using linear consensus algorithms. In: Bemporad A, Heemels M, and Johansson M, editors. Networked Control Systems. Lecture Notes in Control and Information Sciences, vol. 406. London: Springer; 2010. p. 75–107.
[21] Godsil C, and Royle G. Algebraic graph theory. In: Graduate Texts in Mathematics. Berlin, Germany: Springer-Verlag; 2001.
[22] Monajemi Nejad B, Ahmed Attia S, and Raisch J. Max-consensus in a max-plus algebraic setting: the case of fixed communication topologies. In: Proc. Int. Symp. Inf. Commun. Automation Technol.; 2009.
[23] Lucchese R, and Varagnolo D. Average consensus via max consensus. IFAC-PapersOnLine. 2015;48(22):58–63. 5th IFAC Workshop Distrib. Estimation and Control in Netw. Syst.
[24] Xiao L, and Boyd S. Fast linear iterations for distributed averaging. Syst Control Lett. 2004;53:65–78.
[25] Polyak BT. Introduction to optimization. In: Translations Series in Mathematics and Engineering. New York, NY, USA: Optimization Software; 1987.
[26] Gu D. Distributed EM algorithm for Gaussian mixtures in sensor networks. IEEE Trans Neural Netw. 2008;19(7):1154–1166.
[27] Hardy GH. Divergent series. Providence, RI, USA: American Mathematical Society; 1992.
[28] Kar S, and Moura JMF. Consensus + innovations distributed inference over networks. IEEE Signal Process Mag. 2013;30(3):99–109.
[29] Silva Pereira S, Pagès-Zamora A, and López-Valcarce R. Distributed TLS estimation under random data faults. In: Proc. IEEE Int. Conf. Acoust. Speech Signal Process.; Brisbane, Australia; 2015. p. 2949–2943.
[30] Kowalczyk W, Vlassis N. and Newscast EM. In: Proc. Adv. Neural Inf. Process. Syst.; 2005 Dec 13–18; Vancouver, BC, Canada; 2005. p. 713–720.

[31] Forero PA, Cano A, and Giannakis GB. Consensus-based distributed expectation–maximization algorithm for density estimation and classification using wireless sensor networks. In: Proc. IEEE Int. Conf. Acoust., Speech, Signal Process.; 2008 Mar 31–Apr 4; Las Vegas, NV; 2008. p. 1989–1992.
[32] Forero PA, Cano A, and Giannakis GB. Distributed clustering using wireless sensor networks. IEEE J Sel Top Signal Process. 2011;5(4):707–724.
[33] Giménez-Febrer P, Pagès-Zamora A, and López-Valcarce R. Online EM-based distributed estimation in sensor networks with faulty nodes. In: Proc. 24th Eur. Signal Process. Conf.; Budapest, Hungary; 2016. p. 175–179.
[34] Aysal TC, Coates MJ, and Rabbat MG. Distributed average consensus with dithered quantization. IEEE Trans Signal Process. 2008;56(10):4905–4918.
[35] Kar S, and Moura JMF. Distributed consensus algorithms in sensor networks with imperfect communication: link Failures and channel noise. IEEE Trans Signal Process. 2009;57(1):355–369.
[36] Li T, and Zhang JF. Consensus conditions of multi-agent systems with time-varying topologies and stochastic communication noises. IEEE Trans Autom Control. 2010;55(9):2043–2057.

Part IV

Cross-layer issues

Chapter 10

Distributed estimation in energy harvesting wireless sensor networks

Y.-W. Peter Hong[1,2] *and Pradeep Chennakesavula*[1]

In this chapter, distributed estimation is examined for energy-harvesting wireless sensor networks (WSNs), where the energy available at the sensors is converted entirely from ambient sources. In this application, each sensor takes a local measurement of the common parameter of interest and forwards it to the fusion center, where the final estimate is performed. Due to the randomness of the energy arrival, the transmission energy and status of the energy-harvesting sensors are unknown and, thus, the final maximum likelihood estimate at the fusion center can be computed using the expectation–maximization (EM) algorithm. Furthermore, by taking into consideration the spatial heterogeneity of the energy arrival, the sensor deployment problem is also examined for the purpose of reconstructing the entire random field. Numerical simulations are provided to demonstrate the effectiveness of the proposed schemes.

10.1 Introduction

WSNs or the modern-day Internet-of-Things consist of spatially distributed devices that have the ability to sense the physical environment, process the gathered information, and communicate through the wireless interface. In recent years, WSNs have been adopted in a wide range of applications, such as environmental monitoring, disaster recovery, and battlefield surveillance, etc. [1]. In these applications, sensors are often deployed in large-scale and in hostile areas making human maintenance and battery-replacement impractical. Due to these reasons, energy-harvesting techniques have been adopted to prolong the lifetime of sensor nodes [2–4]. This technology enables the conversion of ambient energy (such as solar [5], vibrational [6], or thermal energy [7]) to electricity. However, in these cases, the characteristics of the energy arrival and the efficiency of energy usage will have a significant impact on the sensing performance. These issues will be examined in this chapter in the context of distributed estimation, which underlies many WSN applications.

[1] Institute of Communications Engineering, National Tsing Hua University, Taiwan
[2] MOST Joint Research Center for AI Technology and All Vista Healthcare, Taiwan

Distributed estimation problems have been examined extensively in the literature for systems without energy-harvesting capabilities, e.g., in [8,9]. In these problems, sensors first take local measurements of the common parameter of interest and forward them to the fusion center using analog or digital forwarding (DF) schemes. Without energy harvesting considerations, sensors' transmissions need only take into consideration average power constraints, regardless of the available energy in the battery. In energy-harvesting systems, one must take into account both the energy causality and the battery-storage constraints. The former reflects the fact that energy can only be used after it has been harvested, and the latter sets a constraint on the maximum amount of energy that can be harvested in each time slot. These constraints along with the randomness and intermittence of the energy arrival pose significant challenges on the system design. In this chapter, we first examine in Section 10.3 the impact of energy harvesting on the distributed estimation performance and then propose in Section 10.4 an optimized sensor-deployment policy that takes into consideration the spatial inhomogeneity of the energy arrival rates at different locations. The contents of this chapter are summarized from the works in [10,11].[1]

Specifically, in Section 10.3, we devise the optimal analog-forwarding (AF) transmission policy and the optimal fusion strategy for distributed estimation in WSNs. We consider both cases where sensors may or may not be equipped with batteries to regulate the energy usage over time. In AF systems, each sensor forwards a scaled version of its local observation to the fusion center if its available energy is sufficient for transmission in the given time slot. Due to random variations of the magnitude of the scaled observation, the required transmit energy may sometimes exceed the available battery energy, preventing the sensor from transmitting. This is different from conventional studies that consider only average power constraints and allow sensors to transmit as long as the average power is satisfied in the long term. In this work, these issues are taken into consideration in the estimation at the fusion center and also in the choice of the scaling factors at the sensors. However, even if a sensor is not able to transmit, the fact that it remains silent conveys information about the magnitude of the observation, similar to the concept of censoring sensors [12], and can be incorporated into the estimation at the fusion center. The maximum likelihood estimator is adopted at the fusion center and is derived based on statistics of the energy arrivals at the sensors. Similar issues can also be examined for DF transmission systems, as in [13] , but is not discussed here.

Then, in Section 10.4, we examine the impact of energy harvesting from a more system-wide perspective. In particular, we study the optimal deployment of energy-harvesting sensors for the purpose of sensing and reconstruction of a spatially correlated Gaussian random field. Here, we consider the notion of random

[1]Parts of this chapter are reprinted with permission by ©2014 IEEE, from Y.-W. P. Hong, "Distributed estimation with analog forwarding in energy-harvesting wireless sensor networks," in *Proceedings of IEEE International Conference on Communication Systems*, pp. 142–146 (2014) [10], and by ©2015 IEEE, from T. C. Hsu, Y. W. P. Hong and T. Y. Wang, "Optimized random deployment of energy harvesting sensors for field reconstruction in analog and digital forwarding systems," in *IEEE Transactions on Signal Processing*, vol. 63, no. 19, pp. 5194–5209, Oct. 1, 2015 [11].

deployment where sensors are assumed to be deployed randomly and in large scale according to a spatially nonhomogeneous Poisson point process (NHPPP) [14]. In this case, the issue of sensor deployment is reduced to finding the optimal sensor density at different locations. Here, we assume that the sensors' operations are supported solely by ambient energy (i.e., energy harvested from the environment) and, thus, their transmit powers and probabilities depend strongly on the characteristics of their energy arrival at their respective locations. To reduce the field reconstruction error, the sensors should, on the one hand, be more spread out over the field to gather more informative samples, but should, on the other hand, be concentrated more at locations with large energy arrivals or with large channel gains toward the closest data-gathering node. This trade-off is exploited in this chapter for the AF case to determine the optimal random sensor deployment but can also be examined for the DF case as well [11]. In particular, given the locations of data-gathering nodes, the statistics of the energy arrival at different locations and a constraint on the average number of sensors, we determine the spatially dependent sensor densities and the energy-aware transmission thresholds at the sensors by minimizing an upper bound on the average mean-square reconstruction error. The energy-aware transmission policy allows a sensor to transmit only when its accumulated energy (i.e., its battery level) is beyond a certain threshold and remains silent, otherwise, so that energy can be preserved for use in later time slots.

10.2 Related works and further readings

Field reconstruction and decentralized parameter estimation are essential applications in WSNs and have been studied extensively in the literature for WSNs without energy-harvesting capabilities, e.g., [8,9]. These works focused mainly on the design of sensor transmission schemes under various local constraints and centralized fusion rules. In particular, the amplifying gains in AF systems and the source quantizer and signal modulation in DF systems can be designed to minimize the field reconstruction error. The readers are referred to the above articles for further background on the subject. In this work, we focus on the impact of energy harvesting on the above problems.

Specifically, the impact of energy harvesting on distributed estimation was examined recently in, e.g., [15–20]. Specifically, in [15], the optimal power allocation among sensors was derived by minimizing the mean-square error (MSE) of the estimate at the fusion center. The proposed scheme assumes noncausal knowledge of the energy arrival and, thus, may be impractical. In [16], the optimal power allocation over time was determined for both cases with causal and noncausal energy information. However, only a single sensor was considered and, thus, data fusion was not discussed. In both [15] and [16], the instantaneous energy consumption in each time slot was assumed to be equal to the average transmit energy. This approximation may not be accurate, especially when AF transmissions are adopted, since the transmit power may vary drastically depending on the magnitude of the scaled signal. In [17], a similar remote estimation problem was also considered for the case with a single sensor and causal energy information. The optimal transmission policy was derived using dynamic programming. In [18], an optimal energy allocation problem was considered by minimizing distortion over a finite horizon in a multi-sensor estimation system subject to energy harvesting constraints. The optimal energy allocation policy was also derived by dynamic programming. More recently, [19] studied power allocation for distributed estimation of an unknown scalar random source in sensor networks with a multiple-antenna fusion center, where sensors are equipped with radio-frequency-based energy-harvesting technology; [20] studied an energy allocation problem for distributed estimation with collaborating energy-harvesting sensors. This work utilized the sharing of measurements among neighboring sensors to improve the final estimate at the fusion center.

In the past, sensor deployment problems have been examined mostly for sensors without energy-harvesting capabilities (see [21] for a survey on these topics). In particular, without energy-harvesting considerations, sensor deployment policies have been proposed with the goal of minimizing the field reconstruction error in [22], of guaranteeing connectivity in [23], and of maximizing sensor coverage in [24]. In these works, sensors were only allowed to be placed among a predetermined set of locations and, thus, the optimal sensor-placement problem often results in choosing the best combination of locations within the set that optimizes the system performance. This problem is often NP hard [21] due to its combinatorial nature. Therefore, heuristic or approximate solutions were proposed to reduce the computational complexity. Due to advances in energy-harvesting technology [2–4], similar problems have also been examined recently, e.g., in [25,26], for energy harvesting wireless systems with considerations on the stochastic nature of the energy arrival at each node. In particular, in [25,26], the deployment of energy harvesting relay nodes in sensor networks were examined with the goal of enhancing network throughput and of guaranteeing connectivity, respectively. However, they did not consider the cross-layer impact of the spatial dependencies of the energy arrivals and sensor measurements on the field reconstruction performance. In our work, the optimal random sensor-deployment strategy for field reconstruction was examined by taking into consideration the spatial correlation between the sensor measurements and the energy arrivals. Rather than designing a deterministic deployment scheme which places sensors at precise locations, we consider the random deployment of sensors and determine the optimal sensor densities at different locations. We argue that random deployment [27,28] is more practical when the number of sensors to be deployed is large or when the sensors are to be deployed in a hostile environment.

10.3 Distributed estimation in energy-harvesting wireless sensor networks

Let us consider a WSN with K energy-harvesting sensors and a fusion center as illustrated in Figure 10.1. During each observation period, each sensor takes a local measurement of a common parameter of interest and forwards the observation to the fusion center, where a final estimate is computed. The value of the common parameter

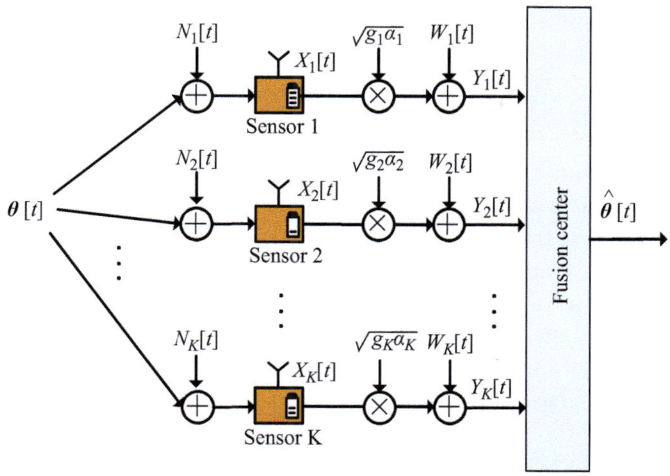

Figure 10.1 Illustration of the sensing and analog-forwarding model

of interest in the tth observation period (or time slot) is denoted by $\theta[t]$ and the noisy observation at sensor k is given by:

$$X_k[t] = \theta[t] + N_k[t], \tag{10.1}$$

where $N_k[t]$ is the additive Gaussian noise with mean 0 and variance σ_k^2, i.e., $N_k[t] \sim \mathcal{N}(0, \sigma_k^2)$. The observation noise is assumed to be independent over time and among sensors. Two sensor forwarding schemes have been adopted the most in the literature, the AF and the DF schemes. In the AF scheme, sensors directly transmit a scaled version of their local observations to the fusion center whereas, in the DF scheme, sensors first quantize and digitally modulate the information to be transmitted. Here, we utilize the AF case as a representative example to demonstrate the key concepts and trade-offs of distributed estimation in energy-harvesting systems. Readers are referred to [13] for a similar study on the DF case.

Specifically, in AF systems, sensors simply transmit scaled versions of their local observations to the fusion center over orthogonal channels. In this case, the signal transmitted by sensor k in the tth observation period can be written as $\sqrt{\alpha_k}X_k[t]$, where α_k is the scaling factor at sensor k. We assume that the sensors' transmit energies are supported solely by ambient sources. Moreover, the sensors' operational energy is assumed to be negligible and, thus, the harvested energy can be used entirely for transmission. Here, we consider two cases: (i) the case where no batteries are available at the sensors (and, thus, the energy gathered in each observation period is used only for the transmission in that period) and (ii) the case where the sensors are equipped with batteries that have capacity $B_{\max}$. Suppose that $E_k[t]$ is the energy arrival at sensor k in the tth observation period and $B_k[t]$ is the battery energy available for transmission at sensor k in slot t. Then, in case (i), we have $B_k[t] = E_k[t]$, for all k and t, and, in case (ii), we have:

$$B_k[t] = (B_k[t-1] - \alpha_k X_k[t-1]^2 O_k[t-1]\Delta)^+ + E_k[t], \tag{10.2}$$

where Δ is the duration of each sensor's transmission and $O_k[t] \in \{0, 1\}$ represents the transmission status of sensor k in slot t. That is, $O_k[t] = 1$ indicates that sensor k transmits and $O_k[t] = 0$ indicates that it remains silent. Without loss of generality, we shall assume in the following that $\Delta = 1$. Due to the randomness of the energy arrivals and the sensors' observation values, the available battery energy at each sensor may not always be sufficient for transmission.

By assuming that a sensor transmits only if its available battery energy is sufficient for transmission, we have:

$$O_k[t] = \mathbf{1}_{\{\alpha_k X_k[t]^2 \le B_k[t]\}}, \tag{10.3}$$

where $\mathbf{1}_{\{\cdot\}}$ is the indicator function. Then, the signal received from sensor k at the fusion center can be expressed as:

$$Y_k[t] = \sqrt{g_k[t]\,\alpha_k} X_k[t] O_k[t] + W_k[t], \tag{10.4}$$

where $g_k[t]$ is the channel gain from sensor k to the fusion center and $W_k[t] \sim \mathcal{N}(0, \sigma_0^2)$ is the additive noise at the fusion center during the reception of sensor k's signal. Hence, there exists a trade-off between the number of signals received at the fusion center and the quality of each received signal. This can be exploited through a proper choice of the scaling factors $\{\alpha_k\}_{k=1}^K$. We assume that the channel gains of the sensors are known at the fusion center, and the transmission status of the sensors can be accurately estimated from the observation. Hence, the information obtained by the fusion center from sensor k at time t can be represented by the observation pair $(Y_k[t], O_k[t])$. The case where only $Y_k[t]$ is observed at the fusion center can be derived similarly, but is not discussed here to avoid redundancy.

10.3.1 Maximum likelihood estimation for the case with no batteries at the sensors

In this section, we consider the case where no batteries are available at the sensors. In this case, the energy harvested by each sensor is used only in the time slot that it was received and is discarded if it is not utilized before the next time slot. Here, the estimation is performed separately in each time slot and, thus, the time index t can be omitted for brevity.

Suppose that the observations at the fusion center are given by $\{(Y_k, O_k)\}_{k=1}^K = \{(y_k, o_k)\}_{k=1}^K$. Then, the log-likelihood function can be written as:

$$L(\theta; \{y_k, o_k\}_{k=1}^K) = \sum_{k=1}^{K} \left[\log f_{Y_k|O_k=o_k}(y_k; \theta) + \log \Pr(O_k = o_k; \theta)\right], \tag{10.5}$$

where:

$$\Pr(O_k = 0;\theta) = \Pr(E_k < \alpha_k X_k^2;\theta) = \mathbf{E}_{E_k}\left[Q\left(\frac{\sqrt{E_k/\alpha_k}-\theta}{\sigma_k}\right)+Q\left(\frac{\sqrt{E_k/\alpha_k}+\theta}{\sigma_k}\right)\right] \tag{10.6}$$

and $\Pr(O_k = 1;\theta) = 1 - \Pr(O_k = 0;\theta)$. It is interesting to observe that the probability that sensor k remains silent is nondecreasing with respect to $|\theta|$ and α. Moreover, given $O_k = 0$, we have $Y_k = W_k$ and, thus,

$$f_{Y_k|O_k=0}(y_k) = \frac{1}{\sqrt{2\pi\sigma_0^2}} e^{-(1/2\sigma_0^2)y_k^2};$$

and, given $O_k = 1$, it can be shown that:

$$f_{Y_k|O_k=1}(y_k;\theta) = \frac{\beta/\sqrt{2\pi\sigma_0^2\sigma_k^2}}{\Pr(O_k = 1;\theta)} \cdot e^{(\beta^2/2)\Gamma(y_k,\theta)^2-(y_k^2/2\sigma_0^2)-(\theta^2/2\sigma_k^2)}$$
$$\cdot \mathbf{E}_{E_k}\left[1-Q\left(\sqrt{\frac{E_k}{\alpha_k\beta^2}}-\Gamma(y_k,\theta)\beta\right)-Q\left(\sqrt{\frac{E_k}{\alpha_k\beta^2}}+\Gamma(y_k,\theta)\beta\right)\right],$$

where $\beta \triangleq (g_k\alpha_k/\sigma_0^2 + 1/\sigma_k^2)^{-(1/2)}$ and $\Gamma(y_k,\theta) \triangleq (\sqrt{g_k\alpha_k}/\sigma_0^2)y_k + \theta/\sigma_k^2$. The maximum likelihood estimate (MLE) is then given by:

$$\hat{\theta}_{\text{MLE}} = \arg\max_{\theta\in\Theta} L(\theta;\{y_k,o_k\}_{k=1}^K), \tag{10.7}$$

where Θ is the set of all possible values of θ. The MLE above takes into consideration the statistics of the energy arrival at the sensors. It is worthwhile to remark that, since Y_k does not contain information about θ when $O_k = 0$, an intuitive approach is to take the MLE based only on the signals received from sensors that actually transmit, i.e., $\{y_k\}_{k\in\Omega_1}$, where $\Omega_1 \triangleq \{k : O_k = 1\}$, and without considerations on the energy arrival statistics. In this case, the MLE is given by:

$$\tilde{\theta}_{\text{BL-MLE}} = \arg\max_{\theta\in\Theta} L(\theta;\{y_k\}_{k\in\Omega_1}) \tag{10.8}$$

$$= \arg\min_{\theta\in\Theta} \sum_{k\in\Omega_1} \frac{(y_k - \sqrt{g_k\alpha_k}\theta)^2}{g_k\alpha_k\sigma_k^2 + \sigma_0^2} \tag{10.9}$$

However, this estimator does not perform well when the probability that a sensor does not transmit is high (e.g., when $|\theta|$ and/or α_k is large). This estimator will be referred to as the baseline MLE (BL-MLE).

Moreover, we can also examine impact of the choice of the scaling factors $\{\alpha_k\}_{k=1}^K$ on the signal quality and the number of sensors that transmit in each time slot. However, these factors (and, thus, the estimation performance at the fusion center) also depend on the deterministic and unknown parameter θ and, thus, it is necessary to acquire preliminary knowledge of θ before the optimal values of $\{\alpha_k\}_{k=1}^K$ can be determined. If θ is known, it would be desirable to choose the scaling factors $\{\alpha_k\}_{k=1}^K$ so that the average estimation error is minimized. However, the estimation error is difficult to evaluate exactly in this case since there is no closed form of the MLE. Hence, due to the asymptotic efficiency of the MLE, we instead adopt Fisher information as the optimizing criterion, which yields:

$$\alpha_k^* = \arg\max_{\alpha_k} \mathbf{E}\left[\left(\frac{\partial}{\partial\theta}\log f(Y_k, O_k; \theta)\right)^2 ; \theta\right]. \tag{10.10}$$

Notice that the computation of $\{\alpha_k^*\}_{k=1}^K$ requires knowledge of θ. In practice, this value can be replaced by a preliminary estimate of θ obtained by polling J sensors, where $J \ll K$. The preliminary estimate can then be broadcast by the fusion center to all sensors, who then compute their gain based on this information. The same parameters are also computed at the fusion center since they are required to compute the MLE. Similar operations have also been adopted in [12].

10.3.2 *Maximum likelihood estimation for the case with battery-enabled sensors*

In this section, we assume that sensors are equipped with finite capacity batteries and, thus, are capable of regulating the energy usage over time. In this case, each sensor's transmission status will become correlated over time since the transmissions (or lack of transmissions) in previous time slots will impact the amount of energy available for transmissions in later time slots. The MLE must take this temporal dependence into consideration in its derivation.

Suppose that sensors feedback their residual energy level at the end of every T observation periods and that the parameter of interest, i.e., θ, remains constant in between two consecutive feedbacks. Specifically, given the feedback at the end of slot mT, the fusion center can compute the MLE based on the statistics of the energy arrival and the signals received in the most recent τ time slots. The estimation is performed separately at the end of every T observation periods and, thus, the index mT is omitted in the following for convenience. Also, for ease of exposition, we shall assume that the sensors' residual energy levels at the end of the previous T observation periods are 0. In this case, the MLE at time τ is given by:

$$\hat{\theta}_{\text{MLE}} = \arg\max_{\theta\in\Theta} L(\theta; \{\mathbf{y}_k^{1:\tau}, \mathbf{o}_k^{1:\tau}\}_{k=1}^K), \tag{10.11}$$

where $\mathbf{y}_k^{1:\tau} \triangleq [y_k[1], \ldots, y_k[\tau]]$ and $\mathbf{o}_k^{1:\tau} \triangleq [o_k[1], \ldots, o_k[\tau]]$. However, the distributions of $\mathbf{y}_k^{1:\tau}$ and $\mathbf{o}_k^{1:\tau}$ depend on the sensors' observations and battery levels and cannot be easily derived. To compute the MLE, we model the sensors' observations and battery levels as missing data and adopt the EM algorithm as described below.

10.3.2.1 Sensor fusion using the EM algorithm

The EM algorithm is an iterative method for finding ML estimates with incomplete datasets. Each iteration of the EM algorithm consists of two steps, i.e., the E-step and the M-step. The E-step computes the expectation of the log-likelihood function over the missing data using the current estimate of the parameter and the M-step computes the parameter that maximizes the expected log-likelihood function obtained in the

E-step. The two steps are performed iteratively until the sequence of parameter estimates converges.

Specifically, let us consider $\mathbf{X}_k^{1:\tau} \triangleq [X_k[1], \dots, X_k[\tau]]$ be the vectors of sensor observations and the vectors of available battery energies $\mathbf{B}_k^{1:\tau} \triangleq [B_k[1], \dots, B_k[\tau]]$, for $k = 1, \dots, K$, as the missing data and take $\{\mathbf{y}_k^{1:\tau}, \mathbf{o}_k^{1:\tau}, \mathbf{X}_k^{1:\tau}, \mathbf{B}_k^{1:\tau}\}_{k=1}^K$ as the complete dataset. Moreover, to simplify notations, let us define $S_k[t] \triangleq [X_k[t], B_k[t]]$ and $z_k \triangleq [y_k[t], o_k[t]]$ as the missing data and observed data of sensor k at time t, and let $\mathbf{S}_k^{1:\tau} \triangleq [S_k[1], \dots, S_k[\tau]]$ and $\mathbf{z}_k^{1:\tau} \triangleq [z_k[1], \dots, z_k[\tau]]$. In this case, the log-likelihood function of the complete dataset is:

$$L(\theta; \{\mathbf{z}_k^{1:\tau}, \mathbf{S}_k^{1:\tau}\}_{k=1}^K) = \sum_{k=1}^{K} \ell_k(\theta; \mathbf{z}_k^{1:\tau}, \mathbf{S}_k^{1:\tau}), \tag{10.12}$$

where

$$\begin{aligned}\ell_k(\theta; \mathbf{z}_k^{1:\tau}, \mathbf{S}_k^{1:\tau}) &= \sum_{t=1}^{\tau} \log f_{B_k[t]|S_k[t-1]}(B_k[t]) + \sum_{t=1}^{\tau} \log f_{X_k[t]}(X_k[t]; \theta) \\ &\quad + \sum_{t=1}^{\tau} \log f_{Y_k[t]|S_k[t]}(y_k[t]) + \sum_{t=1}^{\tau} \log \Pr(O_k = o_k | S_k[t]).\end{aligned} \tag{10.13}$$

In the above, we utilized the fact that, given $S_k[t]$, the observations $y_k[t]$ and $o_k[t]$ no longer depend on θ and are independent; and the fact that, given $S_k[t-1]$, the battery level $B_k[t]$ also does not depend on θ.

Suppose that $\hat{\theta}^{(i-1)}$ is the estimate obtained in the $(i-1)$th iteration of the EM algorithm. Then, the E-step and the M-step in the ith iteration can be described as follows:

E-Step: Compute the expected log-likelihood function:

$$Q(\theta; \hat{\theta}^{(i-1)}) = \sum_{k=1}^{K} \mathbf{E}_{\mathbf{S}_k^{1:\tau}|\mathbf{z}_k^{1:\tau}} \left[\ell_k(\theta; \mathbf{z}_k^{1:\tau}, \mathbf{S}_k^{1:\tau}); \hat{\theta}^{(i-1)} \right] \tag{10.14}$$

M-Step: Update the estimate by taking:

$$\hat{\theta}^{(i)} = \arg\max_{\theta \in \Theta} Q(\theta; \hat{\theta}^{(i-1)}). \tag{10.15}$$

From (10.13), we can see that only the second term (i.e., the term corresponding to the density of $X_k[t]$, for $t = 1, \dots, \tau$) depends on θ. Hence, it is sufficient to consider in the M-step the optimization over the expectation of this term only, i.e.,

$$\hat{\theta}^{(i)} = \arg\max_{\theta \in \Theta} Q'(\theta; \hat{\theta}^{(i-1)}), \tag{10.16}$$

where

$$Q'(\theta; \hat{\theta}^{(i-1)}) \triangleq \sum_{k=1}^{K} \mathbf{E}_{\mathbf{S}_k^{1:\tau}|\mathbf{z}_k^{1:\tau}} \left[\sum_{t=1}^{\tau} \log f_{X_k[t]}(X_k[t]; \theta); \hat{\theta}^{(i-1)} \right].$$

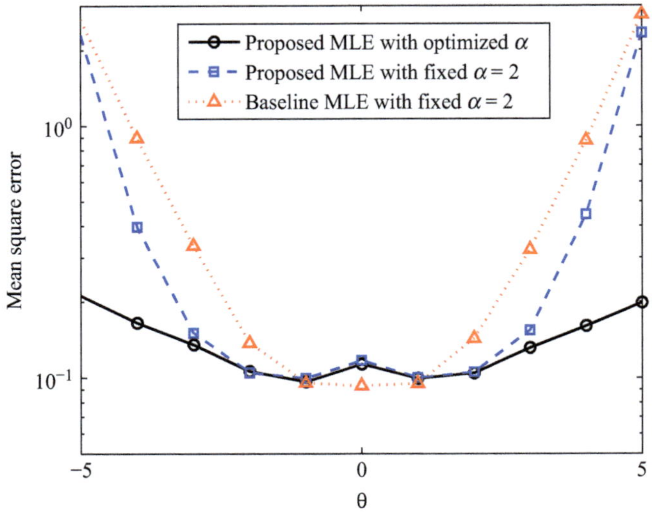

Figure 10.2 For the case without batteries, we show the MSE of the proposed MLE with optimized $\{\alpha_k\}_{k=1}^K$*, the proposed MLE with fixed* $\alpha_k = 2$*,* $\forall k$*, and the baseline MLE with fixed* $\alpha_k = 2$*,* $\forall k$*. ©2014 IEEE. Reprinted, with permission, from Reference [10]*

Notice that the computation of Q' still requires, for each k, integration over 2τ variables whose distributions are nontrivial. The complexity of this integration may rapidly become intractable as τ increases. Interestingly, by exploiting the temporal relation between our observations, we can simplify the computations using particle filtering [29]. Details of the implementation can be found in [10].

10.3.2.2 Computer simulations and performance comparisons

In this section, computer simulations are provided to demonstrate the effectiveness of our proposed schemes. We consider a network with $K = 15$ sensors and let $\sigma_0^2 = \sigma_k^2 = 1$, for all k. The energy arrival is assumed to be independent and identically distributed (i.i.d.) over time and across sensors according to a Rayleigh distribution with parameter 20. The units are normalized such that one unit of transmit energy consumes one unit of battery energy. The number of sensors polled for preliminary estimates is $J = 3$.

In Figure 10.2, we consider the case without batteries and compare the MSE for three schemes: the proposed MLE with optimized scaling factors (according to (10.10)), the proposed MLE with fixed scaling factors and the BL-MLE given in (10.8) with fixed scaling factors. In the latter two cases, we set $\alpha_k = 2$, $\forall k$. We can see that, even with fixed scaling factors, the proposed MLE outperforms the BL-MLE, whose estimation does not utilize knowledge of the sensors' transmission status. The case with optimized scaling factors provides even more advantages, especially when $|\theta|$ is large. However, the performance is sensitive to the choice of α when θ is close to 0.

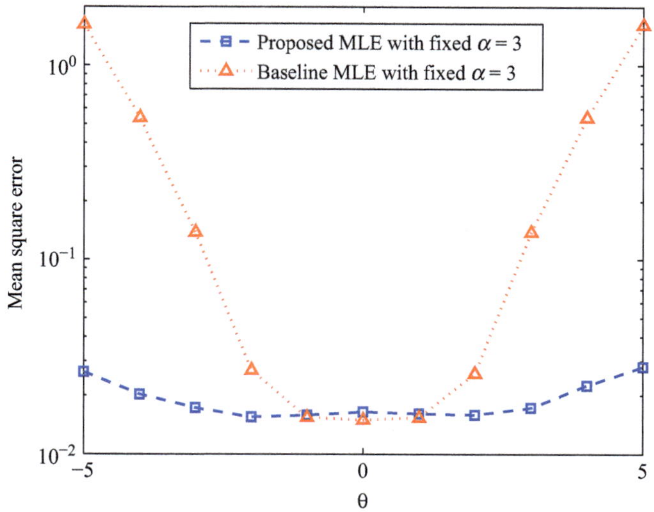

Figure 10.3 For the case with batteries, we show the MSE of the proposed MLE and the baseline MLE, both with fixed $\alpha_k = 3$, $\forall k$. ©2014 IEEE. Reprinted, with permission, from Reference [10]

In Figure 10.3, we consider the case with batteries and compare the MSE of the proposed MLE and the BL-MLE, both with fixed scaling factors $\alpha_k = 3$, for all k. Here, we set the feedback period as $T = 6$ time slots and the maximum battery capacity as $B_{\max} = 60$ units. The number of particles is $M = 2{,}000$. We can see that the MLE approximated by the particle-filtering-based EM algorithm indeed performs well and achieves significantly lower MSE than that of the BL-MLE, especially when θ is large and less sensors transmit. This is due to the fact that the proposed MLE takes into account the information of the sensors' transmission status.

10.4 Deployment of energy harvesting sensors for distributed estimation

In the previous section, we focused on the impact of the randomness and intermittence of energy arrivals on the sensor transmission and the estimation fusion operations. However, from a system perspective, the randomness of the energy arrivals can vary over different locations, causing a system-wide impact on the performance. In this section, we exploit this impact through studies of the optimal sensor deployment strategies previously presented in [11]. In particular, we examine the optimal spatial densities of energy harvesting sensors for the purpose of sensing and estimation (or reconstruction) of a spatially correlated Gaussian random field. We show that, with spatially inhomogeneous energy arrival rates, sensors should, on the one hand, be more spread out over the field to gather more informative samples, but should, on the

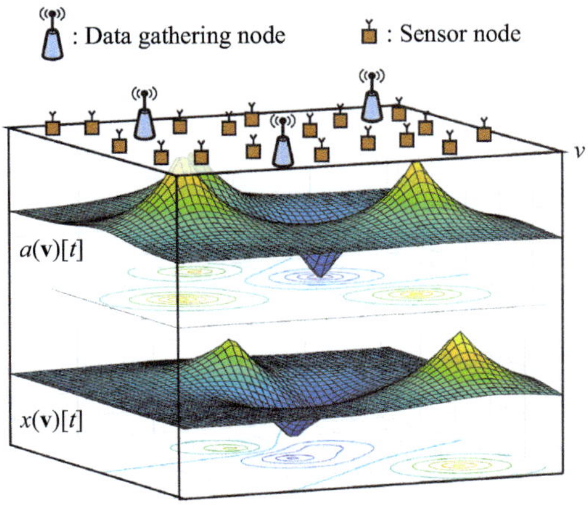

Figure 10.4 Illustration of the sensor deployment, the energy arrival distribution, and the field values at a specific time instant

other hand, be concentrated more at locations with large energy arrivals or with large channel gains toward the closest data-gathering node. Here, we focus only on the AF case, but a similar study for the DF case can also be found in [11].

Suppose that a large number of sensors are deployed randomly in the region $\mathcal{V}$ according to a spatially NHPPP [14] with deterministic densities $\{\lambda(\mathbf{v})\}_{\mathbf{v}\in\mathcal{V}}$, where $\lambda(\mathbf{v})$ is the density at location $\mathbf{v} \in \mathcal{V}$, as illustrated in Figure 10.4. In an observation period, each sensor takes a local sample of the random field and forwards it to the closest data-gathering node. The locations of data-gathering nodes can be arbitrary but are assumed to be fixed over time. The random field is denoted by $\{x(\mathbf{v})[t]\}_{\mathbf{v}\in\mathcal{V}}$, where $x(\mathbf{v})[t]$ is the value of the field at location $\mathbf{v} \in \mathcal{V}$ in the tth observation period. Following [22,30], we assume that the set of random field values $\{x(\mathbf{v})[t]\}_{\mathbf{v}\in\mathcal{V}}$ is jointly Gaussian with zero mean and with covariances $E[x(\mathbf{v}_i)[t]x(\mathbf{v}_j)[t]] = \sigma_x^2\rho(\mathbf{v}_i, \mathbf{v}_j)[t]$, for all $\mathbf{v}_i, \mathbf{v}_j \in \mathcal{V}$, where σ_x^2 is the variance of $x(\mathbf{v})[t]$, for all $\mathbf{v} \in \mathcal{V}$, and $\rho(\mathbf{v}_i, \mathbf{v}_j)[t]$ is the correlation coefficient between $x(\mathbf{v}_i)[t]$ and $x(\mathbf{v}_j)[t]$. The random field is also assumed to be stationary over time. Moreover, let $\{a(\mathbf{v})[t]\}_{\mathbf{v}\in\mathcal{V}}$ be the set of random energy arrivals, where $a(\mathbf{v})[t]$ is the energy arrival at location $\mathbf{v} \in \mathcal{V}$ in the tth observation period. In practice, the set of sensor densities $\{\lambda(\mathbf{v})\}_{\mathbf{v}\in\mathcal{V}}$ to be derived in this work provides a guideline for the numbers of sensors that should be scattered at different locations, e.g., from an air vehicle. To simplify our computations, let us partition the region of interest, i.e., $\mathcal{V}$, into M disjoint subregions $\mathcal{V}_1, \ldots, \mathcal{V}_M$ with equal size. That is, we choose $\mathcal{V}_1, \ldots, \mathcal{V}_M$ such that $\mathcal{V} = \bigcup_{i=1}^{M} \mathcal{V}_i$ with areas $|\mathcal{V}_i| = \Delta$, for all i, and with $\mathcal{V}_i \bigcap \mathcal{V}_j = \phi$, for all $i \neq j$. We assume that Δ is sufficiently small (compared to the variations of the random field) so that the sensor densities remain approximately constant in each subregion (i.e., $\lambda(\mathbf{v}) \approx \lambda_i$, for all $\mathbf{v} \in \mathcal{V}_i$ and for all i). Then, by assuming that $\lambda_i\Delta \ll 1$, the probability that exactly one sensor exists in subregion $\mathcal{V}_i$ can be approximated as $((e^{-\lambda_i\Delta}\lambda_i\Delta)/1!) \approx \lambda_i\Delta$ and the probability that more than one sensor exists in $\mathcal{V}_i$ is negligible. Moreover, with Δ sufficiently small, we can also approximate the field value and the energy arrival at a sensor in subregion $\mathcal{V}_i$ by their respective values at the center of the subregion, which are denoted by $x_i[t]$ and $a_i[t]$, respectively. Therefore, if a sensor exists in subregion $\mathcal{V}_i$, its local observation can be expressed as:

$$\tilde{x}_i[t] = x_i[t] + n_i[t], \tag{10.17}$$

where $n_i[t]$ is the observation noise at the sensor located at the subregion $\mathcal{V}_i$. The observation noise is assumed to be i.i.d. across sensors and over time according to a Gaussian distribution with mean zero and variance σ_n^2, denoted by $\mathcal{N}(0, \sigma_n^2)$.

Here, we adopt an *integrate-and-fire* [31,32] transmission policy where a sensor transmits only if its accumulated energy exceeds a certain threshold but exhausts all its energy when doing so. The energy threshold at the sensor in subregion $\mathcal{V}_i$ is denoted by γ_i and is chosen to minimize the field reconstruction error in later sections. The accumulated energy at each sensor, say, the sensor in subregion $\mathcal{V}_i$, varies over time t and can be expressed as:

$$e_i[t] = \begin{cases} e_i[t-1] + a_i[t], & \text{if } e_i[t-1] \leq \gamma_i \\ a_i[t], & \text{if } e_i[t-1] > \gamma_i, \end{cases} \tag{10.18}$$

at the beginning of the tth observation period. By assuming that the process $\{e_i[t]\}_{t=0}^{\infty}$ is stationary over time, the probability that the sensor transmits in a given time slot is $\bar{F}_{e_i}(\gamma_i) \triangleq \Pr(e_i[t] > \gamma_i)$, which is the complementary cumulative density function of $e_i[t]$. An example of the case where $\{e_i[t]\}_{t=0}^{\infty}$ is stationary is given in the following sections by considering a Bernoulli energy arrival process. By the stationarity of the random field $\{x_i[t]\}_{t=0}^{\infty}$ and of the accumulated energy process $\{e_i[t]\}_{t=0}^{\infty}$, for all i, we shall omit the time index t in later discussions and focus on the field reconstruction performance at a particular instant in time.

Let o_i be a binary random variable defined such that $o_i = 1$ if a sensor exists in subregion $\mathcal{V}_i$ and has energy above the threshold γ_i, and $o_i = 0$, otherwise. In this case, o_i can be viewed as a Bernoulli random variable with probability $\Pr(o_i = 1) = \lambda_i \Delta \bar{F}_{e_i}(\gamma_i)$ and $\Pr(o_i = 0) = 1 - \lambda_i \Delta \bar{F}_{e_i}(\gamma_i)$. By adopting the AF transmission scheme, we assume that each sensor transmits an amplified version of its received signal to the closest data-gathering node, as illustrated in Figure 10.5. The transmissions from the data-gathering nodes to the fusion center are assumed to be noiseless. In this case, the signal transmitted by a sensor in subregion $\mathcal{V}_i$ can be written as:

$$s_i = \frac{\sqrt{\kappa_i e_i}}{\sqrt{\sigma_x^2 + \sigma_n^2}} \tilde{x}_i, \tag{10.19}$$

where $\kappa_i > 0$ is the amplifying factor, e_i is the energy available at the sensor in $\mathcal{V}_i$, and the received signal at the closest data-gathering node is:

$$y_i = \left(h_i \frac{\sqrt{\kappa_i e_i}}{\sqrt{\sigma_x^2 + \sigma_n^2}} \tilde{x}_i + w_i \right) o_i, \tag{10.20}$$

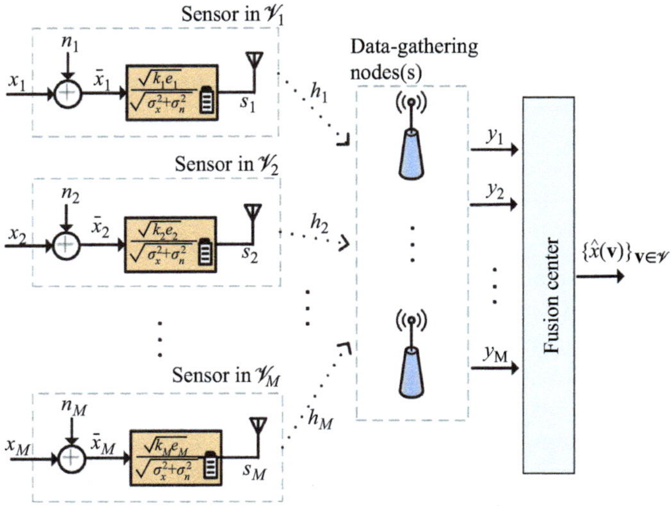

Figure 10.5 Illustration of the analog-forwarding data-gathering procedure

where h_i is the channel coefficient between the sensor and its closest data-gathering node, and $w_i \sim \mathcal{N}(0, \sigma_w^2)$ is the additive white Gaussian noise. Here, we do not consider the effects of short-term fading and assume that h_i is a constant that depends only on the distance between the sensor and its data-gathering node. For notational simplicity, we define $g_i \triangleq (h_i\sqrt{\kappa_i e_i}/(\sqrt{\sigma_x^2+\sigma_n^2}))$ as the effective channel gain experienced by the sensor in $\mathcal{V}_i$. The received signals $\{y_i\}_{i=1}^M$ are then utilized to perform the estimation of the sensor field values $\{x(\mathbf{v})\}_{\mathbf{v}\in\mathcal{V}}$. For simplicity, we assume that the sensor depletes its battery every time it transmits (i.e., any residual energy is dumped after each transmission). Notice that, since the observation $\tilde{x}_i$ is random, the energy required to transmit the signal in (10.19) may exceed the accumulated energy e_i with a certain probability. In this case, a saturation effect may occur, causing additional distortion on the received signal. This effect was examined in the previous section but is not considered here for simplicity. However, by choosing κ_i to be sufficiently small, the probability that saturation occurs is small and, thus, is omitted for simplicity.

Given the received signals $\{y_i\}_{i=1}^M$, the fusion center computes a linear minimum mean square error (LMMSE) estimate of the field value $x(\mathbf{v})$ at location $\mathbf{v}$, which can be written as:

$$\hat{x}(\mathbf{v}) = \mathbf{C}^{\mathbf{o}}_{x(\mathbf{v})\mathbf{y}}\mathbf{C}^{\mathbf{o}\,\dagger}_{\mathbf{yy}}\mathbf{y}, \tag{10.21}$$

where $\mathbf{o} \triangleq [o_1, \ldots, o_M]^T$ and is the vector representing the sensors' transmission status, $\mathbf{C}^{\mathbf{o}}_{\mathbf{yy}} \triangleq E[\mathbf{y}\mathbf{y}^T|\mathbf{o}]$, $\mathbf{C}^{\mathbf{o}}_{x(\mathbf{v})\mathbf{y}} \triangleq E[x(\mathbf{v})\mathbf{y}^T|\mathbf{o}]$, and † represents the Moore–Penrose pseudoinverse. The resulting MSE of the estimate on $x(\mathbf{v})$ is then given by:

$$\xi(\mathbf{v}|\mathbf{o}) = \sigma_x^2 - \mathrm{tr}\left(\mathbf{C}^{\mathbf{o}}_{x(\mathbf{v})\mathbf{y}}\mathbf{C}^{\mathbf{o}\,\dagger}_{\mathbf{yy}}\mathbf{C}^{\mathbf{o}\;T}_{x(\mathbf{v})\mathbf{y}}\right), \tag{10.22}$$

and the average MSE over the entire sensor field is defined as:

$$\bar{\xi} \triangleq \frac{1}{|\mathscr{V}|} \int\limits_{\mathbf{v}\in\mathscr{V}} \sum_{\mathbf{o}\in\mathscr{O}} \xi(\mathbf{v}|\mathbf{o}) \Pr(\mathbf{o}) d\mathbf{v}, \tag{10.23}$$

where $\mathscr{O}$ is the set of all possible realizations of $\mathbf{o}$. Notice that the above expression involves the summation over all possible realizations of $\mathbf{o}$, which can be intractable in practice. To simplify our computations, we consider instead an upper bound of the average MSE given as follows:

$$\bar{\xi} = \frac{1}{|\mathscr{V}|} \int\limits_{\mathbf{v}\in\mathscr{V}} E_{\mathbf{o}} \left[\min_{\mathbf{k}\in\mathbb{R}^M} E\left[|x(\mathbf{v}) - \mathbf{k}^T\mathbf{y}|^2 \, |\mathbf{o} \right] \right] d\mathbf{v} \tag{10.24}$$

$$\leq \frac{1}{|\mathscr{V}|} \int\limits_{\mathbf{v}\in\mathscr{V}} \min_{\mathbf{k}\in\mathbb{R}^M} E_{\mathbf{o}} \left[E\left[|x(\mathbf{v}) - \mathbf{k}^T\mathbf{y}|^2 \, |\mathbf{o} \right] \right] d\mathbf{v} \tag{10.25}$$

$$= \frac{1}{|\mathscr{V}|} \int\limits_{\mathbf{v}\in\mathscr{V}} \min_{\mathbf{k}\in\mathbb{R}^M} E\left[|x(\mathbf{v}) - \mathbf{k}^T\mathbf{y}|^2 \right] d\mathbf{v} \tag{10.26}$$

$$= \frac{1}{|\mathscr{V}|} \int\limits_{\mathbf{v}\in\mathscr{V}} \sigma_x^2 - \mathrm{tr}\left(\mathbf{C}_{x(\mathbf{v})\mathbf{y}} \mathbf{C}_{\mathbf{yy}}^{-1} \mathbf{C}_{x(\mathbf{v})\mathbf{y}}^T \right) d\mathbf{v} \triangleq \bar{\xi}_{\text{upper}}, \tag{10.27}$$

where $\mathbf{C}_{x(\mathbf{v})\mathbf{y}} \triangleq E[x(\mathbf{v})\mathbf{y}^T]$ and $\mathbf{C}_{\mathbf{yy}} \triangleq E[\mathbf{y}\mathbf{y}^T]$ are computed by taking the expectation over all possible realizations of $\mathbf{o}$, and thus, do not depend on the actual subset of transmitting sensors. The vector $\mathbf{k}$ inside the expectation in (10.24) can be viewed as a linear estimator that should be optimized separately for each given value of $\mathbf{v}$ and $\mathbf{o}$. The inequality in (10.25) follows from the fact that the minimization over each term inside the expectation must be smaller than the minimization over the entire expectation (which amounts to using the same $\mathbf{k}$ for each $\mathbf{o}$). The latter can be interpreted as the minimum MSE attainable when the fusion center does not have knowledge of the sensors' locations and transmission status.

Let $\alpha_i \triangleq \Pr(o_i = 1) = \bar{F}_{e_i}(\gamma_i)\lambda_i\Delta$ be the probability that a sensor exists in $\mathscr{V}_i$ and transmits in the given observation period, and let $\bar{g}_i(\gamma_i) \triangleq E[g_i(\gamma_i)|e_i > \gamma_i]$ and $\overline{g_i^2}(\gamma_i) \triangleq E[g_i^2(\gamma_i)|e_i > \gamma_i]$ be the conditional first and second moments of the effective channel gain given that the sensor transmits. In this case, we have $\mathbf{C}_{x(\mathbf{v})\mathbf{y}} = \mathbf{C}_{x(\mathbf{v})\mathbf{x}}\mathbf{D}_{\bar{g}}\mathbf{D}_{\alpha}$, where $\mathbf{D}_{\bar{g}} \triangleq \mathrm{diag}(\bar{g}_1(\gamma_1), \ldots, \bar{g}_M(\gamma_M))$ and $\mathbf{D}_{\alpha} = \mathrm{diag}(\alpha_1, \ldots, \alpha_M)$; thus, the MSE upper bound in (10.27) can be written as:

$$\bar{\xi}_{\text{upper,AF}} = \sigma_x^2 - \mathrm{tr}\left(\mathbf{\Phi} \mathbf{D}_{\bar{g}} \mathbf{D}_{\alpha} \mathbf{C}_{\mathbf{yy}}^{-1} \mathbf{D}_{\alpha} \mathbf{D}_{\bar{g}} \right), \tag{10.28}$$

where $\mathbf{\Phi} \triangleq (1/|\mathscr{V}|) \int_{\mathbf{v}\in\mathscr{V}} \mathbf{C}_{x(\mathbf{v})\mathbf{x}}^T \mathbf{C}_{x(\mathbf{v})\mathbf{x}} d\mathbf{v}$ is defined such that:

$$\{\mathbf{\Phi}\}_{i,j} = \frac{\sigma_x^4}{|\mathscr{V}|} \int\limits_{\mathbf{v}\in\mathscr{V}} \rho(\mathbf{v}, \mathbf{v}_i)\rho(\mathbf{v}, \mathbf{v}_j) d\mathbf{v} \triangleq \phi_{i,j}, \tag{10.29}$$

for all i,j, and the (i,j)th element of $\mathbf{C}_{\mathbf{yy}}$ can be derived as:

$$\{\mathbf{C}_{\mathbf{yy}}\}_{i,j} = \begin{cases} [\overline{g_i^2}(\gamma_i)(\sigma_x^2+\sigma_n^2)+\sigma_w^2]\alpha_i, & \text{for } i=j \\ \bar{g}_i(\gamma_i)\bar{g}_j(\gamma_j)\sigma_x^2\rho(\mathbf{v}_i,\mathbf{v}_j)\alpha_i\alpha_j, & \text{for } i\neq j, \end{cases} \tag{10.30}$$

for all i,j. Notice that Φ is a matrix that depends only on the correlation of the sensor field and not on the optimizing parameters $\{\lambda_i\}_{i=1}^M$ and $\{\gamma_i\}_{i=1}^M$. Let us take the eigenvalue decomposition of $\mathbf{\Phi}$ so that $\mathbf{\Phi} = \mathbf{U}\mathbf{\Sigma}\mathbf{U}^T$, where $\mathbf{\Sigma}$ is a diagonal matrix consisting of the eigenvalues of $\mathbf{\Phi}$, and $\mathbf{U}$ is a unitary matrix consisting of the corresponding eigenvectors. In this case, the term in (10.28) can be written as:

$$\bar{\xi}_{\text{upper,AF}} = \sigma_x^2 - \text{tr}\left(\mathbf{\Sigma}^{1/2}\mathbf{U}^T\mathbf{D}_{\bar{g}}\mathbf{D}_\alpha\mathbf{C}_{\mathbf{yy}}^{-1}\mathbf{D}_\alpha\mathbf{D}_{\bar{g}}\mathbf{U}\mathbf{\Sigma}^{1/2}\right). \tag{10.31}$$

It follows by Cauchy–Schwarz inequality that:

$$\bar{\xi}_{\text{upper,AF}} \leq \sigma_x^2 - \frac{\left[\text{tr}\left(\mathbf{\Sigma}^{1/2}\mathbf{U}^T(\mathbf{D}_{\bar{g}}\mathbf{D}_\alpha)^2\mathbf{U}\mathbf{\Sigma}^{1/2}\right)\right]^2}{\text{tr}\left(\mathbf{\Sigma}^{1/2}\mathbf{U}^T\mathbf{D}_{\bar{g}}\mathbf{D}_\alpha\mathbf{C}_{\mathbf{yy}}\mathbf{D}_\alpha\mathbf{D}_{\bar{g}}\mathbf{U}\mathbf{\Sigma}^{1/2}\right)} \tag{10.32}$$

$$= \sigma_x^2 - \frac{\left[\text{tr}\left(\mathbf{\Phi}(\mathbf{D}_{\bar{g}}\mathbf{D}_\alpha)^2\right)\right]^2}{\text{tr}\left(\mathbf{\Phi}\mathbf{D}_{\bar{g}}\mathbf{D}_\alpha\mathbf{C}_{\mathbf{yy}}\mathbf{D}_\alpha\mathbf{D}_{\bar{g}}\right)} \triangleq \bar{\xi}_{\text{obj,AF}}(\lambda,\gamma), \tag{10.33}$$

where $\lambda = [\lambda_1, \ldots, \lambda_M]^T$ is the vector of sensor densities and $\gamma = [\gamma_1, \ldots, \gamma_M]^T$ is the vector of energy thresholds. The upper bound $\bar{\xi}_{\text{obj,AF}}(\lambda,\gamma)$ is then utilized as the objective function for optimizing λ and γ in the AF case. The bound captures the essential behaviors of the MSE with respect to the sensor densities and transmission thresholds and, thus, allows us to obtain a tractable solution to an otherwise intractable problem.

10.4.1 Optimization of sensor densities and energy thresholds

In this section, we determine the optimal sensor densities $\{\lambda_i\}_{i=1}^M$ and energy thresholds $\{\gamma_i\}_{i=1}^M$ in AF systems based on the minimization of the average MSE upper bound in (10.27).

Specifically, let us consider the optimization problem where the MSE upper bound $\bar{\xi}_{\text{obj,AF}}$ in (10.33) is minimized subject to a constraint on the average total number of sensors, i.e.,

$$\min_{\Lambda,\gamma} \quad \bar{\xi}_{\text{obj,AF}}(\lambda,\gamma) \tag{10.34a}$$

$$\text{subject to} \quad \sum_{i=1}^M \lambda_i\Delta \leq \bar{\Lambda}, \tag{10.34b}$$

$$0 < \lambda_i\Delta \leq \varepsilon_\Lambda, \quad \text{for } i=1,\ldots,M, \tag{10.34c}$$

where $\bar{\Lambda}$ is the constraint on the average number of sensors in the network. Here, the approximate probability that a sensor exists in subregion $\mathscr{V}_i$, i.e., $\lambda_i\Delta$, is restricted within $[0,\varepsilon_\Lambda]$ and $\varepsilon_\Lambda < 1$ should be small enough to make the approximation accurate enough. The value of λ_i is assumed to be positive to ensure that $\mathbf{C}_{\mathbf{yy}}$ is invertible. Following the relation in (10.18), the energy threshold γ determines the distribution of the accumulated energy process and, thus, affects the objective function through the probabilities α_i, for all i, which involve the event that a sensor exists and transmits in each subregion. Since the first term in (10.33) is a constant, the optimization problem can be equivalently formulated as the maximization over the second term in (10.33). By the change of variable $\Lambda_i = \lambda_i \Delta$ and by defining $\mathbf{\Lambda} = [\Lambda_1, \ldots, \Lambda_M]^T$, the optimization problem can be written explicitly as follows:

$$\max_{\mathbf{\Lambda},\gamma} \quad J_{\mathrm{AF}}(\mathbf{\Lambda}, \boldsymbol{\gamma}) = \frac{J_{\mathrm{AF}}^{\mathrm{num}}(\mathbf{\Lambda}, \boldsymbol{\gamma})}{J_{\mathrm{AF}}^{\mathrm{den}}(\mathbf{\Lambda}, \boldsymbol{\gamma})} \tag{10.35a}$$

$$\text{subject to} \quad \sum_{i=1}^{M} \Lambda_i \le \bar{\Lambda}, \tag{10.35b}$$

$$0 < \Lambda_i \le \varepsilon_\Lambda, \quad \text{for } i = 1, \ldots, M. \tag{10.35c}$$

The term in the numerator of the objective function is:

$$J_{\mathrm{AF}}^{\mathrm{num}}(\mathbf{\Lambda}, \boldsymbol{\gamma}) = \left(\sum_{i=1}^{M} \phi_{i,i} [\bar{g}_i(\gamma_i)]^2 \bar{F}_{e_i}^2(\gamma_i) \Lambda_i^2 \right)^2 \tag{10.36}$$

and that in the denominator is:

$$J_{\mathrm{AF}}^{\mathrm{den}}(\mathbf{\Lambda}, \boldsymbol{\gamma}) = \sum_{i=1}^{M} \phi_{i,i} \left[\overline{g_i^2}(\gamma_i) \sigma_{\tilde{x}}^2 + \sigma_w^2 \right] \bar{F}_{e_i}^3(\gamma_i) \Lambda_i^3 \bar{g}_i^2(\gamma_i)$$
$$+ \sum_{i=1}^{M} \sum_{j \ne i} \phi_{i,j} \sigma_x^2 \rho(\mathbf{v}_i, \mathbf{v}_j) \bar{F}_{e_i}^2(\gamma_i) \Lambda_i^2 \bar{g}_i^2(\gamma_i) \bar{F}_{e_j}^2(\gamma_j) \Lambda_j^2 \bar{g}_j^2(\gamma_j), \tag{10.37}$$

where $\sigma_{\tilde{x}}^2 = \sigma_x^2 + \sigma_n^2$.

Notice that the solution to the optimization problem in (10.35) depends on the statistics of the energy arrival. In the following, we provide an example of how such a problem can be solved under the Bernoulli energy arrival model.

10.4.2 Solution for Bernoulli energy arrival case

In this subsection, we apply the above techniques to cases where the energy arrival can be modeled as an i.i.d. Bernoulli process. This model has been widely adopted in the literature, e.g., in [33–36], due to its tractability and because of its ability to model sources with sporadic energy arrivals in discrete time. Examples of such energy sources may include vibrational energy, RF energy, or wind, etc. It can also be used to approximate more stable energy sources, such as solar, by choosing each time slot to be sufficiently small so that energy will arrive approximately at a constant rate when viewed from a larger timescale. Other energy arrival models can also fit into our framework but may require different optimization techniques from that to be introduced in the following.

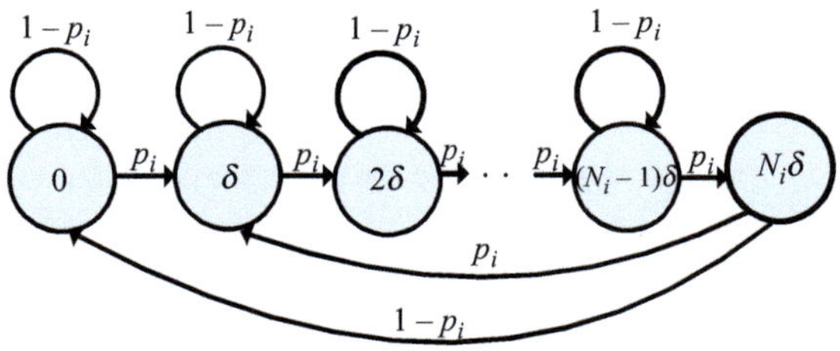

Figure 10.6 Energy harvesting model of Bernoulli energy arrival

Suppose that the energy-arrival process $\{a_i[t]\}_{t=0}^{\infty}$ consists of a sequence of i.i.d. Bernoulli random variables with $\Pr(a_i[t] = \delta) = p_i$ and $\Pr(a_i[t] = 0) = 1 - p_i$, $\forall t$. That is, the sensor in $\mathcal{V}_i$ harvests energy δ in each observation period with probability p_i and fails to harvest energy with probability $1 - p_i$. Since the sensor expends all its energy once the energy threshold γ_i is exceeded,[2] the accumulated energy $\{e_i[t]\}_{t=0}^{\infty}$ forms a time-invariant $(N_i + 1)$-state Markov process, as illustrated in Figure 10.6, where $N_i \triangleq \min\{n : n\delta \geq \gamma_i\}$. This Markov chain is irreducible and aperiodic, and all of its states are positive recurrent. As a result, the stationary (or steady-state) distribution is unique [37] and is given by $\Pr(e_i[t] = 0) = ((1 - p_i)/N_i)$, $\Pr(e_i[t] = N_i\delta) = (p_i/N_i)$, and $\Pr(e_i[t] = k\delta) = (1/N_i)$ for $k = 1, \ldots, N_i - 1$. By assuming that $e_i[0]$ follows the stationary distribution, the distribution of the accumulated energy will remain the same after any number of transitions, and the process will indeed be stationary. In this case, the probability that the sensor in subregion $\mathcal{V}_i$ transmits (if it exists) is $\bar{F}_{e_i}(\gamma_i) = \Pr(e_i[t] = N_i\delta)$. In the Bernoulli energy arrival model, the accumulated energies are integer multiples of δ and, thus, the energy threshold γ_i can also be set as a multiple of δ, in which case, we have $\gamma_i = N_i\delta$. Consequently, we have $\bar{F}_{e_i}(\gamma_i) = (p_i\delta/\gamma_i)$, $\bar{g}_i(\gamma_i) = (\tilde{h}_i\sqrt{\gamma_i}/\sigma_{\tilde{x}})$, and $\overline{g_i^2}(\gamma_i) = (\tilde{h}_i^2\gamma_i/\sigma_{\tilde{x}}^2)$, where $\tilde{h}_i \triangleq \sqrt{\kappa_i}h_i$. Notice that the stationary distribution as well as these values change as the energy threshold γ_i is adjusted. By substituting the above into (10.36) and (10.37), we get:

$$J_{\mathrm{AF}}^{\mathrm{num}}(\mathbf{\Lambda}, \boldsymbol{\gamma}) = \left(\sum_{i=1}^{M} \phi_{i,i}\tilde{h}_i^2 p_i^2 \delta^2 \Lambda_i^2 \gamma_i^{-1}\right)^2 \qquad (10.38)$$

and

$$\begin{aligned} J_{\mathrm{AF}}^{\mathrm{den}}(\mathbf{\Lambda}, \boldsymbol{\gamma}) &= \sum_{i=1}^{M} \phi_{i,i}(\tilde{h}_i^2\gamma_i + \sigma_w^2)\tilde{h}_i^2 \sigma_{\tilde{x}}^2 p_i^3 \delta^3 \Lambda_i^3 \gamma_i^{-2} \\ &\quad + \sum_{i=1}^{M}\sum_{j\neq i} \phi_{i,j}\sigma_x^2 \rho(\mathbf{v}_i, \mathbf{v}_j) p_i^2 \delta^2 \tilde{h}_i^2 \Lambda_i^2 \gamma_i^{-1} p_j^2 \delta^2 \tilde{h}_j^2 \Lambda_j^2 \gamma_j^{-1}, \end{aligned} \qquad (10.39)$$

[2]In practice, the accumulated energy may not be depleted after each transmission in the AF case since the required transmit power is random. However, we assume for simplicity that the remaining energy is omitted after each transmission.

and, thus, the optimization problem in (10.35) can be written as:

$$\max_{\Lambda,\gamma} \frac{\sum_{i=1}^{M}\sum_{j=1}^{M} A_{i,j}(\Lambda_i^2\Lambda_j^2/\gamma_i\gamma_j)}{\sum_{i=1}^{M} C_i(\Lambda_i^3/\gamma_i)+\sum_{i=1}^{M} D_i(\Lambda_i^3/\gamma_i^2)+\sum_{i=1}^{M}\sum_{j\neq i} G_{i,j}(\Lambda_i^2\Lambda_j^2/\gamma_i\gamma_j)} \tag{10.40a}$$

$$\text{subject to} \sum_{i=1}^{M} \Lambda_i \leq \bar{\Lambda}, \tag{10.40b}$$

$$0 < \Lambda_i \leq \varepsilon_\Lambda, \qquad i = 1, \ldots, M, \tag{10.40c}$$

$$\gamma_i = Z_i\delta,\ Z_i \in \mathbb{N}, \qquad i = 1, \ldots, M, \tag{10.40d}$$

where $A_{i,j} \triangleq \phi_{i,i}\phi_{j,j}\tilde{h}_i^2\tilde{h}_j^2 p_i^2 p_j^2\delta^4$, $C_i \triangleq \phi_{i,i}\tilde{h}_i^4 p_i^3\delta^3\sigma_{\tilde{x}}^2$, $D_i \triangleq \phi_{i,i}\sigma_{wi}^2 p_i^3\delta^3\tilde{h}_i^2\sigma_{\tilde{x}}^2$, and $G_{i,j} \triangleq \phi_{i,j}\sigma_x^2\,\rho(\mathbf{v}_i, \mathbf{v}_j)p_i^2 p_j^2\tilde{h}_i^2\tilde{h}_j^2\delta^4$. Notice that this problem is a mixed integer nonlinear programming problem, which is difficult to solve in general. Therefore, we consider a relaxation where the integer constraint on γ_i is replaced with the inequality constraint $\gamma_i \geq \delta$. Then, by defining $f(\Lambda,\gamma) \triangleq \sum_{i=1}^{M}\sum_{j=1}^{M} A_{i,j}(\Lambda_i^2\Lambda_j^2/\gamma_i\gamma_j)$, the optimization problem can be written equivalently as:

$$\min_{\Lambda,\gamma} \frac{\sum_{i=1}^{M} C_i(\Lambda_i^3/\gamma_i)+\sum_{i=1}^{M} D_i(\Lambda_i^3/\gamma_i^2)+\sum_{i=1}^{M}\sum_{j\neq i} G_{i,j}(\Lambda_i^2\Lambda_j^2/\gamma_i\gamma_j)}{f(\Lambda,\gamma)} \tag{10.41a}$$

$$\text{subject to} \sum_{i=1}^{M} \Lambda_i \leq \bar{\Lambda}, \tag{10.41b}$$

$$0 < \Lambda_i \leq \varepsilon_\Lambda \text{ and } \gamma_i \geq \delta, \forall i. \tag{10.41c}$$

Notice that the relaxed problem is still non-convex, but can be approximated by a series of geometric programming (GP) problems using the condensation method [38].

Specifically, in each iteration of the condensation method, the function $f(\Lambda,\gamma)$ is replaced by its monomial approximation so that the objective function in (10.41a) becomes a posynomial and that the problem in (10.41) can be formulated as a standard GP problem. Precisely, more specifically, let $(\tilde{\Lambda}^{(\ell)}, \tilde{\gamma}^{(\ell)})$ be the solution obtained in the ℓth iteration of the condensation method. Then, based on the inequality between arithmetic and geometric means [39], it follows that:

$$f(\Lambda,\gamma) = \sum_{i=1}^{M}\sum_{j=1}^{M} \frac{A_{i,j}\Lambda_i^2\Lambda_j^2}{\gamma_i\gamma_j} \geq \prod_{i=1}^{M}\prod_{j=1}^{M}\left(\frac{A_{i,j}\Lambda_i^2\Lambda_j^2}{\beta_{i,j}^{(\ell)}\gamma_i\gamma_j}\right)^{\beta_{i,j}^{(\ell)}} \triangleq \hat{f}^{(\ell)}(\Lambda,\gamma), \tag{10.42}$$

Algorithm 1: Optimized sensor densities and optimized energy thresholds (OI-OET) scheme in the AF system

Initialization: Set $\ell = 0$ and a solution accuracy $\varepsilon > 0$. Find a feasible solution $\left(\tilde{\Lambda}^{(0)}, \tilde{\gamma}^{(0)}\right)$.

Iteration:

1. Compute $\beta_{i,j}^{(\ell)}$, for $i,j = 1, \ldots, M$, by (10.43).
2. Replace $f(\Lambda, \gamma)$ with $\hat{f}^{(\ell)}(\Lambda, \gamma)$ and solve (10.41). Let the solution be $\left(\tilde{\Lambda}^{(\ell+1)}, \tilde{\gamma}^{(\ell+1)}\right)$.
3. Repeat (1) and (2) until $\frac{|J_{\mathrm{AF}}(\tilde{\Lambda}^{(\ell+1)},\tilde{\gamma}^{(\ell+1)})-J_{\mathrm{AF}}(\tilde{\Lambda}^{(\ell)},\tilde{\gamma}^{(\ell)})|}{J_{\mathrm{AF}}(\tilde{\Lambda}^{(\ell)},\tilde{\gamma}^{(\ell)})} \leq \varepsilon$.

where

$$\beta_{i,j}^{(\ell)} = A_{i,j} \frac{((\tilde{\Lambda}_i^{(\ell)})^2 (\tilde{\Lambda}_j^{(\ell)})^2 / \tilde{\gamma}_i^{(\ell)} \tilde{\gamma}_j^{(\ell)})}{f\left(\tilde{\Lambda}^{(\ell)}, \tilde{\gamma}^{(\ell)}\right)}. \tag{10.43}$$

By approximating $f(\Lambda, \gamma)$ with $\hat{f}^{(\ell)}(\Lambda, \gamma)$, the optimization problem in (10.41) becomes a standard GP, which can be converted to a convex optimization problem and solved using the interior point method. Note that solving the problem given in (10.41) by replacing $f(\Lambda, \gamma)$ with $\hat{f}^{(\ell)}(\Lambda, \gamma)$ yields a solution that is also feasible in (10.41) and can be used to find the solution in the next iteration of the condensation method. By initiating with a feasible solution $(\tilde{\Lambda}^{(0)}, \tilde{\gamma}^{(0)})$, the above process can be repeated until the objective value converges. The procedure is summarized in Algorithm 1. An approximated solution to the original problem in (10.40) is thus obtained, and the optimized energy threshold is further rounded to the nearest multiple of δ. The effectiveness of this scheme is demonstrated through Monte Carlo simulations in Section 10.4.3.

10.4.2.1 Complexity reduction

It is necessary to note that, even though the above approach can yield good solutions to the random sensor deployment problem, the complexity can be high when the number of subregions, M, is large. This is not a problem in most cases since the sensor densities and energy thresholds need only be computed offline. However, if it is necessary to reduce the complexity without reducing the resolution of the subregions, one can further reduce the number of parameters by assuming that the parameters in neighboring subregions are the same. This assumption is reasonable since the dimension of a subregion is assumed to be much smaller than the spatial variations of the sensor field and the energy arrivals.

Suppose that the M subregions are combined into N clusters $\mathscr{C}_1, \mathscr{C}_2, \ldots, \mathscr{C}_N$, each consisting of $M_C = M/N$ subregions. Let $\Lambda_C = [\Lambda_{(1)}, \Lambda_{(2)}, \ldots, \Lambda_{(N)}]$ and $\gamma_C = [\gamma_{(1)}, \gamma_{(2)}, \ldots, \gamma_{(N)}]$, where $\Lambda_{(n)}$ and $\gamma_{(n)}$ are the values of Λ_i and γ_i, respectively, for all $i \in \mathscr{C}_n$. Then, by letting $A_{(n),(m)} \triangleq \sum_{i \in \mathscr{C}_n} \sum_{j \in \mathscr{C}_m} A_{i,j}$, $C_{(n)} \triangleq \sum_{i \in \mathscr{C}_n} C_i$,

$D_{(n)} \triangleq \sum_{i \in \mathscr{C}_n} D_i$, and $G_{(n),(m)} \triangleq \sum_{i \in \mathscr{C}_n} \sum_{j \in \mathscr{C}_{m}, i \neq j} G_{i,j}$, the optimization problem in (10.41) can be reduced:

$$\min_{\Lambda_C, \gamma_C} \frac{\sum_{n=1}^{N} \left[C_{(n)}(\Lambda_{(n)}^3/\gamma_{(n)}) + D_{(n)}(\Lambda_{(n)}^3/\gamma_{(n)}^2) + \sum_{m=1}^{N} G_{(n),(m)}(\Lambda_{(n)}^2 \Lambda_{(m)}^2/\gamma_{(n)}\gamma_{(m)}) \right]}{f_C(\Lambda_C, \gamma_C)} \tag{10.44a}$$

$$\text{subject to} \sum_{n=1}^{N} M_C \Lambda_{(n)} \leq \bar{\Lambda}, \tag{10.44b}$$

$$0 < \Lambda_{(n)} \leq \varepsilon_{\Lambda}, \ \gamma_{(n)} \geq \delta, \ n = 1, \ldots, N. \tag{10.44c}$$

The problem can then be solved using the condensation method, similar to that in Algorithm 1, and the complexity is reduced to $O((M/M_C)^7)$.

10.4.3 Computer simulations and performance comparisons

In this section, Monte Carlo simulations are performed to demonstrate the effectiveness of the proposed sensor deployment and transmission control schemes. In the experiments, the sensors are deployed in a square area, denoted by $\mathscr{V} = [0, L) \times [0, L)$ with $L = 100$, which is further partitioned into $M = 100$ square subregions of equal area. By adopting the exponential field correlation model [40], the correlation coefficient between sensor measurements $x(\mathbf{v}_i)$ and $x(\mathbf{v}_j)$ is $\rho(\mathbf{v}_i, \mathbf{v}_j) = \exp(-(\|\mathbf{v}_i - \mathbf{v}_j\|/\theta_1)^{\theta_2})$. Here, $\theta_1 > 0$ is the parameter which controls the decay with respect to distance and $\theta_2 = 2$ is chosen as the smoothness parameter [40]. The variances of the random field and the observation noise are set as $\sigma_x^2 = 1$ and $\sigma_n^2 = 0.01$, respectively. The energy harvested at the sensors are provided by $K_E = 4$ radio frequency (RF) energy sources that are located at points $\mathbf{v}_{E_1} = ((1/10)L, (1/10)L)$, $\mathbf{v}_{E_2} = ((1/10)L, (9/10)L)$, $\mathbf{v}_{E_3} = ((7/10)L, (7/10)L)$, and $\mathbf{v}_{E_4} = ((5/10)L, (2/10)L)$, and the probability of energy arrival in subregion $\mathscr{V}_i$ is modeled as $p_i = \min\{\sum_{j=1}^{K_E} \exp(-\theta_e \|\mathbf{v}_i - \mathbf{v}_{E_j}\|), 1\}$, which decays exponentially with rate $\theta_e > 0$ with respect to distance. The unit energy is set as the scavenged energy from an electric field of 1 V/m, i.e., $\delta = 2 \times 10^{-4}$ mW/cm^2 [7]. Moreover, we assume that there are $K_D = 9$ data-gathering nodes located at $\mathbf{v}_{D_1} = ((1/10)L, (1/10)L)$, $\mathbf{v}_{D_2} = ((1/10)L, (3/10)L)$, $\mathbf{v}_{D_3} = ((1/10)L, (7/10)L)$, $\mathbf{v}_{D_4} = ((1/10)L, (9/10)L)$, $\mathbf{v}_{D_5} = ((4/10)L, (6/10)L)$, $\mathbf{v}_{D_6} = ((6/10)L, (4/10)L)$, $\mathbf{v}_{D_7} = ((6/10)L, (6/10)L)$, $\mathbf{v}_{D_8} = ((7/10)L, (7/10)L)$, and $\mathbf{v}_{D_9} = ((7/10)L, (9/10)L)$. By assuming that each sensor transmits only to its closest data-gathering node and by the path loss model, the channel coefficient experienced by a sensor in $\mathscr{V}_i$ is given by $h_i = \min\{\max_{j=1,\ldots,K_D} \|\mathbf{v}_i - \mathbf{v}_{D_j}\|^{-1}, 1\}$. Finally, we set $\varepsilon_{\Lambda} = 0.35$.

In the following, we evaluate the average MSE $\bar{\xi}$ under the parameters specified by five different schemes: (i) the proposed scheme with optimized sensor densities and optimized energy thresholds (OD-OET); (ii) the scheme with only optimized sensor densities but fixed energy thresholds (OD-FET), i.e., the solutions to (10.41) with $\gamma_i =$

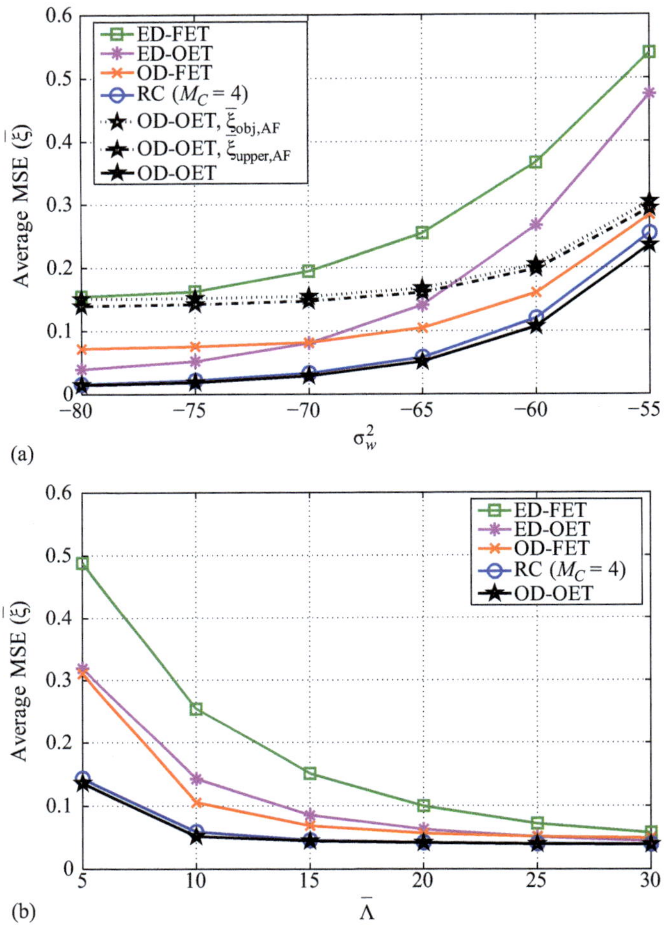

Figure 10.7 Average MSE in the AF system with $\sigma_w^2 = -65$ dB m, $\theta_l = 400$, $\theta_e = 0.06$, and $\bar{\Lambda} = 10$, unless the parameters are varied: (a) MSE vs channel noise variance σ_w^2 in dB m and (b) MSE vs the constraint on the average number of sensors $\bar{\Lambda}$. ©2015 IEEE. Reprinted, with permission, from Reference [11]

2δ, $\forall i$; (iii) the scheme with equal sensor densities but with optimized energy thresholds (ED-OET), i.e., the solutions to (10.41) with $\Lambda_i = (\bar{\Lambda}/M)$, $\forall i$; (iv) the scheme with equal sensor densities $\Lambda_i = (\bar{\Lambda}/M)$ and fixed energy thresholds (ED-FET) $\gamma_i = 2\delta$, $\forall i$; and (v) the reduced-complexity (RC) scheme, i.e., the solution to (10.44) for the AF case. In the schemes with FET, we choose $\gamma_i = 2\delta$ as a trade-off between the transmit probability and the transmission quality. The amplifying factor is set as $\kappa_i = 0.6$, which limits the probability of saturation below 0.1. In all figures, the proposed OD-OET scheme is shown to outperform all other schemes due to the optimization of both the sensor densities and the energy thresholds. Moreover, the proposed RC

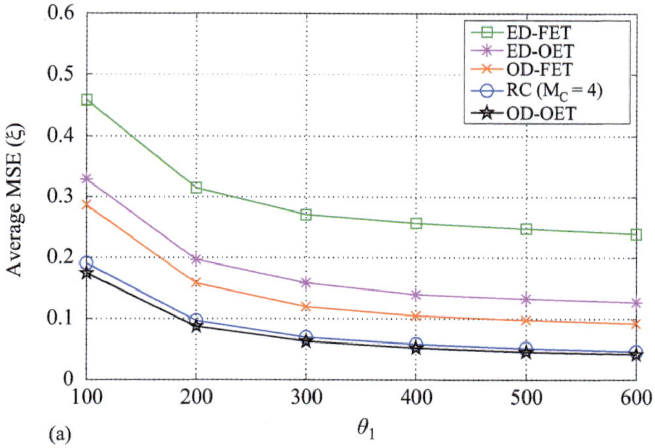

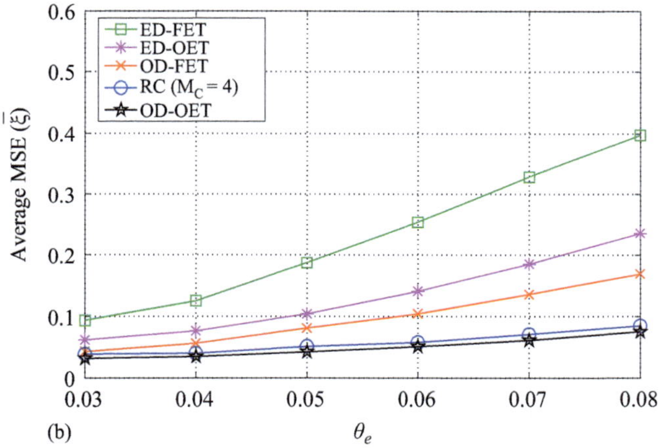

Figure 10.8 Average MSE in the AF system with $\sigma_w^2 = -65\,\text{dB m}$, $\theta_l = 400$, $\theta_e = 0.06$ and $\bar{\Lambda} = 10$, unless the parameters are varied: (a) MSE vs field correlation parameter θ_l and (b) MSE vs energy arrival probability exponent θ_e. ©2015 IEEE. Reprinted, with permission, from Reference [11]

scheme also closely approximates the performance of the OD-OET scheme in most cases, which further validates the effectiveness of the proposed approach.

In Figure 10.7(a), we observe that the amount of MSE improvement provided by the proposed scheme increases with the noise variance. This is because, as the noise variance increases, the sensors must be more carefully placed to obtain more energy or achieve higher channel gains so as to combat the effect of noise. The upper bounds $\bar{\xi}_{\text{upper,AF}}$ and $\bar{\xi}_{\text{obj,AF}}$ of the proposed OD-OET scheme, as given in (10.28) and (10.33), respectively, are also illustrated in the figure for comparison. The latter is the objective of our optimization problem. Even though the bounds may not be tight in general, they demonstrate the same behaviors as the actual MSE and, thus, the solutions obtained by minimizing the upper bounds can indeed provide significant MSE improvement. Moreover, in Figure 10.7(b), we can see that the MSE decreases as the constraint $\bar{\Lambda}$ on the average number of sensors increases and that the MSE improvement is particularly significant when $\bar{\Lambda}$ is small. This is reasonable since the optimality of the sensor deployment would be more critical when the average number of sensors is small.

Furthermore, in Figure 10.8(a), we can see that the MSE decreases as θ_1 increases and that a steady MSE improvement is provided by the OD-OET scheme. This is reasonable since the increase of θ_1 causes the correlation between sensor measurements to increase and, thus, allows each point of the random field to be more accurately estimated by observations in other subregions. Also, in Figure 10.8(b), we can see that the MSE increases with the decay rate θ_e of the energy arrival probability. We can see that the MSE improvement provided by the schemes with either optimized sensor densities or optimized energy thresholds increases significantly as θ_e increases since, in this case, the energy arrival probability decays rapidly over space and, thus, sensors must be deployed more carefully to obtain sufficient energy for transmission.

10.5 Conclusion

In this chapter, we examined the impact of energy harvesting on the distributed estimation performance in WSNs. First, we examined the optimal AF transmission policy at the sensors and the optimal estimation scheme at the fusion center taking into consideration the randomness and intermittence of the energy arrivals. Both cases with and without batteries are considered. Then, we further examined the impact of the spatial heterogeneity of the energy arrivals in the deployment of energy harvesting sensors. These works are introduced as representative examples to illustrate the need to redesign many communications and networking protocols when the system relies on energy harvesting as the main energy source.

References

[1] Akyildiz IF, Su W, Sankarasubramaniam Y, *et al.* Wireless sensor networks: a survey. Computer Networks. 2002;38(4):393–422.

[2] Raghunathan V, Ganeriwal S, and Srivastava M. Emerging techniques for long lived wireless sensor networks. IEEE Communications Magazine. 2006;44(4): 108–114.

[3] Seah WK, Eu ZA, and Tan HP. Wireless sensor networks powered by ambient energy harvesting (WSN-HEAP)-survey and challenges. In: Proceedings of Wireless Communication, Vehicular Technology, Information Theory and Aerospace & Electronic Systems Technology. IEEE; 2009. p. 1–5.

[4] Sudevalayam S, and Kulkarni P. Energy harvesting sensor nodes: survey and implications. IEEE Communications Surveys Tutorials. 2011;13(3):443–461.

[5] Taneja J, Jeong J, and Culler D. Design, modeling, and capacity planning for micro-solar power sensor networks. In: Proceedings of International Conference on Information Processing in Sensor Networks (IPSN); 2008. p. 407–418.
[6] Zhu Q, Guan M, and He Y. Vibration energy harvesting in automobiles to power wireless sensors. In: Proceedings of IEEE International Conference on Information and Automation; 2012. p. 349–354.
[7] Paradiso JA, and Starner T. Energy scavenging for mobile and wireless electronics. IEEE Pervasive Computing. 2005;4(1):18–27.
[8] Ribeiro A, and Giannakis GB. Bandwidth-constrained distributed estimation for wireless sensor networks-part I: Gaussian case. IEEE Transactions on Signal Processing. 2006;54(3):1131–1143.
[9] Cui S, Xiao JJ, Goldsmith AJ, *et al.* Estimation diversity and energy efficiency in distributed sensing. IEEE Transactions on Signal Processing. 2007;55(9):4683–4695.
[10] Hong YWP. Distributed estimation with analog forwarding in energy-harvesting wireless sensor networks. In: Proceedings of IEEE International Conference on Communication Systems (ICCS). IEEE; 2014. p. 142–146.
[11] Hsu TC, Hong YWP, and Wang TY. Optimized random deployment of energy harvesting sensors for field reconstruction in analog and digital forwarding systems. IEEE Transactions on Signal Processing. 2015;63(19): 5194–5209.
[12] Msechu EJ, and Giannakis GB. Sensor-centric data reduction for estimation with WSNs via censoring and quantization. IEEE Transactions on Signal Processing. 2012;60(1):400–414.
[13] Wang J. Distributed estimation with digital forwarding in energy-harvesting wireless sensor networks. Master Thesis, National Tsing Hua University. Taiwan; 2016.
[14] Baccelli F, and Błaszczyszyn B, Stochastic geometry and wireless networks: volume-I. Foundations and Trends® in Networking. 2010;3(3–4):249–449.
[15] Huang C, Zhou Y, Jiang T, *et al.* Power allocation for joint estimation with energy harvesting constraints. In: Proceedings of IEEE International Conference on Acoustics, Speech and Signal Processing; 2013. p. 4804–4808.
[16] Zhao Y, Chen B, and Zhang R. Optimal power allocation for an energy harvesting estimation system. In: Proceedings of IEEE International Conference on Acoustics, Speech and Signal Processing; 2013. p. 4549–4553.
[17] Nayyar A, Basar T, Teneketzis D, *et al.* Optimal strategies for communication and remote estimation with an energy harvesting sensor. IEEE Transactions on Automatic Control. 2013;58(9):2246–2260.
[18] Nourian M, Dey S, and Ahlén A. Distortion minimization in multi-sensor estimation with energy harvesting. IEEE Journal on Selected Areas in Communications. 2015;33(3):524–539.
[19] Mai VV, Shin WY, and Ishibashi K. Wireless power transfer for distributed estimation in sensor networks. IEEE Journal of Selected Topics in Signal Processing. 2017;11(3):549–562.

[20] Liu S, Sharma V, and Varshney PK. Towards an online energy allocation policy for distributed estimation with sensor collaboration using energy harvesting sensors. In: Proceedings of IEEE Global Conference on Signal and Information Processing (GlobalSIP); 2016. p. 500–504.

[21] Younis M, and Akkaya K. Strategies and techniques for node placement in wireless sensor networks: a survey. Ad Hoc Networks. 2008;6(4):621–655.

[22] Yang Y, and Blum RS. Sensor placement in Gaussian random field via discrete simulation optimization. IEEE Signal Processing Letters. 2008;15:729–732.

[23] Ishizuka M, and Aida M. Performance study of node placement in sensor networks. In: Proceedings of 24th International Conference on Distributed Computing Systems Workshops; 2004. p. 598–603.

[24] Huang CF, and Tseng YC. The Coverage problem in a wireless sensor network. Mobile Networks and Applications. 2005;10(4):519–528.

[25] Eu ZA, Tan HP, and Seah WKG. Routing and relay node placement in wireless sensor networks powered by ambient energy harvesting. In: Proceedings of IEEE Wireless Communications and Networking Conference; 2009. p. 1–6.

[26] Misra S, Majd NE, and Huang H. Constrained relay node placement in energy harvesting wireless sensor networks. In: Proceedings of IEEE Eighth International Conference on Mobile Ad-Hoc and Sensor Systems; 2011. p. 25–34.

[27] Kwon T, and Cioffi JM. Random deployment of data collectors for serving randomly-located sensors. IEEE Transactions on Wireless Communications. 2013;12(6):2556–2565.

[28] Tang J, Hao B, and Sen A. Relay node placement in large scale wireless sensor networks. Computer Communications. 2006;29(4):490–501. Current areas of interest in wireless sensor networks designs.

[29] Djuric PM, Kotecha JH, Zhang J, *et al.* Particle filtering. IEEE Signal Processing Magazine. 2003;20(5):19–38.

[30] Nevat I, Peters GW, and Collings IB. Random field reconstruction with quantization in wireless sensor networks. IEEE Transactions on Signal Processing. 2013;61(23):6020–6033.

[31] Scaglione A, and Hong YW. Opportunistic large arrays: cooperative transmission in wireless multihop ad hoc networks to reach far distances. IEEE Transactions on Signal Processing. 2003;51(8):2082–2092.

[32] Huang K. Spatial throughput of mobile ad Hoc networks powered by energy harvesting. IEEE Transactions on Information Theory. 2013;59(11):7597–7612.

[33] Medepally B, Mehta NB, and Murthy CR. Implications of energy profile and storage on energy harvesting sensor link performance. In: Proceedings of IEEE Global Telecommunications Conference; 2009. p. 1–6.

[34] Michelusi N, Stamatiou K, and Zorzi M. On optimal transmission policies for energy harvesting devices. In: Proceedings of Information Theory and Applications Workshop; 2012. p. 249–254.

[35] Kashef M, and Ephremides A. Optimal packet scheduling for energy harvesting sources on time varying wireless channels. Journal of Communications and Networks. 2012;14(2):121–129.

[36] Michelusi N, Stamatiou K, and Zorzi M. Transmission policies for energy harvesting sensors with time-correlated energy supply. IEEE Transactions on Communications. 2013;61(7):2988–3001.

[37] Ross SM. Stochastic process, 2nd Ed. New York: John Wiley & Sons Inc.; 1996.

[38] Boyd S, Kim SJ, Vandenberghe L, *et al.* A tutorial on geometric programming. Optimization and Engineering. 2007;8(1):67.

[39] Chiang M, Tan CW, Palomar DP, *et al.* Power control by geometric programming. IEEE Transactions on Wireless Communications. 2007;6(7):2640–2651.

[40] Vuran MC, Akan ÖB, and Akyildiz IF. Spatio-temporal correlation: theory and applications for wireless sensor networks. Computer Networks. 2004;45(3):245–259. In Memory of Olga Casals.

Chapter 11

Secure estimation in wireless sensor networks in the presence of an eavesdropper

Xiaoxi Guo[1], Alex S. Leong[2], and Subhrakanti Dey[3]

In this chapter, we investigate the performance of distributed estimation schemes in wireless sensor networks (WSNs), in the presence of an eavesdropper. The sensors transmit observations to the fusion center (FC), which at the same time are overheard by the eavesdropper. Both the FC and the eavesdropper reconstruct a minimum mean squared error (MSE) (MMSE) estimate of the physical quantity observed. We address the problem of transmit power allocation for system performance optimization subject to a total average power constraint on the sensor(s), and a security/secrecy constraint on the eavesdropper. We introduce a notion of *security in estimation* that aims to keep the eavesdropper's MSE or distortion above a certain level. We study this notion in (1) an expected sense, (2) a short-term sense and (3) a probabilistic sense via the concept of secrecy outage. Various system scenarios will be considered such as multiple sensors, multiple transmit antennas and full and partial channel state information (CSI).

11.1 Introduction

WSNs are networks consisting of some small, inexpensive and low-power sensors, which are deployed over a region and communicate with a remote processor over wireless links. Due to their low cost, robustness and high flexibility, WSNs are widely employed in many military and civilian applications such as environmental monitoring, traffic control, battlefield surveillance, etc. [1]. In distributed estimation, sensors independently collect data about some phenomena and send the measurements to an FC, which then attempts to reconstruct the phenomena.

One crucial issue in WSNs is the limited battery life of the sensors. As sensors are normally geographically widespread, replacing batteries can be costly. The problem of power allocation for distributed estimation has been well studied, see, e.g., [2–4].

[1]ANZ Group, Australia
[2]Department of Electrical Engineering (EIM-E), Paderborn University, Germany
[3]Institute for Telecommunications Research, University of South Australia, Australia

In the context of communications and information theory, the idea of *information outage* probability minimization was introduced in [5] for block-fading channels and has been further extended in [6,7]. An analogous concept of *estimation outage* probability for distributed estimation was introduced by the authors in [2], which is defined as the probability that the estimation distortion exceeds a certain threshold, see also [8–10].

Due to the broadcast nature of wireless communications, security and privacy issues have become one of the biggest challenges in WSNs. Traditional encryption schemes (or cryptography) might be vulnerable because of the problems in the secret key distribution and management. In addition, if an eavesdropper has sufficiently large computational power, cryptographic schemes with small key size may provide little secrecy. As an alternative, the notion of perfect secrecy,[1] introduced by Shannon [11], provides a different perspective on the data confidentiality. Later, in the 1970s, Wyner introduced the concept of wiretap channel [12] and showed that if the adversary's channel is a degraded version of the legitimate receiver's, reliable information can be received at the legitimate receiver without the eavesdropper being able to extract any useful information. From an information theoretic perspective, the authors in [13–15] studied the secrecy capacity in the case of full CSI or partial CSI, and investigated multiple-input multiple-output (MIMO) channels in [16–18]. Multiterminal source coding or CEO problems with secrecy constraints were also considered in [19–22].

In this chapter, we investigate the secure estimation problem from a "signal processing" viewpoint, where sensors employ simple uncoded analog-forwarding techniques [23] to transmit their observations to the FC. Indeed, various secrecy schemes from a signal processing rather than information theoretic point of view have also been studied and discussed in [24–30], where different performance metrics, such as Bayesian detection-operational privacy metric, bit-error-rate, signal-interference-to-noise ratio, Ali-Silvey distances or error probability were used to measure secrecy in a system. Related techniques based on cooperating relays, artificial noise generation or beamforming were also proposed in [31–34] to secure a system.

Moreover, it is known that mutual information and minimum MSE, fundamental quantities in information theory and estimation theory respectively, are closely related, see, e.g., [20,35,36]. Therefore, in this chapter, we consider analog uncoded transmission at the sensors and introduce the MSE as a security metric to secure the system at the physical layer.

We consider the estimation of a single point Gaussian source by a sensor network in the presence of an eavesdropper, where the analog amplify and forward technique over a slow-fading orthogonal MAC is used. We assume the same observed signal passes through another orthogonal MAC before reaching the eavesdropper, and both the FC and the eavesdropper attempt to obtain an MMSE estimation of the observations. To secure the system, a minimum distortion threshold is set for the eavesdropper to ensure that the distortion at the eavesdropper is no smaller than this threshold.

[1] Perfect secrecy was first introduced in 1949 by Shannon. In this model, it is assumed that the confidential message W is encrypted and then transmitted over a noiseless channel [11]. Perfect secrecy then requires the mutual information between W and the signal Z received by the eavesdropper to be equal to zero.

We first consider this notion in expected and short-term senses (Section 11.2), i.e., ensuring the condition is satisfied for the expected distortion or at every time instance. However, for real-time applications, it may be more appropriate to maintain a target distortion level throughout the fading process and minimize a *distortion outage probability*[2] at the FC and a *secrecy outage* constraint at the eavesdropper. This is the subject of the second part of this chapter (Section 11.3).

Notation: Throughout this chapter, we will use ^{T} to denote transposition, * to denote complex conjugate, ^{H} to denote conjugate transposition, $^{-1}$ to denote matrix inversion, $\mathbf{I}_M$ to denote the $M \times M$ identity matrix, $\|\mathbf{x}\|$ to denote the Euclidean norm of the vector $\mathbf{x}$, $|z|$ to denote the absolute value of z.

11.2 Estimation with security constraints

11.2.1 Multiple antennas scenario

Consider a scenario with one sensor equipped with N_t transmit antennas observing a single point Gaussian source, denoted by $\theta[t]$, $t = 0, 1, 2, \ldots$, which has zero mean and variance σ_θ^2. The measurement received by the sensor at time t is given as:

$$x[t] = \theta[t] + \omega[t], \tag{11.1}$$

where we assume $\omega[t]$ is independent and identically distributed (i.i.d.) Gaussian noise over time t, with zero mean and variance σ_ω^2.

The analog amplify and forward technique [23] is employed, where the sensor transmits over fading channels a scaled version of the analog measurements to the FC. In our model, the sensor amplifies the signal with a beamforming vector $\mathbf{p}[t] \in \mathbb{C}^{N_t \times 1}$ before transmitting it to the FC in the presence of an eavesdropper, as illustrated in Figure 11.1. We assume both channels experience block fading, i.e., the channels

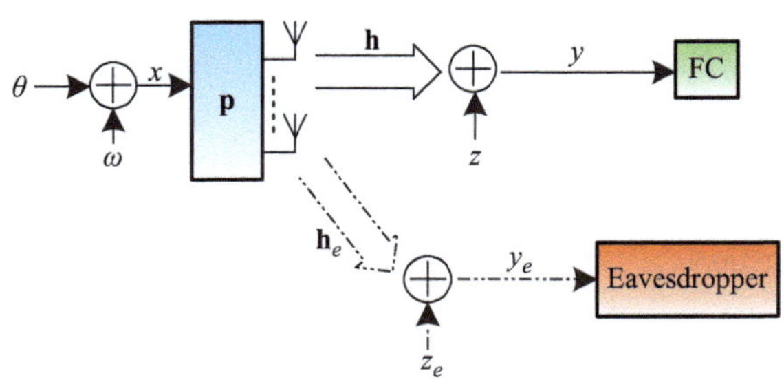

Figure 11.1 Diagram of a multiple-antenna single sensor system with the presence of an eavesdropper

[2]This is analogous to the situation in wireless communication where the ergodic capacity describes the maximum achievable long-term average rate without a delay constraint, while in real-time applications it may be more suitable to adopt the notion of outage capacity which determines the maximum achievable rate with an outage probability less than ε [5].

remain constant during each coherence time interval and are i.i.d. over different time intervals [5]. We further assume that the CSI of the FC is available, while the eavesdropper's CSI may or may not be available to the FC. The FC designs the optimal power allocation strategy based on the available CSI and then sends $\mathbf{p}[t]$ back to the sensor via a secure feedback link.

The signals received by the FC and the eavesdropper are given by, respectively:

$$y[t] = \mathbf{h}^{\mathrm{T}}[t]\mathbf{p}[t]\theta[t] + \mathbf{h}^{\mathrm{T}}[t]\mathbf{p}[t]\omega[t] + z[t], \tag{11.2a}$$

$$y_e[t] = \mathbf{h}_e^{\mathrm{T}}[t]\mathbf{p}[t]\theta[t] + \mathbf{h}_e^{\mathrm{T}}[t]\mathbf{p}[t]\omega[t] + z_e[t], \tag{11.2b}$$

where both $z[t]$ and $z_e[t]$ are i.i.d. zero mean complex Gaussian channel noises at the FC and the eavesdropper with variances σ_n^2 and σ_e^2, respectively, and $\mathbf{h}[t] = \left[h_1[t], \ldots, h_n[t], \ldots, h_{N_t}[t]\right]^{\mathrm{T}}$ and $\mathbf{h}_e[t] = \left[h_{e1}[t], \ldots, h_{en}[t], \ldots, h_{eN_t}[t]\right]^{\mathrm{T}}$ are, respectively, the channels from the sensor to the FC and to the eavesdropper. We assume that $\{h_n[t]\}$ are i.i.d. complex Gaussian with zero mean and variance σ_h^2, and the elements in $\mathbf{h}_e[t]$ are also i.i.d. complex Gaussian, with zero mean and variance $\sigma_{h_e}^2$.

In order to minimize the MSE or *distortion* at the FC, the minimum mean squared error (MMSE) estimator is used to estimate θ under the model (11.2a)–(11.2b) [37]. At time t, the distortion at the FC and the eavesdropper can be shown to be, respectively:

$$D[t] = \sigma_\theta^2 - \frac{\sigma_\theta^4 \left(\mathbf{h}^{\mathrm{T}}[t]\mathbf{p}[t]\right)^{\mathrm{H}} \mathbf{h}^{\mathrm{T}}[t]\mathbf{p}[t]}{\sigma_n^2 + \left(\sigma_\theta^2 + \sigma_\omega^2\right)\left(\mathbf{h}^{\mathrm{T}}[t]\mathbf{p}[t]\right)^{\mathrm{H}} \mathbf{h}^{\mathrm{T}}[t]\mathbf{p}[t]}, \tag{11.3a}$$

$$D_e[t] = \sigma_\theta^2 - \frac{\sigma_\theta^4 \left(\mathbf{h}_e^{\mathrm{T}}[t]\mathbf{p}[t]\right)^{\mathrm{H}} \mathbf{h}_e^{\mathrm{T}}[t]\mathbf{p}[t]}{\sigma_e^2 + \left(\sigma_\theta^2 + \sigma_\omega^2\right)\left(\mathbf{h}_e^{\mathrm{T}}[t]\mathbf{p}[t]\right)^{\mathrm{H}} \mathbf{h}_e^{\mathrm{T}}[t]\mathbf{p}[t]}. \tag{11.3b}$$

For a limited transmission power budget $\mathscr{P}_{\mathrm{tot}}$, we would like to minimize the distortion at the FC by adapting $\mathbf{p}[t]$, while maintaining a certain level of security of the transmission. In information theoretic security, the secrecy capacity is defined as the maximum transmission rate at which the mutual information between the confidential message and the signal received by the eavesdropper is less than a threshold [14]. Motivated by this idea, plus the close relationships between MMSE and mutual information [20,35,36], we consider a notion of *secrecy in estimation* by requiring the distortion at the eavesdropper to be greater than a threshold $D_{\min}$. In this way, some level of confidentiality can be achieved at the FC. We will refer to the minimum distortion threshold $D_{\min}$ as the *secrecy threshold* in the following.

Due to the assumption of system independence over time t, we will drop the time index t for the rest of this section.

11.2.1.1 Full CSI

In the case of full CSI, where we assume the FC can also acquire the channel information between the sensor and the eavesdropper, the power control policies can be derived such that the sensor is able to adjust the antenna transmission power depending on both the FC's and the eavesdropper's channel information. While the requirement of full CSI of the eavesdropper channels is difficult to achieve in practice, this case is nevertheless instructive as well as a useful benchmark for the distortion performance with partial CSI of the eavesdropper channels, to be analyzed subsequently.

Long-term optimal power allocation

In long-term power allocation, we assume that the crucial information lies in the long-term behavior of the estimates, such as long-term trends in the physical process observed, hence the FC is interested in the estimation over multiple fading blocks. We would then like to minimize the long-term average distortion at the FC by adapting $\mathbf{p}$, where the average is across coherence time intervals, while keeping the long-term average sum of sensor transmission powers, defined as:

$$\mathbb{E}\left[\mathbf{p}^{\mathrm{H}}\mathbf{p}\mathbb{E}\left[x_k^2\right]\right] = \mathbb{E}\left[\mathbf{p}^{\mathrm{H}}\mathbf{p}(\sigma_\theta^2 + \sigma_{\omega k}^2)\right], \tag{11.4}$$

to be less than the power budget $\mathscr{P}_{\mathrm{tot}}$. We also seek to maintain the average distortion at the eavesdropper to be greater than the threshold $D_{\min}$, i.e., $\mathbb{E}[D_e] \geq D_{\min}$, to ensure that a level of confidentiality can be achieved at the FC over the long-term. Furthermore, an additional constraint ensuring the average estimation quality being better at the FC is considered for more meaningful solutions. Therefore, the power control problem can be formulated as:

$$\begin{aligned} &\min_{\mathbf{p}} \quad \mathbb{E}\left[D\right] \\ &\text{s.t.} \quad \mathbb{E}\left[\left(\sigma_\theta^2 + \sigma_\omega^2\right)\mathbf{p}^{\mathrm{H}}\mathbf{p}\right] \leq \mathscr{P}_{\mathrm{tot}}, \quad \mathbb{E}\left[D_e\right] \geq D_{\min}, \quad \mathbb{E}\left[D_e\right] \geq \mathbb{E}\left[D\right]. \end{aligned} \tag{11.5}$$

Given the distortion expressions in (11.3a) and (11.3b), we can rewrite problem (11.5) as:

$$\begin{aligned} &\min_{\mathbf{p}} \quad \mathbb{E}\left[\left(\alpha + \left(\mathbf{h}^{\mathrm{T}}\mathbf{p}\right)^{\mathrm{H}}\mathbf{h}^{\mathrm{T}}\mathbf{p}\right)^{-1}\right] \\ &\text{s.t.} \quad \mathbb{E}\left[\mathbf{p}^{\mathrm{H}}\mathbf{p}\right] \leq \frac{\mathscr{P}_{\mathrm{tot}}}{\sigma_\theta^2 + \sigma_\omega^2}, \quad \mathbb{E}\left[\left(\alpha_e + \left(\mathbf{h}_e^{\mathrm{T}}\mathbf{p}\right)^{\mathrm{H}}\mathbf{h}_e^{\mathrm{T}}\mathbf{p}\right)^{-1}\right] \geq D_{\mathrm{ma\_L}}, \end{aligned} \tag{11.6a}$$

$$\mathbb{E}\left[\frac{\alpha_e}{\alpha_e + \left(\mathbf{h}_e^{\mathrm{T}}\mathbf{p}\right)^{\mathrm{H}}\mathbf{h}_e^{\mathrm{T}}\mathbf{p}}\right] \geq \mathbb{E}\left[\frac{\alpha}{\alpha + \left(\mathbf{h}^{\mathrm{T}}\mathbf{p}\right)^{\mathrm{H}}\mathbf{h}^{\mathrm{T}}\mathbf{p}}\right], \tag{11.6b}$$

where $D_{\mathrm{ma\_L}} \triangleq \left(\left(D_{\min}/(\sigma_\theta^2\sigma_\omega^2/(\sigma_\theta^2 + \sigma_\omega^2))\right) - 1\right) / \left(\sigma_e^2\sigma_\theta^2/\left(\left(\sigma_\theta^2 + \sigma_\omega^2\right)\sigma_\omega^2\right)\right)$, $\alpha \triangleq (\sigma_n^2/(\sigma_\theta^2 + \sigma_\omega^2))$, and $\alpha_e \triangleq (\sigma_e^2/(\sigma_\theta^2 + \sigma_\omega^2))$.

To solve problem (11.6a)–(11.6b), we will apply the technique of Lagrange multipliers. The dual problem of (11.6a)–(11.6b) is defined as:

$$\max_{\lambda,\nu,\tau} g(\lambda, \nu, \tau), \tag{11.7}$$

where λ, ν and τ are nonnegative Lagrange multipliers, and the dual function $g(\lambda,\nu,\tau)$ associated with problem (11.6a)–(11.6b) is:

$$g(\lambda,\nu,\tau) \;=\; \min_{\mathbf{p}(\mathbf{h},\mathbf{h}_e)} \int_{\mathbf{h}}\int_{\mathbf{h}_e} l(\mathbf{p}(\mathbf{h},\mathbf{h}_e),\lambda,\nu,\tau)\, f_{\mathbf{h}} f_{\mathbf{h}_e}\, d\mathbf{h}\, d\mathbf{h}_e + \nu D_{\text{ma\_L}} - \lambda \frac{\mathscr{P}_{\text{tot}}}{\sigma_\theta^2+\sigma_\omega^2}, \tag{11.8}$$

where $f_{\mathbf{h}} = \prod_{n=1}^{N_t} f(h_n)$ and $f_{\mathbf{h}_e} = \prod_{n=1}^{N_t} f(h_{en})$, with $f(\cdot)$ denoting the probability density function, and $l(\mathbf{p}(\mathbf{h},\mathbf{h}_e),\lambda,\nu,\tau) = ((1+\tau\sigma_n^2/\sigma_e^2)/(\alpha + (\mathbf{h}^{\mathrm{T}}\mathbf{p})^{\mathrm{H}}\mathbf{h}^{\mathrm{T}}\mathbf{p})) + \lambda\mathbf{p}^{\mathrm{H}}\mathbf{p} - ((\tau+\nu)/(\alpha_e + (\mathbf{h}_e^{\mathrm{T}}\mathbf{p})^{\mathrm{H}}\mathbf{h}_e^{\mathrm{T}}\mathbf{p}))$.

Problem (11.6a)–(11.6b) is in general non-convex. Letting $[p_1(\mathbf{h},\mathbf{h}_e),\ldots, p_{N_t}(\mathbf{h},\mathbf{h}_e)]^{\mathrm{T}} = \mathbf{p}(\mathbf{h},\mathbf{h}_e)$ be the complex gains allocated on each antenna, we can obtain a locally optimal solution from the following necessary Karush–Kuhn–Tucker (KKT) conditions [38] for the optimal point:

$$\frac{-h_n^{\mathrm{H}}\left(\mathbf{h}^{\mathrm{T}}\mathbf{p}\right)^{\mathrm{H}}}{\left[\alpha+\left(\mathbf{h}^{\mathrm{T}}\mathbf{p}\right)^{\mathrm{H}}\mathbf{h}^{\mathrm{T}}\mathbf{p}\right]^2} + \frac{h_{en}^{\mathrm{H}}\left(\mathbf{h}_e^{\mathrm{T}}\mathbf{p}\right)^{\mathrm{H}}(\nu+\tau)}{\left(1+\tau\sigma_n^2/\sigma_e^2\right)\left[\alpha_e+\left(\mathbf{h}_e^{\mathrm{T}}\mathbf{p}\right)^{\mathrm{H}}\mathbf{h}_e^{\mathrm{T}}\mathbf{p}\right]^2} + \frac{\lambda}{1+\tau\sigma_n^2/\sigma_e^2} p_n^{\mathrm{H}} = 0,\ \forall n \tag{11.9a}$$

$$\lambda\left(\mathbb{E}\left[\mathbf{p}^{\mathrm{H}}\mathbf{p}\right] - \frac{\mathscr{P}_{\text{tot}}}{\sigma_\theta{}^2+\sigma_\omega{}^2}\right) = 0, \tag{11.9b}$$

$$\nu\left(D_{\text{ma\_L}} - \mathbb{E}\left[\left(\alpha_e + \left(\mathbf{h}_e^{\mathrm{T}}\mathbf{p}\right)^{\mathrm{H}}\mathbf{h}_e^{\mathrm{T}}\mathbf{p}\right)^{-1}\right]\right) = 0, \tag{11.9c}$$

$$\tau\left(\frac{\sigma_n^2}{\sigma_e^2}\mathbb{E}\left[\left(\alpha+\left(\mathbf{h}^{\mathrm{T}}\mathbf{p}\right)^{\mathrm{H}}\mathbf{h}^{\mathrm{T}}\mathbf{p}\right)^{-1}\right] - \mathbb{E}\left[\left(\alpha_e+\left(\mathbf{h}_e^{\mathrm{T}}\mathbf{p}\right)^{\mathrm{H}}\mathbf{h}_e^{\mathrm{T}}\mathbf{p}\right)^{-1}\right]\right) = 0. \tag{11.9d}$$

Specifically, we first assign arbitrary initial values to λ, ν and τ, and then iteratively apply the following steps until (11.9b), (11.9c) and (11.9d) are satisfied.

Step 1: With fixed $\tau^{(i)}$, $\lambda^{(i)}$ and $\nu^{(i)}$, find the optimal solution $\tilde{\mathbf{p}}$ of the Lagrange dual function (11.8), which can be obtained by solving the equations in (11.9a).

Step 2: With the resulting allocated power, apply the subgradient method to update the Lagrange multipliers, i.e.,

$$\begin{aligned}
\lambda^{(i+1)} &= \left[\lambda^{(i)} + \varepsilon\left(\mathbb{E}\left[\tilde{\mathbf{p}}^{\mathrm{H}}\tilde{\mathbf{p}}\right] - \frac{\mathscr{P}_{\text{tot}}}{\sigma_\theta{}^2+\sigma_\omega{}^2}\right)\right]^+, \\
\nu^{(i+1)} &= \left[\nu^{(i)} + \kappa\left(D_{\text{ma_L}} - \mathbb{E}\left[\left(\alpha_e + \left(\mathbf{h}_e^{\mathrm{T}}\tilde{\mathbf{p}}\right)^{\mathrm{H}}\mathbf{h}_e^{\mathrm{T}}\tilde{\mathbf{p}}\right)^{-1}\right]\right)\right]^+ \\
\tau^{(i+1)} &= \left[\tau^{(i)} + \upsilon\left(\frac{\sigma_n^2}{\sigma_e^2}\mathbb{E}\left[\left(\alpha + \left(\mathbf{h}^{\mathrm{T}}\tilde{\mathbf{p}}\right)^{\mathrm{H}}\mathbf{h}^{\mathrm{T}}\tilde{\mathbf{p}}\right)^{-1}\right] - \mathbb{E}\left[\left(\alpha_e + \left(\mathbf{h}_e^{\mathrm{T}}\tilde{\mathbf{p}}\right)^{\mathrm{H}}\mathbf{h}_e^{\mathrm{T}}\tilde{\mathbf{p}}\right)^{-1}\right]\right)\right]^+,
\end{aligned} \tag{11.10}$$

where υ, κ and ε are sufficiently small step-sizes for updating τ, ν and λ, respectively.

Zero information leakage

Other than diversity gain, another advantage with multiple transmit antennas is that we can employ techniques to hide the observation data from the eavesdropper, e.g., by transmitting it onto the null space of the eavesdropper's channel. As a result, the eavesdropper is unable to detect any information about x.

Let the singular value decomposition[3] of $\mathbf{h}_e^{\mathrm{T}}$ be $\mathbf{h}_e^{\mathrm{T}} = \mathbf{USV}^{\mathrm{H}}$. The null space of the eavesdropper's channel can be represented by the span of the orthonormal column vectors of $\bar{\mathbf{V}}$, where $\bar{\mathbf{V}}$ are the last $N_t - 1$ columns of $\mathbf{V}$. Then we can express the eavesdropper's channel null space as $\bar{\mathbf{V}}\bar{\mathbf{V}}^{\mathrm{H}}$ [39].

Next, we define a precoding matrix:

$$\mathbf{W} = \bar{\mathbf{V}}\bar{\mathbf{V}}^{\mathrm{H}}, \tag{11.11}$$

where $\mathbf{W} \in \mathbb{C}^{N_t}$. The sensor sends $\mathbf{W}\mathbf{p}x$. The signal received by the FC and the eavesdropper are then given by, respectively:

$$y = \mathbf{h}^{\mathrm{T}}\mathbf{W}\mathbf{p}x + z = \mathbf{h}^{\mathrm{T}}\mathbf{W}\mathbf{p}\theta + \mathbf{h}^{\mathrm{T}}\mathbf{W}\mathbf{p}\omega + z, \tag{11.12a}$$

$$y_e = \mathbf{h}_e^{\mathrm{T}}\mathbf{W}\mathbf{p}x + z_e = z_e, \tag{11.12b}$$

and the transmission power can be computed as $\left((\mathbf{Wp})^{\mathrm{H}}\,\mathbf{Wp}\right)(\sigma_\theta^2 + \sigma_\omega^2)$. Since the eavesdropper receives only noise, the distortion at the eavesdropper reaches its maximum level of σ_θ^2, and we can remove constraints (11.6a)–(11.6b) in the problem. In addition, we know that the beamforming vector should lineup with $\mathbf{h}^{\mathrm{T}}\mathbf{W}$ to minimize the distortion at the FC; thus, $\mathbf{p} = \sqrt{p_0}(\left(\mathbf{h}^{\mathrm{T}}\mathbf{W}\right)^{\mathrm{H}}/\|\mathbf{hW}\|)$, with p_0 being real-valued. Therefore, problem (11.6a)–(11.6b) can be simplified to:

$$\begin{aligned} &\min_{p_0 \geq 0} \ \mathbb{E}\left[\left(\alpha + p_0\mathbf{h}^{\mathrm{T}}\mathbf{W}\mathbf{h}^*\right)^{-1}\right] \\ &\text{s.t.} \ \ p_o \leq \frac{\mathscr{P}_{\mathrm{tot}}}{\sigma_\theta^2 + \sigma_\omega^2}. \end{aligned} \tag{11.13}$$

It can be easily seen that when $p_0 = (\mathscr{P}_{\mathrm{tot}}/(\sigma_\theta^2 + \sigma_\omega^2))$, the long-term distortion at the FC reaches its minimum.

Short-term optimal power allocation

We can also formulate a power allocation problem that minimizes the distortion at the FC, while satisfying a secrecy constraint at the eavesdropper and meeting the total power budget *in every transmission time slot*. We refer to this as short-term power allocation. Note that for short-term optimal power allocation, we cannot guarantee

[3] Because $\mathbf{h}_e^{\mathrm{T}}$ is a row vector, $\mathbf{U}$ is a complex scalar and $\mathbf{S}$ is a complex row vector.

that the estimation quality is better at the FC at all times. Keeping this in mind, the optimal power allocation problem for a given set of channels can be cast as:

$$\min_{\mathbf{p}} D = \min_{\mathbf{p}} \ \sigma_\theta^2 - \frac{\alpha\sigma_\theta^4}{\sigma_n^2}\left[1 - \frac{\alpha}{\alpha + \left(\mathbf{h}^{\mathrm{T}}\mathbf{p}\right)^{\mathrm{H}}\mathbf{h}^{\mathrm{T}}\mathbf{p}}\right]$$

$$\text{s.t. } \mathbf{p}^{\mathrm{H}}\mathbf{p} \leq \frac{\mathscr{P}_{\text{tot}}}{\sigma_\theta^2 + \sigma_\omega^2}, \quad \sigma_\theta^2 - \frac{\alpha_e\sigma_\theta^4}{\sigma_e^2}\left[1 - \frac{\alpha_e}{\alpha_e + \left(\mathbf{h}_e^{\mathrm{T}}\mathbf{p}\right)^{\mathrm{H}}\mathbf{h}_e^{\mathrm{T}}\mathbf{p}}\right] \geq D_{\min}, \quad (11.14)$$

where $D_{\min}$ and $\mathscr{P}_{\text{tot}}$ are, respectively, the distortion threshold at the eavesdropper and the total transmission power budget. The Lagrange multiplier technique can be applied to solve this non-convex optimization problem in every fading block, with the details omitted to avoid repetition.

11.2.1.2 Partial CSI

Due to the difficulties of perfectly acquiring the eavesdropper's CSI in practice, in this subsection, we will assume that the FC only has statistical knowledge of the eavesdropper's channel information. As it is then not practical to consider short-term constraints that need to be satisfied at every time instant, we only look at the long-term scenario for the partial CSI case.

The power allocation problem is formulated similar to (11.5), but now with $\mathbf{p}$ being a function of $\mathbf{h}$, rather than a function of both $\mathbf{h}$ and $\mathbf{h}_e$ as in the full CSI case. As the problem is again non-convex, a locally optimal solution can be obtained as follows. Similar to (11.8), we define the Lagrange dual function as:

$$g(\lambda, \nu, \tau) = \min_{p_n(\mathbf{h}),\ \forall n} \int_{\mathbf{h}} l(\{p_n(\mathbf{h})\}, \lambda, \nu, \tau)\, f_{\mathbf{h}}\, d\mathbf{h} + \nu D_{\text{ma\_L}} - \lambda \frac{\mathscr{P}_{\text{tot}}}{\sigma_\theta^2 + \sigma_\omega^2},$$

with $l(\{p_n(\mathbf{h})\}, \lambda, \nu, \tau)$ expressed as:

$$l(\{p_n(\mathbf{h})\}, \lambda, \nu, \tau) = \frac{1 + \tau\sigma_n^2/\sigma_e^2}{\alpha + \left(\mathbf{h}^{\mathrm{T}}\mathbf{p}\right)^{\mathrm{H}}\mathbf{h}^{\mathrm{T}}\mathbf{p}} + \lambda\mathbf{p}^{\mathrm{H}}\mathbf{p} - \int_{\mathbf{h}_e} \frac{\tau + \nu}{\alpha_e + \left(\mathbf{h}^{\mathrm{T}}{}_e\mathbf{p}\right)^{\mathrm{H}}\mathbf{h}_e^{\mathrm{T}}\mathbf{p}} f_{\mathbf{h}_e} d\mathbf{h}_e. \quad (11.15)$$

For any set of channels $\mathbf{h}$, the optimal transmission power of the sensor is determined by the stationary points (or KKT points). We can then adapt similar methods as described in Section 11.2.1.1. In **Step 1**, the power policies $\tilde{\mathbf{p}}(\mathbf{h})$ can be derived by applying **Algorithm 1**. For fixed $\tau^{(i)}$, $\lambda^{(i)}$ and $\nu^{(i)}$, **Algorithm 1** sequentially updates the transmit power on each antenna by minimizing the function given in (11.15), until a locally optimal solution is found. In **Step 2**, we update the Lagrange multipliers via the subgradient method.

Algorithm 1:

1: Initialize the iteration index $q = 0$, choose an arbitrary initial value for $\{p_n(\mathbf{h})^{(q)}\}_{n=1}^{N_t}$, and obtain $l^{(q)} = l(\{p_n(\mathbf{h})^{(q)}\}, \lambda, \nu, \tau)$ from (11.15).
2: **repeat**
3: For $j = 1 : N_t$
   1. Find the complex gain $p_j(\mathbf{h})$ on antenna j such that $l\left(\left\{\{p_n(\mathbf{h})^{(q)}\}_{n\neq j}, p_j(\mathbf{h})\right\}, \lambda, \nu, \tau\right)$ is minimized.
   2. Update the transmission power of antenna j by $[p_1(\mathbf{h})^{(q)}, \ldots, p_j'(\mathbf{h})^{(q)}, \ldots, p_K(\mathbf{h})^{(q)}]$.
4: update $l^{(q+1)} = l(\{p_n'(\mathbf{h})^{(q)}\}, \lambda, \nu, \tau)$, and $q = q + 1$.
5: **until** Convergence: $\left(l^{(q+1)} - l^{(q)}\right)/\left(l^{(q+1)}\right) < \zeta$; set $\{\tilde{p}_n(\mathbf{h})\} = \{p_n'(\mathbf{h})^{(q)}\}$.

Remark: In **Step 1**, $\tau^{(i)}$, $\lambda^{(i)}$ and $\nu^{(i)}$ are fixed; hence, we drop the iteration number i in **Algorithm 1**; and ζ is a prespecified convergence criterion. Additionally, **Algorithm 1** only gives a locally optimal solution, as the different initial values of $\mathbf{p}(\mathbf{h})^{(0)}$ may lead l in (11.15) to converge to a different minimum. Thus, in practice, the FC begins with several different initial points and chooses the best resulting powers and forwards them to the sensor.

Artificial noise

To enhance the system performance, we can use the technique of artificial noise [32] to degrade the eavesdropper's channel. The artificial noise is generated by the transmitter (the sensor) in such a way that the additional noise lies in the null space of the intended receiver's (the FC's) channel; as a result, this noise will not cause any damage toward the message received by the FC but will degrade the eavesdropper's channel.

Let the column vectors of $\hat{\mathbf{W}}^{\mathrm{H}} = [\mathbf{w}_1 \mathbf{W}_2]$ form an orthonormal basis of $\mathbb{C}^{N_t}$, with $\mathbf{w}_1 \in \mathbb{C}^{1\times N_t}$ representing the signal space of $\mathbf{h}$. The sensor then transmits:

$$\mathbf{w}_1\sqrt{\beta}x + \mathbf{W}_2\mathbf{v}, \tag{11.16}$$

where $\mathbf{W}_2\mathbf{v}$ is the artificial noise, which is chosen to be a random vector in the null space of $\mathbf{h}^{\mathrm{T}}$, to reduce the possibility of small "noise" being seen by the eavesdropper. Here, $\mathbf{v} \in \mathbb{C}^{(N_t-1)\times 1}$ has $N_t - 1$ i.i.d. complex Gaussian entries with each having zero mean and variance β_a. Hence, the transmit power in each fading block is given as $\beta(\sigma_\theta^2 + \sigma_\omega^2) + \beta_a(N_t - 1)$. The signals received by the FC and the eavesdropper are, respectively:

$$y = \mathbf{h}^{\mathrm{T}}\hat{\mathbf{W}}^{\mathrm{H}}\left[\sqrt{\beta}x,\ \mathbf{v}^{\mathrm{T}}\right]^{\mathrm{T}} + z = \mathbf{h}^{\mathrm{T}}\mathbf{w}_1\sqrt{\beta}x + \mathbf{h}^{\mathrm{T}}\mathbf{W}_2\mathbf{v} + z = \mathbf{h}^{\mathrm{T}}\mathbf{w}_1\sqrt{\beta}x + z, \tag{11.17}$$

$$y_e = \mathbf{h}_e^{\mathrm{T}}\hat{\mathbf{W}}^{\mathrm{H}}\left[\sqrt{\beta}x, \mathbf{v}^{\mathrm{T}}\right]^{\mathrm{T}} + z_e = \mathbf{h}_e^{\mathrm{T}}\mathbf{w}_1\sqrt{\beta}x + \mathbf{h}_e^{\mathrm{T}}\mathbf{W}_2\mathbf{v} + z_e. \tag{11.18}$$

Our objective is to determine the power used to produce artificial noise and forward the observation signal, so that the long-term distortion at the FC is minimized. The functional optimization problem can be written as:

$$\begin{aligned}
\min_{\beta(\mathbf{h}),\ \beta_a(\mathbf{h})} \quad & \mathbb{E}\left[\left(\frac{1}{\sigma_\theta^2} + \frac{\beta(\mathbf{h})\mathbf{h}^{\mathrm{T}}\mathbf{h}^*}{\sigma_n^2 + \beta(\mathbf{h})\mathbf{h}^{\mathrm{T}}\mathbf{h}^*\sigma_\omega^2}\right)^{-1}\right] \\
\text{s.t.} \quad & (\sigma_\theta^2 + \sigma_\omega^2)\mathbb{E}\left[\beta(\mathbf{h})\right] + (N_t - 1)\,\mathbb{E}\left[\beta_a(\mathbf{h})\right] \le \mathscr{P}_{\text{tot}} \\
& \mathbb{E}\left[\left(\frac{1}{\sigma_\theta^2} + \frac{\beta(\mathbf{h})|\mathbf{h}_e^{\mathrm{T}}\mathbf{w}_1|^2}{\sigma_e^2 + \beta(\mathbf{h})|\mathbf{h}_e^{\mathrm{T}}\mathbf{w}_1|^2\sigma_\omega^2 + \mathbf{h}_e^{\mathrm{T}}\mathbf{W}_2\mathbf{W}_2^{\mathrm{H}}\mathbf{h}_e^*\beta_a(\mathbf{h})}\right)^{-1}\right] \ge D_{\min} \\
& \mathbb{E}\left[\left(\frac{1}{\sigma_\theta^2} + \frac{\beta(\mathbf{h})|\mathbf{h}_e^{\mathrm{T}}\mathbf{w}_1|^2}{\sigma_e^2 + \beta(\mathbf{h})|\mathbf{h}_e^{\mathrm{T}}\mathbf{w}_1|^2\sigma_\omega^2 + \mathbf{h}_e^{\mathrm{T}}\mathbf{W}_2\mathbf{W}_2^{\mathrm{H}}\mathbf{h}_e^*\beta_a(\mathbf{h})}\right)^{-1}\right] \\
& \ge \mathbb{E}\left[\left(\frac{1}{\sigma_\theta^2} + \frac{\beta(\mathbf{h})\mathbf{h}^{\mathrm{T}}\mathbf{h}^*}{\sigma_n^2 + \beta(\mathbf{h})\mathbf{h}^{\mathrm{T}}\mathbf{h}^*\sigma_\omega^2}\right)^{-1}\right]
\end{aligned} \tag{11.19}$$

To solve problem (11.19), we apply the technique of Lagrange multipliers and use two steps similar to those described in Section 11.2.1.1, where in **Step 1**, with fixed Lagrange multipliers, we need to sequentially find $\beta(\mathbf{h})$ and $\beta_a(\mathbf{h})$.

11.2.2 Multiple sensors scenario

If the use of multiple antennas on a sensor is not an option, an alternative is to employ multiple sensors. Therefore, in this subsection, we investigate the behavior of a multiple-sensor signal antenna system.

A schematic diagram of the wireless system model is shown in Figure 11.2. We assume that the same single point Gaussian source θ as defined in Section 11.2.1 is observed by K sensors. The measurement received by the kth sensor is corrupted with noise ω_k and given as:

$$x_k = \theta + \omega_k, \tag{11.20}$$

where we assume ω_k is i.i.d. Gaussian noise over time, with zero mean and variance $\sigma_{\omega k}^2$. The sensors employ the analog amplify and forward technique to scale the signal with $p_k \in \mathbb{C}$ before sending it to the FC via a set of orthogonal channels $[h_1, \ldots, h_K]$. The observation $\{x_k\}$ is also listened to by the eavesdropper via another set of orthogonal channels $[h_{e1}, \ldots, h_{ek}]$.

The signals received by the FC and the eavesdropper from the kth sensor are given by, respectively:

$$y_k = h_k p_k \theta + h_k p_k \omega_k + z_k, \tag{11.21a}$$

$$y_{ek} = h_{ek} p_k \theta + h_{ek} p_k \omega_k + z_{ek}, \tag{11.21b}$$

where both h_k and h_{ek} are zero mean i.i.d. complex Gaussian channels (Rayleigh fading) from sensor k to the FC and the eavesdropper with variances $\sigma_{h_k}^2$ and $\sigma_{h_{e}k}^2$,

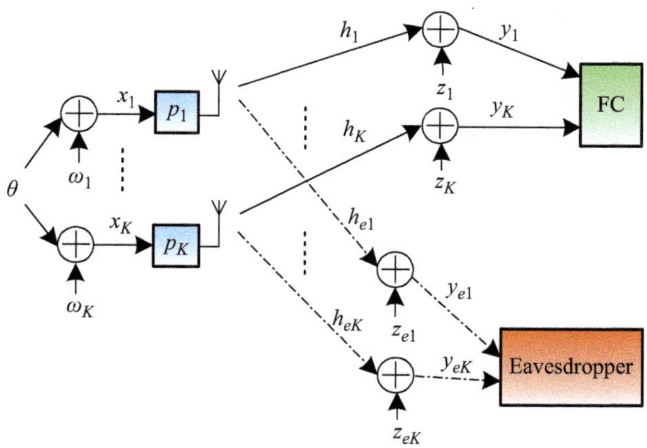

Figure 11.2 Diagram of the wireless sensor network using orthogonal MAC scheme with the presence of an eavesdropper

respectively, and z_k and z_{ek} represent i.i.d. complex Gaussian noises with zero mean and variances σ_{nk}^2 at the FC and σ_{ek}^2 at the eavesdropper respectively.

The MMSE estimator is used at both the FC and the eavesdropper to estimate θ. At each channel instant, the distortion at the FC and the eavesdropper can be shown to be, respectively:

$$D = \left(\frac{1}{\sigma_\theta^2} + \sum_{k=1}^{K} \frac{(h_k p_k)^{\mathrm{H}} h_k p_k}{(h_k p_k)^{\mathrm{H}} h_k p_k \sigma_{\omega k}^2 + \sigma_{nk}^2}\right)^{-1} = \left(\frac{1}{\sigma_\theta^2} + \sum_{k=1}^{K} \frac{g_k \beta_k}{g_k \beta_k \sigma_{\omega k}^2 + \sigma_{nk}^2}\right)^{-1}, \tag{11.22a}$$

$$D_e = \left(\frac{1}{\sigma_\theta^2} + \sum_{k=1}^{K} \frac{(h_{ek} p_k)^{\mathrm{H}} h_{ek} p_k}{(h_{ek} p_k)^{\mathrm{H}} h_{ek} p_k \sigma_{\omega k}^2 + \sigma_{ek}^2}\right)^{-1} = \left(\frac{1}{\sigma_\theta^2} + \sum_{k=1}^{K} \frac{g_{ek} \beta_k}{g_{ek} \beta_k \sigma_{\omega k}^2 + \sigma_{ek}^2}\right)^{-1}, \tag{11.22b}$$

where $g_k = h_k^{\mathrm{H}} h_k \in \mathbb{R}$ and $g_{ek} = {h_{ek}}^{\mathrm{H}} h_{ek} \in \mathbb{R}$ are respectively the channel power gains from sensor k to the FC and the eavesdropper, and $\beta_k = p_k^{\mathrm{H}} p_k \in \mathbb{R}$ is the power allocated on the kth sensor. This means that for a given set of $\{\beta_k\}$, any $\{p_k\}$ satisfying $p_k^{\mathrm{H}} p_k = \beta_k, \forall k$ would result in the same distortion, which implies p_k does not necessarily need to lineup with sensor k's channel direction; hence, we focus on $\{\beta_k\}$ in the multiple-sensor scenario.

11.2.2.1 Full CSI

We would like to minimize the distortion at the FC by adapting the sensors' transmit powers while satisfying the same constraints as considered in the multiple-antenna scenario. With full knowledge of the eavesdropper's channel information, the power control problem can be formulated as:

$$\begin{aligned}
&\min_{\beta_k(g_k, g_{ek})\geq 0,\ \forall k} \quad \mathbb{E}\left[\left(\frac{1}{\sigma_\theta^2}+\sum_{k=1}^{K}\frac{g_k\beta_k}{g_k\beta_k\sigma_{\omega k}^2+\sigma_{nk}^2}\right)^{-1}\right]\\
&\text{s.t.}\ \mathbb{E}\left[\sum_{k=1}^{K}(\sigma_\theta^2+\sigma_{\omega k}^2)\,\beta_k\right]\leq \mathscr{P}_{\text{tot}},\quad \mathbb{E}\left[\left(\frac{1}{\sigma_\theta^2}+\sum_{k=1}^{K}\frac{g_{ek}\beta_k}{g_{ek}\beta_k\sigma_{\omega k}^2+\sigma_{ek}^2}\right)^{-1}\right]\geq D_{\min}\\
&\mathbb{E}\left[\left(\frac{1}{\sigma_\theta^2}+\sum_{k=1}^{K}\frac{g_{ek}\beta_k}{g_{ek}\beta_k\sigma_{\omega k}^2+\sigma_{ek}^2}\right)^{-1}\right]\geq \mathbb{E}\left[\left(\frac{1}{\sigma_\theta^2}+\sum_{k=1}^{K}\frac{g_k\beta_k}{g_k\beta_k\sigma_{\omega k}^2+\sigma_{nk}^2}\right)^{-1}\right].
\end{aligned} \tag{11.23}$$

A similar setup was considered in [40]; however, here an additional constraint guaranteeing a larger distortion at the eavesdropper is included. Despite this, one can still use the same methods as described in [40] to numerically obtain locally optimal solutions for problem (11.23).

Next, we explore the short-term distortion performance at the FC while satisfying a secrecy constraint at the eavesdropper and a total power constraint at the sensor *in every transmission instant*. We can formulate the optimization problem as:

$$\begin{aligned}
&\min_{\beta_k\geq 0,\ \forall k}\ \sum_{k=1}^{K}\frac{-g_k\beta_k}{g_k\beta_k\sigma_{\omega k}^2+\sigma_{nk}^2}\\
&\text{s.t.}\ \sum_{k=1}^{K}\left(\sigma_\theta^2+\sigma_{\omega k}^2\right)\beta_k\leq \mathscr{P}_{\text{tot}},\quad \sum_{k=1}^{K}\frac{g_{ek}\beta_k}{g_{ek}\beta_k\sigma_{\omega k}^2+\sigma_{ek}^2}\leq I_{\text{ms}},
\end{aligned} \tag{11.24}$$

where $I_{\text{ms}} \triangleq (1/D_{\min}) - (1/\sigma_\theta^2)$. Similar techniques as in Section 11.2.1.1 can be used to find a locally optimal solution, we omit details to avoid repetition.

11.2.2.2 Partial CSI

In this scenario, optimal long-term power allocation problem is formulated similar to problem (11.23), but with the power scheme $\{\beta_k\}$ only being a function of the FC's channel information. A locally optimal solution can be then derived by applying similar techniques as used in [40].

The artificial noise technique can also be used with multiple sensors, by allowing some of the sensors to act as relay nodes [32], see [41] for details.

11.2.3 Numerical results

We first show the performance of a multiple-antenna single sensor system via numerical simulations. We consider the source θ to be Gaussian distributed with zero mean and variance $\sigma_\theta^2 = 1$ mW. The sensor measurement sensitivity is set to $\sigma_\omega^2 = 10^{-3}$ mW.

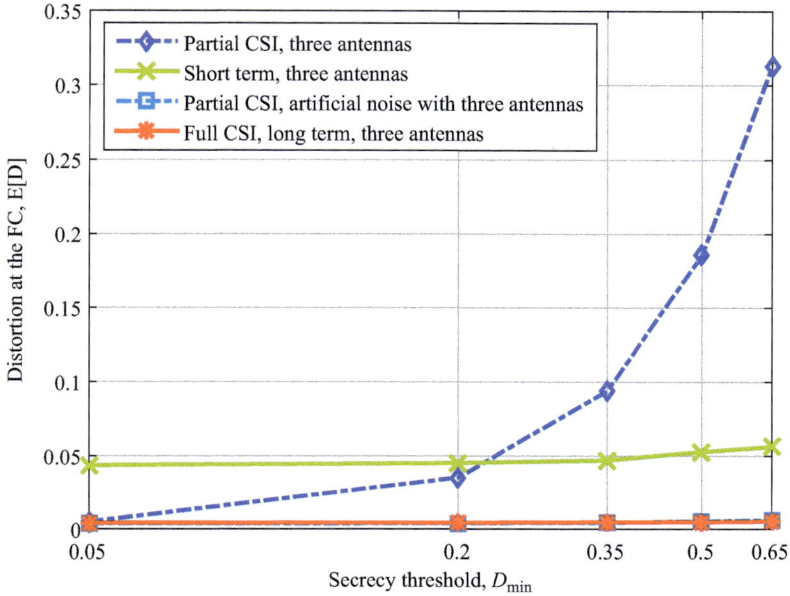

Figure 11.3 Performance comparison between full CSI, partial CSI and artificial noise in a multiple-antenna single sensor system

We assume the same noise level for both the FC and the eavesdropper's channel, where $\sigma_n^2 = \sigma_e^2 = 10^{-8}$ mW. The secrecy threshold is chosen from the range $0.05 \leq D_{\min} \leq 0.65$. Furthermore, we consider the pathloss of signal power, in decibel scale, at the FC and the eavesdropper to follow the free-space pathloss model [42]:

$$PL = 20\log_{10}(Dist) + 20\log_{10}(f) - 27.55, \tag{11.25}$$

where $Dist \in \{d, d_e\}$ is the distance between the sensor and the FC or the eavesdropper in meters, and f is the signal frequency in megahertz (we assume the network uses operation frequency of 800 MHz, and the sensor is closer to the FC than to the eavesdropper with the distance from the sensor to the FC and to the eavesdropper being set to 127 and 130 m, respectively). Thus, the channel power gain follows an exponential distribution with mean of $10^{-(PL/10)}$ mW.

Figure 11.3 depicts the distortion performance at the FC versus the secrecy threshold for a three-antenna single sensor system. For comparison, we plot the system performance under four scenarios: long-term full CSI, partial CSI, partial CSI with artificial noise and short-term full CSI. First, owing to the channel knowledge of both the FC and the eavesdropper, it is not surprising to see that the performance of the full CSI scenario is superior to the performance of partial CSI. Similar performance gains can be seen for the full CSI short-term distortion. We also notice the superior performance of artificial noise in the partial CSI case. This is because in the full CSI scenario, due to the full channel information of both FC and the eavesdropper, the direction of the beamformer can be designed to benefit the FC with little information being leaked to the eavesdropper; and in the case of artificial noise, a small amount of "noise" is deliberately generated to degrade the eavesdropper's channel, which indicates that the secrecy threshold can be easily achieved without sacrificing much transmit power, whereas for the case of partial CSI without artificial noise, some antennas need to be switched off to achieve the secrecy requirements, which is also the case for the short-term scenario.

Next we study the distortion performance at the FC for a multiple-sensor network, where we assume the total transmit power budget is 30 mW and all sensors share the same measurement sensitivity, i.e., $\sigma_{\omega_k}^2 = \sigma_{\omega}^2 = 10^{-3}$, $\forall k$. The distance from each sensor to the FC and to the eavesdropper are 125, 126, 127, 128, 129, 130, 131, 132 m, and 139, 138, 137, 136, 135, 131, 130, 129 m, respectively. We apply the same pathloss model (11.25) and we consider the same noise level for both the FC and the eavesdropper's channel, where $\sigma_n^2 = \sigma_e^2 = 10^{-8}$ mW. In Figure 11.4, the secrecy threshold is chosen from the range $0.05 \leq D_{\min} \leq 0.25$. The short-term distortion result is obtained by averaging over 10,000 channel realizations. Not surprisingly, we can see that the long-term distortion performances are superior to the short-term performance due to a smaller feasibility region for the latter, where the sensors are required to ensure that the power constraint, and the secrecy constraint are satisfied in every transmission slot.

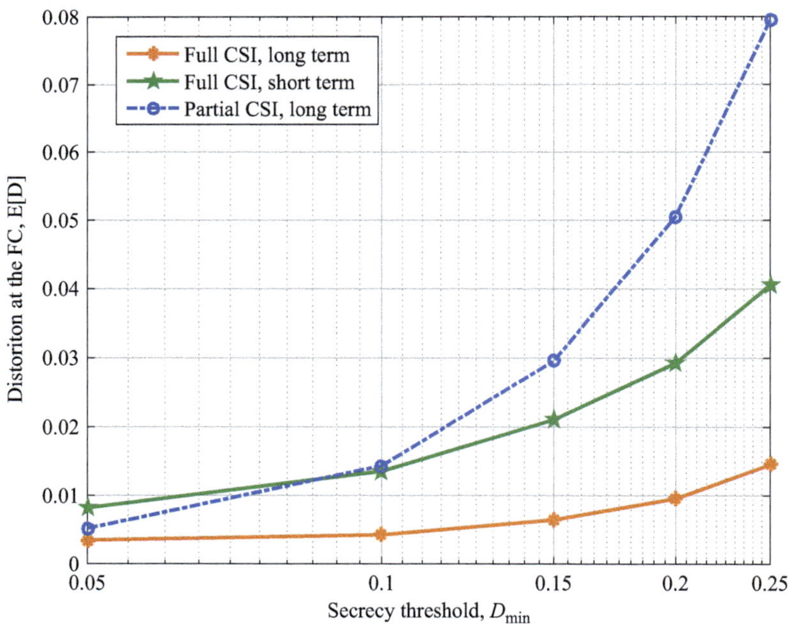

Figure 11.4 Performance comparison in an eight-sensor network

11.3 Estimation with secrecy outage constraints

11.3.1 Multiple sensors scenario

A schematic diagram of the WSN model is shown in Figure 11.5, where we have K sensors observing a single point Gaussian source with zero mean and variance σ_θ^2, denoted by $\theta[t]$, $t = 0, 1, 2, \ldots$. The measurement $x_k[t]$ received by the kth sensor at time t is given by:

$$x_k[t] = \theta[t] + \omega_k[t], \tag{11.26}$$

where $\omega_k[t]$ is the sensor measurement noise which is i.i.d. Gaussian with zero mean and variance $\sigma_{\omega k}^2$.

The sensors are assumed to have a single transmit antenna, see Section 11.3.2 for the case of multiple transmit antennas. Each sensor amplifies and forwards their measurements to a N_r-antenna fusion center (FC) with amplification factor $\beta_k[t] \in \mathbb{C}$ via a slow-fading orthogonal MAC. The transmissions are overheard by an eavesdropper

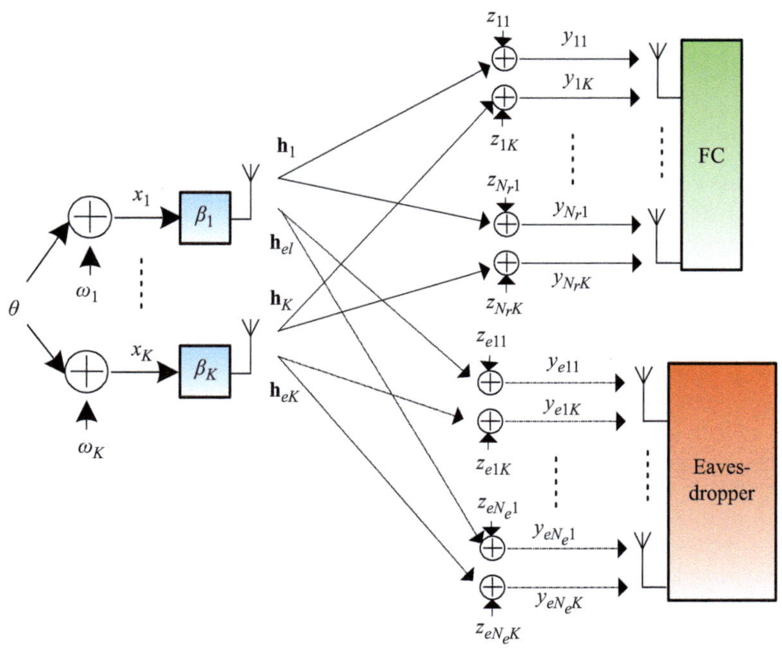

Figure 11.5 Diagram of the wireless sensor network using orthogonal MAC scheme with the presence of an eavesdropper

which is equipped with N_e receive antennas. The signals received by the FC and eavesdropper from the kth sensor are then given by, respectively:

$$\mathbf{y}_k[t] = \theta[t]\beta_k[t]\mathbf{h}_k[t] + \omega_k[t]\beta_k[t]\mathbf{h}_k[t] + \mathbf{z}_k[t], \quad (11.27a)$$

$$\mathbf{y}_{ek}[t] = \theta[t]\beta_k[t]\mathbf{h}_{ek}[t] + \omega_k[t]\beta_k[t]\mathbf{h}_{ek}[t] + \mathbf{z}_{ek}[t], \quad (11.27b)$$

where $\mathbf{y}_k[t] = \left[y_{1k}[t], \ldots, y_{N_rk}[t]\right]^{\mathrm{T}}$ and $\mathbf{y}_{ek}[t] = \left[y_{e1k}[t], \ldots, y_{eN_ek}[t]\right]^{\mathrm{T}}$, the entries of $\mathbf{h}_k[t]$ and $\mathbf{h}_{ek}[t]$ are the instantaneous zero mean i.i.d. complex Gaussian channels from sensor k to the FC and the eavesdropper with variances σ^2_{hk} and $\sigma^2_{h_ek}$, respectively, and $\mathbf{z}_k[t] = \left[z_{1k}[t], \ldots, z_{N_rk}[t]\right]^{\mathrm{T}}$ and $\mathbf{z}_{ek}[t] = \left[z_{e1k}[t], \ldots, z_{eN_ek}[t]\right]^{\mathrm{T}}$ represent i.i.d. additive Gaussian noises with zero mean and covariances $\sigma_k^2\mathbf{I}_{N_r}$ at the FC and $\sigma_{ek}^2\mathbf{I}_{N_e}$ at the eavesdropper, respectively. The set of received signals at the FC from all the sensors can be written as:

$$\begin{aligned}\mathbf{Y}[t] &= \left[\mathbf{y}_1[t], \ldots, \mathbf{y}_K[t]\right]^{\mathrm{T}} \\ &= \theta[t]\left[\beta_1[t]\mathbf{h}_1[t], \ldots, \beta_k[t]\mathbf{h}_k[t]\right]^{\mathrm{T}} + \left[\mathbf{z}_1[t], \ldots, \mathbf{z}_k[t]\right]^{\mathrm{T}} \\ &\quad + \left[\omega_1[t]\beta_1[t]\mathbf{h}_1[t], \ldots, \omega_k[t]\beta_k[t]\mathbf{h}_k[t]\right]^{\mathrm{T}}. \end{aligned} \quad (11.28)$$

Using the fact that each sensor transmits through an orthogonal MAC, the covariance of the noise factor $[\omega_1[t]\beta_1[t]\mathbf{h}_1[t], \ldots, \omega_k[t]\beta_k[t]\mathbf{h}_k[t]]^{\mathrm{T}} + [\mathbf{z}_1[t], \ldots, \mathbf{z}_k[t]]^{\mathrm{T}}$ can be derived as a $KN_r \times KN_r$ matrix:

$$C[t] = \begin{bmatrix} \sigma_{w1}^2\beta_1^2[t]\mathbf{h}_1[t]\mathbf{h}_1^{\mathrm{H}}[t] + \sigma_1^2\mathbf{I}_{N_r} & & 0 \\ & \ddots & \\ 0 & & \sigma_{wK}^2\beta_K^2[t]\mathbf{h}_K[t]\mathbf{h}_K^{\mathrm{H}}[t] + \sigma_K^2\mathbf{I}_{N_r} \end{bmatrix}. \quad (11.29)$$

At time t, the distortion at the FC using the MMSE estimator is:

$$\begin{aligned} D[t] &= \left(\frac{1}{\sigma_\theta^2} + \begin{bmatrix}\beta_1[t]\mathbf{h}_1[t] \\ \vdots \\ \beta_K[t]\mathbf{h}_K[t]\end{bmatrix}^{\mathrm{H}} C[t]^{-1} \begin{bmatrix}\beta_1[t]\mathbf{h}_1[t] \\ \vdots \\ \beta_K[t]\mathbf{h}_K[t]\end{bmatrix}\right)^{-1} \\ &\overset{(a)}{=} \left[\frac{1}{\sigma_\theta^2} + \sum_{k=1}^{K} \beta_k^{\mathrm{H}}[t]\beta_k[t]\Big(\sigma_k^{-2}\mathbf{h}_k^{\mathrm{H}}[t]\mathbf{h}_k[t] - \sigma_k^{-2}\mathbf{h}_k^{\mathrm{H}}[t]\mathbf{h}_k[t] \right. \\ &\quad \left. \times\left(\sigma_{wk}^{-2}\beta_k^{-2}[t] + \sigma_k^{-2}\mathbf{h}_k^{\mathrm{H}}[t]\mathbf{h}_k[t]\right)^{-1}\sigma_k^{-2}\mathbf{h}_k^{\mathrm{H}}[t]\mathbf{h}_k[t]\Big)\right]^{-1} \end{aligned} \quad (11.30)$$

$$= \left(\frac{1}{\sigma_\theta^2} + \sum_{k=1}^{K} \frac{g_k[t]p_k[t]}{\sigma_k^2 + g_k[t]\sigma_{wk}^2 p_k[t]}\right)^{-1}, \quad (11.31)$$

where (a) results from applying the matrix inversion lemma, $p_k[t] \triangleq \beta_k^{\mathrm{H}}[t]\beta_k[t]$ is the power allocated on the kth sensor, and $g_k[t] \triangleq \mathbf{h}_k^{\mathrm{H}}[t]\mathbf{h}_k[t] = \sum_{m=1}^{N_r} h_{mk}^{\mathrm{H}}[t]h_{mk}[t]$ is the sum of channel power gains from the kth sensor to the FC, with $h_{mk}[t]$ being the channel gain from sensor k to the mth antenna at the FC. Similarly, the distortion at the eavesdropper is given as:

$$D_e[t] = \left(\frac{1}{\sigma_\theta^2} + \sum_{k=1}^{K} \frac{g_{ek}[t]p_k[t]}{g_{ek}[t]p_k[t]\sigma_{\omega k}^2 + \sigma_{ek}^2} \right)^{-1}, \tag{11.32}$$

where $g_{ek}[t] \triangleq \mathbf{h}_{ek}^{\mathrm{H}}[t]\mathbf{h}_{ek}[t] = \sum_{n=1}^{N_e} h_{enk}^{\mathrm{H}}[t]h_{enk}[t]$. Note that for a given set of $\{p_k[t]\}$, any $\{\beta_k[t]\}$ satisfying $\beta_k[t]^{\mathrm{H}}\beta_k[t] = p_k[t], \forall k$ would result in the same distortion; hence, our primary focus is on $\{p_k[t]\}$.

In this section, we focus on the *distortion outage minimization* problem. For a given maximum acceptable distortion level $\mathbb{D}$ at the FC, we define a *distortion outage* to be the event that the instantaneous distortion $D[t]$ exceeds $\mathbb{D}$. The *distortion outage probability* at the FC is then given as $\mathrm{Pr}_{\mathrm{outage\_FC}} \triangleq \mathrm{Pr}\left[D[t] > \mathbb{D}\right]$. At the eavesdropper, for a given minimum acceptable distortion level $\mathbb{D}_e$, a *secrecy outage* event is declared if the instantaneous distortion $D_e[t]$ is less than $\mathbb{D}_e$ (which means that the eavesdropper has a good quality estimate), and the *secrecy outage probability* is defined as $\mathrm{Pr}_{\mathrm{outage\_EVE}} \triangleq \mathrm{Pr}\left[D_e[t] < \mathbb{D}_e\right]$.

We wish to minimize the distortion outage probability at the FC by adapting the transmit powers of the sensors at each channel instance, while keeping the secrecy outage probability under a certain threshold, i.e., $\mathrm{Pr}_{\mathrm{outage\_EVE}} \le \delta$, and the long-term average sum of sensor transmission powers, defined as $\mathbb{E}\left[\sum_{k=1}^{K} p_k \mathbb{E}\left[x_k^2[t]\right]\right] = \mathbb{E}\left[\sum_{k=1}^{K} p_k(\sigma_\theta^2 + \sigma_{\omega k}^2)\right]$, to be less than a power budget $\mathscr{P}_{\mathrm{tot}}$.

Due to the assumption of system independence over time t, we will again drop the time index t for the rest of the section.

11.3.1.1 Full CSI

Let the channel states at the FC and the eavesdropper be denoted by $\mathbf{g} = [g_1, \ldots, g_K]$ and $\mathbf{g}_e = \left[g_{e1}, \ldots, g_{ek}\right]$, respectively. The outage minimization problem is:

$$\begin{aligned} &\min_{\mathbf{P}(\mathbf{G})} \ \mathrm{Pr}[D(\mathbf{G}, \mathbf{P}(\mathbf{G})) > \mathbb{D}] \\ &\text{s.t.} \ \ \mathrm{Pr}[D_e(\mathbf{G}, \mathbf{P}(\mathbf{G})) < \mathbb{D}_e] \le \delta, \quad \mathbb{E}_{\mathbf{G},\mathbf{p}}\left[\langle \mathbf{P}(\mathbf{G}) \rangle\right] \le \mathscr{P}_{\mathrm{tot}}, \end{aligned} \tag{11.33}$$

where $\mathbf{G} = \left[\mathbf{g}; \mathbf{g}_e\right]$, and $\langle \mathbf{p}(\mathbf{G}) \rangle \triangleq \sum_{k=1}^{K} \left(\sigma_\theta^2 + \sigma_{\omega k}^2\right) p_k(\mathbf{G})$ is the total power consumption. $\mathbf{P}(\mathbf{G})$ is a vector of random variables with conditional probability density function $f_{\mathbf{P}|\mathbf{G}}\left(\mathbf{p}|\,\mathbf{G}\right)$, where $\mathbf{p}$ is one of the deterministic schemes and $\mathbf{p} = [p_1, \ldots, p_K]$ are the powers allocated across the sensors.

Notice from (11.32) that when zero power is allocated to the sensors, we obtain $D_e|_{\mathbf{p}=\mathbf{0}} = \sigma_\theta^2$, giving the largest possible distortion at the eavesdropper, while if the transmit power on each sensor approaches infinity, we have the smallest possible distortion at the eavesdropper $D_e \to \left((1/\sigma_\theta^2) + \sum_{k=1}^{K}(1/\sigma_{\omega k}^2)\right)^{-1}$. Therefore, in order to produce a meaningful solution to problem (11.33), $\mathbb{D}_e$ should satisfy $\left((1/\sigma_\theta^2) + \sum_{k=1}^{K}(1/\sigma_{\omega k}^2)\right)^{-1} < \kappa < \mathbb{D}_e < \sigma_\theta^2$, where κ is a nonnegative threshold to ensure the first constraint in (11.33) is achievable for a given transmit power budget $\mathscr{P}_{\mathrm{tot}}$ and a secrecy outage probability threshold δ.

In communications theory, it was shown in [5,7] that for information outage minimization problems, the optimal power allocation policy is in general a probabilistic policy, in particular this is often the case for discrete channel distributions. Motivated by these results, we start with a probabilistic power allocation $\mathbf{P}(\mathbf{G})$.

Denote the indicator function by $\mathbb{1}\{x\}$, where $\mathbb{1}\{x\} = 1$ if x is true; otherwise, $\mathbb{1}\{x\} = 0$. With full CSI at the FC, the distortion outage probability at the FC and the secrecy outage probability at the eavesdropper can be expressed as, respectively:

$$\Pr[D(\mathbf{G},\mathbf{P}) > \mathbb{D}] = \iint \mathbb{1}\{D(\mathbf{G},\mathbf{p}) > \mathbb{D}\} f_{\mathbf{P}|\mathbf{G}}(\mathbf{p}|\,\mathbf{G})\, d\mathbf{p}(\mathbf{G})\, dF(\mathbf{G}), \tag{11.34}$$

$$\Pr[D_e(\mathbf{G},\mathbf{P}) < \mathbb{D}_e] = \iint \mathbb{1}\{D_e(\mathbf{G},\mathbf{p}) < \mathbb{D}_e\} f_{\mathbf{P}|\mathbf{G}}(\mathbf{p}|\,\mathbf{G})\, d\mathbf{p}(\mathbf{G})\, dF(\mathbf{G}), \tag{11.35}$$

where $F(\mathbf{G})$ denotes the cumulative distribution function of $\mathbf{G}$.

Using similar techniques as in [7], see [43] for details, one can show that problem (11.33) can be reformulated into the following:

$$\min_{\{\omega_j(\mathbf{G})\},\{\mathbf{p}_j(\mathbf{G})\}} \quad 1 - \mathbb{E}\left[\omega_1(\mathbf{G}) + \omega_2(\mathbf{G})\right]$$

$$\text{s.t.}\ \mathbb{E}\left[\omega_2(\mathbf{G})\right] \le \delta, \tag{11.36a}$$

$$\mathbb{E}\left[\langle\omega_1(\mathbf{G})\,\mathbf{p}_1(\mathbf{G})\rangle + \langle\omega_2(\mathbf{G})\,\mathbf{p}_2(\mathbf{G})\rangle\right] \le \mathscr{P}_{\text{tot}}, \tag{11.36b}$$

$$\omega_1(\mathbf{G})\,D_e\big(\mathbf{G},\mathbf{p}_1(\mathbf{G})\big) - \omega_1(\mathbf{G})\,\sigma_\theta^2 + \omega_2(\mathbf{G})\left(\mathbb{D}_e - \sigma_\theta^2\right) \ge \mathbb{D}_e - \sigma_\theta^2, \tag{11.36c}$$

$$\omega_1(\mathbf{G})\,D\big(\mathbf{G},\mathbf{p}_1(\mathbf{G})\big) + \omega_2(\mathbf{G})\,D\left(\mathbf{G},\mathbf{p}_2(\mathbf{G})\right) - \left(\omega_1(\mathbf{G}) + \omega_2(\mathbf{G})\right)\mathbb{D} \le 0, \tag{11.36d}$$

$$\omega_1(\mathbf{G}) + \omega_2(\mathbf{G}) \le 1, \tag{11.36e}$$

$$0 \le \omega_j(\mathbf{G}) \le 1, \qquad j = 1,2. \tag{11.36f}$$

where the optimal solution to problem (11.33) is of the form

$$\mathbf{P}^*(\mathbf{G}) = \mathbf{p}_i(\mathbf{G}) \text{ w.p. } \omega_i(\mathbf{G}), \quad i = 1,2,3.$$

By analyzing the generalized KKT conditions [44] of the functional optimization problem (11.36a)–(11.36f), we can show that for continuous fading channel distributions, the optimal power allocation policies are deterministic. Furthermore, we have:

Theorem 11.1. *Consider the following optimization problems* (11.37) *and* (11.38)*:*

$$\begin{aligned}&\min_{\mathbf{p}}\ \langle\mathbf{p}(\mathbf{G})\rangle\\ &\text{s.t.}\ \ D_e(\mathbf{G},\mathbf{p}(\mathbf{G})) \ge \mathbb{D}_e, \quad D(\mathbf{G},\mathbf{p}(\mathbf{G})) \le \mathbb{D},\end{aligned} \tag{11.37}$$

and

$$\begin{aligned}&\min_{\mathbf{p}}\ \langle\mathbf{p}(\mathbf{G})\rangle\\ &\text{s.t.}\ \ D(\mathbf{G},\mathbf{p}(\mathbf{G})) = \mathbb{D},\end{aligned} \tag{11.38}$$

with optimal solutions $\mathbf{p}_a^*(\mathbf{G})$ *and* $\mathbf{p}_b^*(\mathbf{G})$*, respectively. Then a locally optimal solution to problem (11.36a)–(11.36f) is given by:*

$$\mathbf{P}^*(\mathbf{G}) = \begin{cases} \mathbf{p}_a^*(\mathbf{G}), & \text{if } \omega_1^*(\mathbf{G}) = 1 \\ \mathbf{p}_b^*(\mathbf{G}), & \text{if } \omega_2^*(\mathbf{G}) = 1 \text{ and } D_e\big(\mathbf{G}, \mathbf{p}_b^*(\mathbf{G})\big) < \mathbb{D}_e \\ \mathbf{0}, & \text{otherwise.} \end{cases} \tag{11.39}$$

The proof can be found in [43].

11.3.1.2 Partial CSI

From the results in Section 11.3.1.1, we know that the optimal transmit power policies are deterministic if both the FC's and eavesdropper's fading channels have continuous distributions, based on which, in this part of the work, we aim to develop deterministic transmit power policies with full knowledge of only the sensor-to-FC channels $\mathbf{g}$. Using a similar setup as problem (11.33), the Lagrangian in the partial CSI case can be constructed as:

$$\begin{aligned} l(\mathbf{g}, \nu, \lambda) = \int_{\mathbf{g}} \Big[& \mathbb{1}\{D(\mathbf{g}, \mathbf{p}(\mathbf{G})) > \mathbb{D}\} + \lambda \langle \mathbf{p}(\mathbf{G}) \rangle \\ & + \nu \int_{\mathbf{g}_e} \mathbb{1}\big\{D_e\big(\mathbf{g}_e, \mathbf{p}(\mathbf{G})\big) < \mathbb{D}_e\big\}\, dF\big(\mathbf{g}_e\big)\Big] dF(\mathbf{G}), \end{aligned} \tag{11.40}$$

where λ and ν are nonnegative Lagrange multipliers satisfying the following equations at the optimal point:

$$\begin{aligned} &\lambda^*(\mathscr{P}_{\text{tot}} - \mathbb{E}[\langle \mathbf{p}^*(\mathbf{G}) \rangle]) = 0, \\ &\nu^*\big(\delta - \Pr\big[D_e\big(\mathbf{g}_e, \mathbf{p}^*(\mathbf{g})\big) < \mathbb{D}_e\big]\big) = 0. \end{aligned} \tag{11.41}$$

To minimize the Lagrangian given in (11.40), we need to find the optimal power allocation for each channel state at the FC such that $\mathbb{1}\{D(\mathbf{g}, \mathbf{p}(\mathbf{G})) > \mathbb{D}\} + \lambda \langle \mathbf{p}(\mathbf{G}) \rangle + \nu \int_{\mathbf{g}_e} \mathbb{1}\big\{D_e\big(\mathbf{g}_e, \mathbf{p}(\mathbf{G})\big) < \mathbb{D}_e\big\}\, dF\big(\mathbf{g}_e\big)$ is minimized.

Lemma 11.1. *Let* $\xi(\mathbf{p}(\mathbf{g})) = \lambda \langle \mathbf{p}(\mathbf{g}) \rangle + \nu \int_{\mathbf{g}_e} \mathbb{1}\{D_e(\mathbf{g}_e, \mathbf{p}(\mathbf{g})) < \mathbb{D}_e\}\, dF(\mathbf{g}_e)$*. Then a globally optimal* $\mathbf{p}^*(\mathbf{g})$ *of problem* (11.33) *in the partial CSI case must satisfy* $0 \le \mathbb{1}\{D(\mathbf{g}, \mathbf{p}^*(\mathbf{g})) > \mathbb{D}\} + \xi(\mathbf{p}^*(\mathbf{g})) \le 1$*, where the optimal power allocation* $\mathbf{p}^*(\mathbf{g})$ *has the form:*

$$\mathbf{p}^*(\mathbf{g}) = \begin{cases} \hat{\mathbf{p}}(\mathbf{g}), & \text{if } \xi\big(\hat{\mathbf{p}}(\mathbf{g})\big) < 1 \\ \mathbf{0}, & \text{otherwise,} \end{cases} \tag{11.42}$$

with $\hat{\mathbf{p}}(\mathbf{g})$ *being a locally optimal solution of the following problem:*

$$\begin{aligned} \min_{\mathbf{p}(\mathbf{g})} \quad & \lambda \langle \mathbf{p}(\mathbf{g}) \rangle + \nu \int \mathbb{1}\{D_e(\mathbf{g}_e, \mathbf{p}(\mathbf{g})) < \mathbb{D}_e\} f(\mathbf{g}_e)\, d\mathbf{g}_e \\ \text{s.t.} \quad & D(\mathbf{g}, \mathbf{p}(\mathbf{g})) \le \mathbb{D}. \end{aligned} \tag{11.43}$$

The proof can be found in [43].

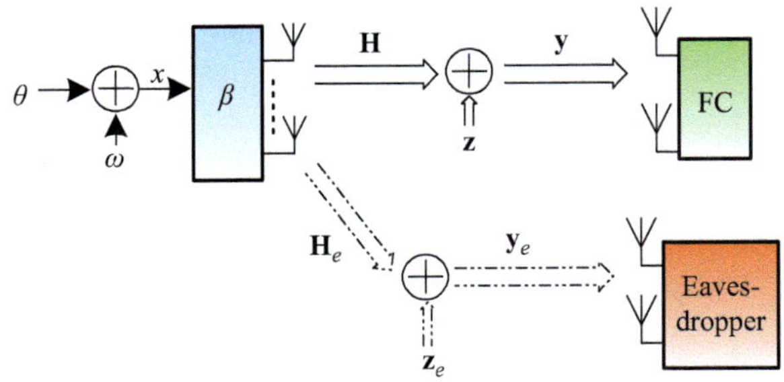

Figure 11.6 Diagram of a single sensor multiple-antenna scenario with the presence of an eavesdropper

11.3.2 Single sensor with multiple antennas scenario

In this subsection, we consider a situation where only one sensor with multiple-antennas is in the network observing the source. As in Section 11.2, with multiple antennas additional techniques, such as artificial noise and beamforming [31,32,34], can be used to further enhance the system performance for both the full CSI and partial CSI cases.

A schematic diagram is shown in Figure 11.6. We assume that the same single point Gaussian source θ as defined in Section 11.3.1 is observed by a sensor with N_t transmit antennas, which employs the analog amplify and forward technique to scale the observed signal with a complex vector $\boldsymbol{\beta} \in \mathbb{C}^{N_t \times 1}$, before sending it to the FC via a set of complex fading channels $\mathbf{H} \in \mathbb{C}^{Nr \times Nt}$. The observed signal x is also listened to by the eavesdropper after passing through another set of channels $\mathbf{H}_e \in \mathbb{C}^{Ne \times Nt}$, where we assume that the FC and the eavesdropper are equipped with N_r and N_e receive antennas, respectively.

The signals received by the FC and the eavesdropper are, respectively:

$$\mathbf{y} = \mathbf{H}\boldsymbol{\beta}\theta + \mathbf{H}\boldsymbol{\beta}\omega + \mathbf{z}, \tag{11.44a}$$

$$\mathbf{y}_e = \mathbf{H}_e\boldsymbol{\beta}\theta + \mathbf{H}_e\boldsymbol{\beta}\omega + \mathbf{z}_e, \tag{11.44b}$$

where $\mathbf{z} \in \mathbb{C}^{Nr \times 1}$ and $\mathbf{z}_e \in \mathbb{C}^{Ne \times 1}$ are complex Gaussian channel noises at the FC and the eavesdropper, with covariances $\sigma^2\mathbf{I}_{N_r}$ and $\sigma_e^2\mathbf{I}_{N_e}$, respectively.

For a given channel instance, the distortion D at the FC can be obtained as:

$$\begin{aligned} D &= \left(\frac{1}{\sigma_\theta^2} + (\mathbf{H}\boldsymbol{\beta})^{\mathrm{H}}\, \Sigma^{-1}\mathbf{H}\boldsymbol{\beta}\right)^{-1}, \\ &\overset{(b)}{=} \sigma_\theta^2\left(1-(\mathbf{H}\boldsymbol{\beta})^{\mathrm{H}}\left[\mathbf{H}\boldsymbol{\beta}(\mathbf{H}\boldsymbol{\beta})^{\mathrm{H}}\left(\sigma_\theta^2+\sigma_\omega^2\right)+\sigma^2\mathbf{I}_{N_r}\right]^{-1}\mathbf{H}\boldsymbol{\beta}\sigma_\theta^2\right) \\ &\overset{(c)}{=} \sigma_\theta^2\left[1 - \frac{\sigma_\theta^2}{\sigma_\theta^2+\sigma_\omega^2}\left(1 - \frac{\alpha}{\alpha + (\mathbf{H}\boldsymbol{\beta})^{\mathrm{H}}\mathbf{H}\boldsymbol{\beta}}\right)\right], \end{aligned} \tag{11.45}$$

where $\Sigma \triangleq \mathbf{H}\boldsymbol{\beta}(\mathbf{H}\boldsymbol{\beta})^{\mathrm{H}}\sigma_\omega^2 + \sigma^2\mathbf{I}_{N_r}$ is the covariance matrix of $\mathbf{H}\boldsymbol{\beta}\omega + \mathbf{z}$, $\alpha \triangleq (\sigma^2/(\sigma_\omega^2+\sigma_\theta^2))$, and (b)–(c) result from applying the matrix inversion lemma. Similarly, the distortion at the eavesdropper is given as:

$$D_e = \sigma_\theta^2\left[1 - \frac{\sigma_\theta^2}{\sigma_\theta^2+\sigma_\omega^2}\left(1 - \frac{\alpha_e}{\alpha_e + (\mathbf{H}_e\boldsymbol{\beta})^{\mathrm{H}}\mathbf{H}_e\boldsymbol{\beta}}\right)\right], \tag{11.46}$$

where $\alpha_e \triangleq (\sigma_e^2/(\sigma_\omega^2+\sigma_\theta^2))$.

We set a maximum acceptable distortion level $\mathbb{D}$ at the FC. Then the *distortion outage probability* is:

$$\Pr_{\text{outage\_FC}} = \Pr\left[D > \mathbb{D}\right] = \Pr\left[\frac{1}{(\mathbf{H}\boldsymbol{\beta})^{\mathrm{H}}\mathbf{H}\boldsymbol{\beta}} > \mathbb{S}\right], \tag{11.47}$$

where $\mathbb{S} \triangleq ((\sigma_\omega^2+\sigma_\theta^2)/\sigma^2)\left[(\sigma^4/(\sigma^4 - \mathbb{D}\left(\sigma_\omega^2+\sigma_\theta^2\right) + \sigma_\omega^2\sigma_\theta^2)) - 1\right]$. We also set a minimum acceptable distortion level $\mathbb{D}_e$ at the eavesdropper. Letting $\mathbb{S}_e \triangleq ((\sigma_\omega^2+\sigma_\theta^2)/\sigma_e^2)\left[(\sigma_e^4/(\sigma_e^4 - \mathbb{D}_e\left(\sigma_\omega^2+\sigma_\theta^2\right) + \sigma_\omega^2\sigma_\theta^2)) - 1\right]$, the *secrecy outage probability* at the eavesdropper can be expressed as:

$$\Pr_{\text{outage\_EVE}} = \Pr[D_e < \mathbb{D}_e] = \Pr\left[\frac{1}{(\mathbf{H}_e\boldsymbol{\beta})^{\mathrm{H}}\mathbf{H}_e\boldsymbol{\beta}} < \mathbb{S}_e\right].$$

With a given power budget at the sensor, our objective is to minimize the distortion outage probability at the FC, while keeping the secrecy outage probability at the eavesdropper below δ. Hence the optimization problem can be formulated as:

$$\begin{aligned} &\min_{\boldsymbol{\beta}} \ \Pr\left[\frac{1}{(\mathbf{H}\boldsymbol{\beta})^{\mathrm{H}}\mathbf{H}\boldsymbol{\beta}} > \mathbb{S}\right] \\ &\text{s.t.} \ \Pr\left[\frac{1}{(\mathbf{H}_e\boldsymbol{\beta})^{\mathrm{H}}\mathbf{H}_e\boldsymbol{\beta}} < \mathbb{S}_e\right] \le \delta, \quad \mathbb{E}\left[\boldsymbol{\beta}^{\mathrm{H}}\boldsymbol{\beta}\right] \le \frac{\mathscr{P}_{\text{tot}}}{\sigma_\theta^2+\sigma_\omega^2}. \end{aligned} \tag{11.48}$$

11.3.2.1 Full CSI

Using techniques from [43], we can show that problem (11.48) can be recast as:

$$\min_{\{\boldsymbol{\beta}_j(\mathbf{H},\mathbf{H}_e)\},\ \{\omega_j(\mathbf{H},\mathbf{H}_e)\}} \quad 1 - \mathbb{E}[\omega_1 + \omega_2]$$

$$\text{s.t.} \quad \mathbb{E}[\omega_2] \le \delta, \tag{11.49a}$$

$$\mathbb{E}\left[\sum_{j=1}^{2}\omega_j\boldsymbol{\beta}_j^{\mathrm{H}}\boldsymbol{\beta}_j\right] \le \frac{\mathscr{P}_{\text{tot}}}{\sigma_\theta^2+\sigma_\omega^2}, \tag{11.49b}$$

$$\frac{\omega_1}{(\mathbf{H}_e\boldsymbol{\beta}_1)^{\mathrm{H}}\mathbf{H}_e\boldsymbol{\beta}_1} \ge (1-\omega_2)\,\mathbb{S}_e, \tag{11.49c}$$

$$\frac{\omega_1}{(\mathbf{H}\boldsymbol{\beta}_1)^{\mathrm{H}}\mathbf{H}\boldsymbol{\beta}_1} + \frac{\omega_2}{(\mathbf{H}\boldsymbol{\beta}_2)^{\mathrm{H}}\mathbf{H}\boldsymbol{\beta}_2} \le (\omega_1+\omega_2)\,\mathbb{S}, \tag{11.49d}$$

$$0 \le \omega_1 + \omega_2 \le 1, \quad 0 \le \omega_1, \omega_2 \le 1, \tag{11.49e}$$

where the optimal power allocation randomises among the three deterministic schemes $\boldsymbol{\beta}_1$, $\boldsymbol{\beta}_2$, and $\boldsymbol{\beta}_3$ with corresponding weighting factors $\{\omega_i\}_{i=1}^3$. With the assumption that both $\mathbf{H}$ and $\mathbf{H}_e$ are continuously distributed, a locally optimal solution of problem (11.48) in the case of full CSI is given as:

$$\boldsymbol{\beta}^*(\mathbf{H},\mathbf{H}_e) = \begin{cases} \boldsymbol{\beta}_1^*(\mathbf{H},\mathbf{H}_e), & \text{if } \lambda^* \boldsymbol{\beta}_1^{*\mathrm{H}} \boldsymbol{\beta}_1^* \le 1, \\ \boldsymbol{\beta}_2^*(\mathbf{H},\mathbf{H}_e), & \text{if } \lambda^* \boldsymbol{\beta}_2^{*\mathrm{H}} \boldsymbol{\beta}_2^* \le 1-\gamma^* \text{ and } \left(\mathbf{H}_e\boldsymbol{\beta}_2^*\right)^{\mathrm{H}} \mathbf{H}_e\boldsymbol{\beta}_2^* > \mathbb{S}_e^{-1}, \\ \mathbf{0}, & \text{otherwise}, \end{cases} \tag{11.50}$$

where λ^* and $\gamma^*(\mathbf{H},\mathbf{H}_e)$ are the optimal Lagrange multipliers chosen to satisfy the constraints $\mathbb{E}\left[\sum_{j=1}^2 \omega_j^* \boldsymbol{\beta}_j^{\mathrm{H}*} \boldsymbol{\beta}_j^*\right] \le (\mathscr{P}_{\mathrm{tot}}/(\sigma_\theta^2+\sigma_\omega^2))$ and $\mathbb{E}\left[\omega_2^*\right] \le \delta$, respectively, and $\boldsymbol{\beta}_1^*$ and $\boldsymbol{\beta}_2^*$ are, respectively, the optimal solutions of the following two problems:

$$\begin{aligned} \min_{\boldsymbol{\beta}_1(\mathbf{H},\,\mathbf{H}_e)} \quad & \boldsymbol{\beta}_1^{\mathrm{H}} \boldsymbol{\beta}_1 \\ \text{s.t.} \quad & \left(\mathbf{H}\boldsymbol{\beta}_1\right)^{\mathrm{H}} \mathbf{H}\boldsymbol{\beta}_1 \ge \mathbb{S}^{-1}, \quad \left(\mathbf{H}_e\boldsymbol{\beta}_1\right)^{\mathrm{H}} \mathbf{H}_e\boldsymbol{\beta}_1 \le \mathbb{S}_e^{-1}; \end{aligned} \tag{11.51}$$

and

$$\begin{aligned} \min_{\boldsymbol{\beta}_2(\mathbf{H})} \quad & \boldsymbol{\beta}_2^{\mathrm{H}} \boldsymbol{\beta}_2 \\ \text{s.t.} \quad & \left(\mathbf{H}\boldsymbol{\beta}_2\right)^{\mathrm{H}} \mathbf{H}\boldsymbol{\beta}_2 \ge \mathbb{S}^{-1}. \end{aligned} \tag{11.52}$$

Note that problems (11.51) and (11.52) can be solved via the Lagrangian multiplier techniques. However, as all the constraints and objective functions in problem (11.51) and (11.52) are real-valued over the complex field, we need to consider both the real and imaginary parts when applying the KKT conditions for the optimal points [45,46]. Furthermore, because problem (11.51) is a non-convex optimization problem, while (11.52) is a convex problem, we obtain a locally optimal solution $\boldsymbol{\beta}_1^*$ of the problem (11.51) and a globally optimal solution $\boldsymbol{\beta}_2^*$ of the problem (11.52).

Zero outage probability at the eavesdropper

If the sensor has more transmit antennas than the number of receive antennas at the eavesdropper, i.e., $N_t > N_e$, then it can transmit the observation signal x onto the null space of the eavesdropper's channel, thus leaking no useful information to the eavesdropper. To be more specific, let the singular value decomposition of $\mathbf{H}_e$ be $\mathbf{H}_e = \mathbf{USV}^{\mathrm{H}}$. Then we can express the eavesdropper's channel null space as $\tilde{\mathbf{V}}\tilde{\mathbf{V}}^{\mathrm{H}}$, where $\tilde{\mathbf{V}}$ contains the last $N_t - N_e$ columns of $\mathbf{V}$ [39]. Define a precoding matrix $\mathbf{W} = \tilde{\mathbf{V}}\tilde{\mathbf{V}}^{\mathrm{H}} \in \mathbb{C}^{N_t}$. The signals received by the FC and the eavesdropper are then given by, respectively:

$$\mathbf{y} = \mathbf{HW}\boldsymbol{\beta}\theta + \mathbf{HW}\boldsymbol{\beta}\omega + \mathbf{z}, \tag{11.53}$$

$$\mathbf{y}_e = \mathbf{H}_e\mathbf{W}\boldsymbol{\beta}\theta + \mathbf{H}_e\mathbf{W}\boldsymbol{\beta}\omega + \mathbf{z}_e = \mathbf{z}_e. \tag{11.54}$$

On the eavesdropper side, as no information about x is received, we obtain the secrecy outage probability $\Pr_{\mathrm{outage\_EVE}} = 0$.

The outage minimization problem can then be given as:

$$\begin{aligned} \min_{\boldsymbol{\beta}(\mathbf{H})} \quad & \Pr\left[\frac{1}{(\mathbf{HW}\boldsymbol{\beta})^{\mathrm{H}}\,\mathbf{HW}\boldsymbol{\beta}} > \mathbb{S}\right] \\ \text{s.t.} \quad & \mathbb{E}\left[\boldsymbol{\beta}^{\mathrm{H}}\boldsymbol{\beta}\right] \leq \frac{\mathscr{P}_{\mathrm{tot}}}{\sigma_\theta^2+\sigma_\omega^2}. \end{aligned} \tag{11.55}$$

Similar techniques as used in [43] can be employed to solve problem (11.55), and it can be shown that the globally optimal $\boldsymbol{\beta}^*$ is constructed by randomizing among two deterministic power schemes $\boldsymbol{\beta}_1$ and $\boldsymbol{\beta}_2 = \mathbf{0}$ with corresponding weighting factors ω and $1 - \omega$. Furthermore, problem (11.55) can be reformulated into the following problem:

$$\begin{aligned} \min_{\boldsymbol{\beta}_1(\mathbf{H}),\, \omega(\mathbf{H})} \quad & 1 - \mathbb{E}[\omega] \\ \text{s.t.} \quad & \mathbb{E}\left[\omega\boldsymbol{\beta}_1^{\mathrm{H}}\boldsymbol{\beta}_1\right] \leq \frac{\mathscr{P}_{\mathrm{tot}}}{\sigma_\theta^2+\sigma_\omega^2}, \quad (\mathbf{HW}\boldsymbol{\beta}_1)^{\mathrm{H}}\,\mathbf{HW}\boldsymbol{\beta}_1 \geq \mathbb{S}^{-1}. \end{aligned} \tag{11.56}$$

The solution is given as

$$\boldsymbol{\beta}^*(\mathbf{H}) = \begin{cases} \boldsymbol{\beta}_1^*(\mathbf{H}), & \text{if } \lambda^*\left(\sigma_\theta^2+\sigma_\omega^2\right)\boldsymbol{\beta}_1^{\mathrm{H}*}\boldsymbol{\beta}_1^* < 1 \\ \mathbf{0}, & \text{otherwise,} \end{cases} \tag{11.57}$$

where λ^* is the optimal Lagrange multiplier associated with the power constraint in (11.56) which is obtained numerically, and $\boldsymbol{\beta}_1^*$ is the globally optimal solution of the problem:

$$\begin{aligned} \min_{\boldsymbol{\beta}_1(\mathbf{H})} \quad & \left(\sigma_\theta^2+\sigma_\omega^2\right)\boldsymbol{\beta}_1^{\mathrm{H}}\boldsymbol{\beta}_1, \\ \text{s.t.} \quad & (\mathbf{HW}\boldsymbol{\beta}_1)^{\mathrm{H}}\,\mathbf{HW}\boldsymbol{\beta}_1 \leq \mathbb{S}^{-1}. \end{aligned} \tag{11.58}$$

11.3.2.2 Partial CSI

From the full CSI case, we know that a deterministic power allocation is optimal for continuously distributed fading channels. Therefore, applying the results derived in Section 11.3.1.2, we can obtain a locally optimal $\boldsymbol{\beta}^*$ at each FC channel instance as:

$$\boldsymbol{\beta}^*(\mathbf{H}) = \begin{cases} \hat{\boldsymbol{\beta}}(\mathbf{H}), & \text{if } \nu(\mathbf{H})\displaystyle\int_{\mathbf{H}_e} \mathbb{1}\left\{\left(\mathbf{H}_e\hat{\boldsymbol{\beta}}\right)^{\mathrm{H}}\mathbf{H}_e\hat{\boldsymbol{\beta}} > \tfrac{1}{\mathbb{S}_e}\right\} dF(\mathbf{H}_e) + \lambda\hat{\boldsymbol{\beta}}^{\mathrm{H}}\hat{\boldsymbol{\beta}} < 1 \\ \mathbf{0}, & \text{otherwise,} \end{cases}$$

where $\hat{\boldsymbol{\beta}}$ is a locally optimal solution to the problem:

$$\begin{aligned} \min_{\boldsymbol{\beta}(\mathbf{H})} \quad & \lambda\boldsymbol{\beta}^{\mathrm{H}}\boldsymbol{\beta} + \nu(\mathbf{H})\int_{\mathbf{H}_e} \mathbb{1}\left\{(\mathbf{H}_e\boldsymbol{\beta})^{\mathrm{H}}\,\mathbf{H}_e\boldsymbol{\beta} > \mathbb{S}_e^{-1}\right\} dF(\mathbf{H}_e) \\ \text{s.t.} \quad & (\mathbf{H}\boldsymbol{\beta})^{\mathrm{H}}\,\mathbf{H}\boldsymbol{\beta} \geq \mathbb{S}^{-1}, \end{aligned} \tag{11.59}$$

with λ and $\nu(\mathbf{H})$ being nonnegative Lagrange multipliers corresponding to respectively the power constraint and the secrecy outage constraint at the eavesdropper.

Note that solving the subproblem (11.59) requires high computational costs owing to the difficulties of explicitly expressing $\nu(\mathbf{H}) \int_{\mathbf{H}_e} \mathbb{1}\left\{(\mathbf{H}_e\boldsymbol{\beta})^{\mathrm{H}}\mathbf{H}_e\boldsymbol{\beta} > \mathbb{S}_e^{-1}\right\} dF(\mathbf{H}_e)$, but a locally optimal solution can still be found using the Lagrangian multiplier techniques.

Artificial noise

Let $[\mathbf{W}_1, \mathbf{W}_2]$ be an orthonormal basis of $\mathbb{C}^{N_t}$ with $\mathbf{W}_1 \in \mathbb{C}^{N_t \times N_r}$ and $\mathbf{W}_2 \in \mathbb{C}^{N_t \times (N_t - N_r)}$ representing, respectively, the signal space and the null space of $\mathbf{H}$. The signals received by the FC and the eavesdropper are, respectively:

$$\mathbf{y} = \mathbf{H}\mathbf{W}_1\boldsymbol{\beta}x + \mathbf{H}\mathbf{W}_2\mathbf{v} + \mathbf{z} = \mathbf{H}\mathbf{W}_1\boldsymbol{\beta}\theta + \mathbf{H}\mathbf{W}_1\boldsymbol{\beta}\omega + \mathbf{z}, \tag{11.60a}$$

$$\begin{aligned}\mathbf{y}_e &= \mathbf{H}_e\mathbf{W}_1\boldsymbol{\beta}x + \mathbf{H}_e\mathbf{W}_2\mathbf{v} + \mathbf{z}_e \\ &= \mathbf{H}_e\mathbf{W}_1\boldsymbol{\beta}\theta + \mathbf{H}_e\mathbf{W}_1\omega + \mathbf{H}_e\mathbf{W}_2\mathbf{v} + \mathbf{z}_e,\end{aligned} \tag{11.60b}$$

where the artificial noise $\mathbf{v} \in \mathbb{C}^{(N_t - N_r)\times 1}$ has $N_t - N_r$ i.i.d. complex Gaussian elements with zero mean and variance p_a. As $[\mathbf{W}_1, \mathbf{W}_2]$ is a unitary matrix, we obtain that $\mathbf{H}_e\mathbf{W}_1$ is independent of $\mathbf{H}_e\mathbf{W}_2$, giving the effective noise at the eavesdropper as $\mathbf{H}_e\mathbf{W}_2\mathbf{v} + \mathbf{z}_e$. The transmit power in each fading block is given as $\left(\sigma_\theta^2 + \sigma_\omega^2\right)\boldsymbol{\beta}^{\mathrm{H}}\boldsymbol{\beta} + (N_t - N_r)p_a$.

We want to minimize the distortion outage probability at the FC, by finding the best $\boldsymbol{\beta}^*(\mathbf{H})$ and $p_a^*(\mathbf{H})$ to meet the long-term power constraint and the secrecy outage constraint at the eavesdropper. The optimization problem can be written as:

$$\begin{aligned}\min_{p_a(\mathbf{H}),\, \boldsymbol{\beta}(\mathbf{H})} \quad & \Pr\left[(\mathbf{H}\mathbf{W}_1\boldsymbol{\beta})^{\mathrm{H}}\mathbf{H}\mathbf{W}_1\boldsymbol{\beta} < \mathbb{S}^{-1}\right] \\ \text{s.t.} \quad & \Pr\left[(\mathbf{H}_e\mathbf{W}_1\boldsymbol{\beta})^{\mathrm{H}}\left(\left(\sigma_\theta^2 + \sigma_\omega^2\right)\mathbf{H}_e\mathbf{W}_1\boldsymbol{\beta}(\mathbf{H}_e\mathbf{W}_1\boldsymbol{\beta})^{\mathrm{H}} + \sigma_e^2\mathbf{I}_{N_e}p_a\mathbf{H}_e\mathbf{W}_2(\mathbf{H}_e\mathbf{W}_2)^{\mathrm{H}}\right)^{-1}\right. \\ & \left. \times\ \mathbf{H}_e\mathbf{W}_1\boldsymbol{\beta} > \frac{\sigma^2 - \mathbb{D}_e}{\sigma^4}\right] \le \delta, \\ & \mathbb{E}\left[\left(\sigma_\theta^2 + \sigma_\omega^2\right)\boldsymbol{\beta}^{\mathrm{H}}\boldsymbol{\beta} + (N_t - N_r)p_a\right] \le \mathscr{P}_{\text{tot}}.\end{aligned} \tag{11.61}$$

In order to solve problem (11.61), we can employ similar techniques as described in Section 11.3.1.2, whose details are omitted for brevity.

11.3.3 Numerical results

We first consider a situation with three sensors. For simplicity, we consider the source σ_θ^2 to be distributed as $N(0, 1)$, and all three sensors share the same measurement sensitivity of $\sigma_{\omega k}^2 = 10^{-3}, \forall k$. We assume that the distances from each sensor to the eavesdropper are 125, 127 and 129 m, whereas it is 125, 130 and 135 m to the FC, respectively. Furthermore, we consider the path-loss of the signal power at the FC and the eavesdropper as following the free-space path-loss model:

$$PL = 20\log_{10}(d) + 20\log_{10}(f) - 27.55, \tag{11.62}$$

where $d \in \{d_k, d_{ek}\}$ is the distance between sensor k and the FC or the eavesdropper in meters, and $f = 800$ MHz. In addition, the total power budget range is set to 1 mW $\le \mathscr{P}_{\text{tot}} \le 11$ mW, to ensure that the secrecy outage probability requirement at the

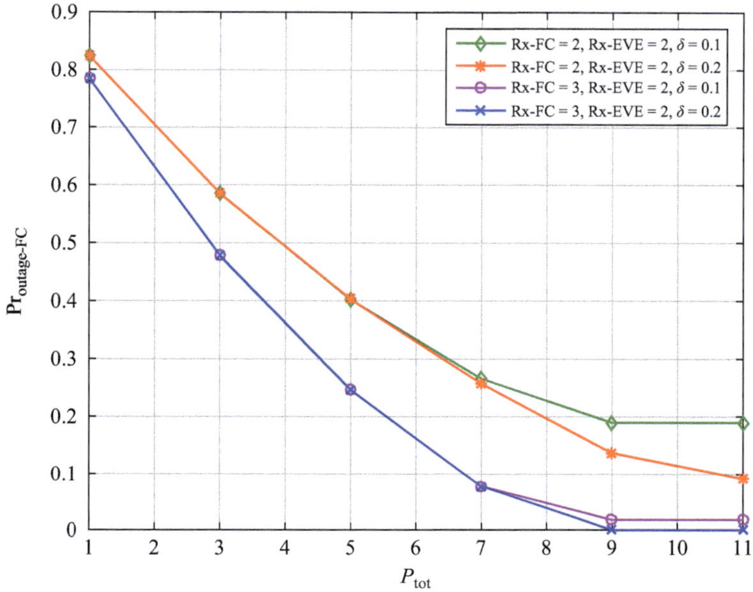

Figure 11.7 Performance comparison in a three-sensor network with $N_e = 2$ and full CSI of the FC and the eavesdropper

eavesdropper is achievable. The maximum acceptable distortion level $\mathbb{D}$ at the FC is set to 0.007, while the required minimum distortion level $\mathbb{D}_e$ at the eavesdropper is 0.01.

Figure 11.7 shows the distortion outage probability at the FC with two antennas at the eavesdropper, under different secrecy outage probability requirements at the eavesdropper, namely 0.1 and 0.2. When the number of receive antennas at the FC is fixed, it is seen that for both sets (i.e., $N_r = 2$ and $N_r = 3$) the outage probability at the FC behaves similarly for the two different outage requirements at the eavesdropper when $\mathscr{P}_{\text{tot}}$ is small. As we increase the total power budget, they start to decrease until saturation. This is because when $\mathscr{P}_{\text{tot}}$ is small, the sensors are more likely to choose small power consumption policies that only guarantee non-outage at the FC, or the sensors would simply stop transmitting to save power. As the transmission power budget increases, sensors begin to transmit in channel states where outage happens neither at the FC nor at the eavesdropper, until a point where more incremental power would lead to the secrecy outage probability at the eavesdropper being greater than the security requirement δ, at which the distortion outage probability at the FC saturates.

Next, we look at the distortion outage probability at the FC for the multiple-antenna single sensor scenario, where we assume that the sensor is 127 m away from the FC, and 130 m away from the eavesdropper. For simplicity, we assume that the sensor is equipped with three antennas, whereas there is only one antenna at the FC and one or two antennas at the eavesdropper. We consider the maximum acceptable distortion level $\mathbb{D}_e$ at the eavesdropper as 0.013, which is twice as large as the required

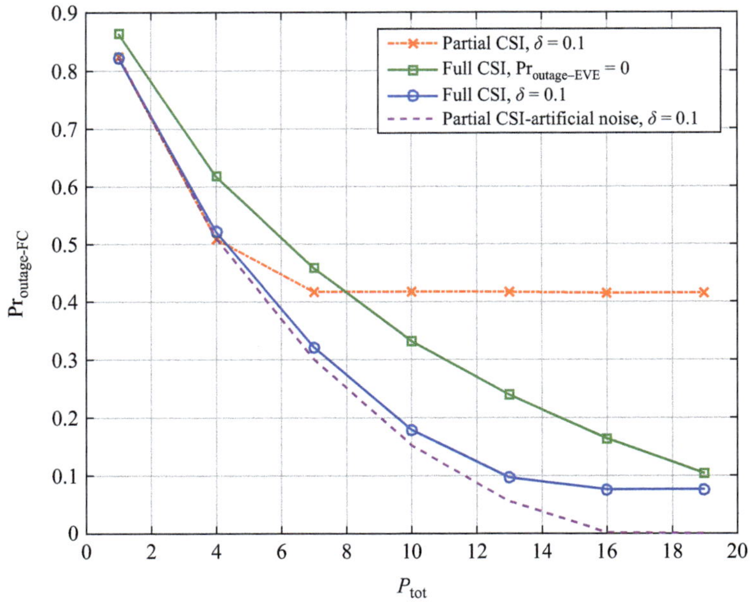

Figure 11.8 Performance comparison for a single sensor multiple-antenna scenario with $N_r = N_e = 1$ and $\delta = 0.1$

minimum distortion level $\mathbb{D}$ at the FC. We assume the same noise level for both the FC and the eavesdropper, where $\sigma^2 = \sigma_e^2 = 10^{-8}$ mW. In Figure 11.8, the distortion outage probability at the FC versus the long-term power budget is plotted for full CSI, full CSI with $\mathrm{Pr}_{\mathrm{outage\_EVE}} = 0$, partial CSI and partial CSI-Artificial Noise schemes. As we can see, the full CSI case outperforms the partial CSI case, and in both cases, the distortion outage probability at the FC saturates. By contrast, the full CSI with $\mathrm{Pr}_{\mathrm{outage\_EVE}} = 0$ and partial CSI-artificial noise schemes perform better when we have a relatively large transmit power budget, where $\mathrm{Pr}_{\mathrm{outage\_FC}}$ keeps decreasing as $\mathscr{P}_{\mathrm{tot}}$ increases. More interestingly, it is seen that the full CSI $\mathrm{Pr}_{\mathrm{outage\_EVE}} = 0$ scheme performs no better than the partial CSI-artificial noise scheme across the entire power range. This is owing to the fact that the effective channel gains of the FC are largely reduced when projecting it onto the eavesdropper's channel null space, whereas in the case of partial CSI-artificial noise, only a small portion of the transmit power is used to generate "noise."

11.4 Conclusion

This chapter has investigated secure estimation in the presence of an eavesdropper. Motivated by information theoretic concepts, and relations between the mutual information and minimum MSE, we introduced a notion of security in estimation that aims to keep the eavesdropper's MSE or distortion above a certain level. Power allocation problems were formulated taking into account security constraints, in expected, short term and probabilistic senses. Various system scenarios were investigated such as multiple sensors, multiple transmit antennas and full and partial CSI. In the case of partial CSI, the use of artificial noise was found to give good performance.

Notes: This chapter describes some of the work in [41,43]. The reader is referred to these original works for additional technical details and problem scenarios. The problem of state estimation of *dynamical systems* in the presence of eavesdroppers has also been recently studied, see [47–51].

References

[1] Akyildiz IF, Su W, Sankarasubramaniam Y, *et al.* Wireless sensor networks: A survey. Computer Networks. 2002;38(4):393–422.

[2] Cui S, Xiao JJ, Goldsmith AJ, *et al.* Estimation diversity and energy efficiency in distributed sensing. IEEE Transactions on Signal Processing. 2007;55(9):4683–4695.

[3] Xiao JJ, Cui S, Luo ZQ, *et al.* Power scheduling of universal decentralized estimation in sensor networks. IEEE Transactions on Signal Processing. 2006;54(2):413–422.

[4] Bahceci I, and Khandani AK. Linear estimation of correlated data in wireless sensor networks with optimum power allocation and analog modulation. IEEE Transactions on Communications. 2008;56(7):1146–1156.

[5] Caire G, Taricco G, and Biglieri E. Optimum power control over fading channels. IEEE Transactions on Information Theory. 1999;45(5):1468–1489.

[6] Negi R, and Cioffi JM. Delay-constrained capacity with causal feedback. IEEE Transactions on Information Theory. 2002;48(9):2478–2494.

[7] Luo J, Yates R, and Spasojevic P. Service outage based power and rate allocation for parallel fading channels. IEEE Transactions on Information Theory. 2005;51(7):2594–2611.

[8] Wang CH, Leong AS, and Dey S. Distortion outage minimization and diversity order analysis for coherent multiaccess. IEEE Transactions on Signal Processing. 2011;59(12):6144–6159.

[9] Wang CH, and Dey S. Distortion Outage Minimization in Rayleigh Fading Using Limited Feedback. In: 2009 IEEE Global Telecommunications Conference. IEEE; 2009. p. 1–8.

[10] Leong AS, Dey S, Nair GN, *et al.* Power allocation for outage minimization in state estimation over fading channels. IEEE Transactions on Signal Processing. 2011;59(7):3382–3397.

[11] Shannon CE. Communication theory of secrecy systems. Bell System Technical Journal. 1949;28(4):656–715.

[12] Wyner AD. The wire-tap channel. Bell System Technical Journal. 1975;54(8):1355–1387.

[13] Khisti A, Tchamkerten A, and Wornell GW. Secure broadcasting over fading channels. IEEE Transactions on Information Theory. 2008;54(6): 2453–2469.
[14] Gopala PK, Lai L, and El-Gamal H. On the secrecy capacity of fading channels. IEEE Transactions on Information Theory. 2008;54(10):4687–4698.
[15] Liang Y, Poor HV, and Shamai S. Secure communication over fading channels. IEEE Transactions on Information Theory. 2008;54(6):2470–2492.
[16] Khisti A, and Wornell GW. Secure transmission with multiple antennas – Part II: The MIMOME wiretap channel. IEEE Transactions on Information Theory. 2010;56(11):5515–5532.
[17] Bustin R, Liu R, Poor HV, *et al.* An MMSE approach to the secrecy capacity of the MIMO Gaussian wiretap channel. EURASIP Journal on Wireless Communications and Networking. 2009;2009:3.
[18] Liu T, and Shamai S. A note on the secrecy capacity of the multiple-antenna wiretap channel. IEEE Transactions on Information Theory. 2009;55(6): 2547–2553.
[19] Ekrem E, and Ulukus S. Secure lossy transmission of vector Gaussian sources. IEEE Transactions on Information Theory. 2013;59(9):5466–5487.
[20] Naghibi F, Salimi S, and Skoglund M. The CEO Problem with Secrecy Constraints. In: IEEE International Symposium on Information Theory (ISIT). IEEE; 2014. p. 756–760.
[21] Bagherikaram G, and Plataniotis KN. Secure Hybrid Digital-analog Wyner-Ziv Coding. In: IEEE 22nd International Symposium on Personal Indoor and Mobile Radio Communications (PIMRC). IEEE; 2011. p. 1161–1166.
[22] Villard J, and Piantanida P. Secure multiterminal source coding with side information at the eavesdropper. IEEE Transactions on Information Theory. 2013;59(6):3668–3692.
[23] Gastpar M, and Vetterli M. Source-Channel Communication in Sensor Networks. In: Information Processing in Sensor Networks. Springer; 2003. p. 162–177.
[24] Marano S, Matta V, and Willett PK. Distributed detection with censoring sensors under physical layer secrecy. IEEE Transactions on Signal Processing. 2009;57(5):1976–1986.
[25] Soosahabi R, and Naraghi-Pour M. Scalable PHY-layer security for distributed detection in wireless sensor networks. IEEE Transactions on Information Forensics and Security. 2012;7(4):1118–1126.
[26] Li Z, and Oechtering TJ. Privacy-aware distributed Bayesian detection. IEEE Journal of Selected Topics in Signal Processing. 2015;9(7):1345–1357.
[27] Kailkhura B, Nadendla VSS, and Varshney PK. Distributed inference in the presence of eavesdroppers: A survey. IEEE Communications Magazine. 2015;53(6):40–46.
[28] Nadendla VSS, and Varshney PK. Design of binary quantizers for distributed detection under secrecy constraints. IEEE Transactions on Signal Processing. 2016;64(10):2636–2648.

[29] Aysal TC, and Barner KE. Sensor data cryptography in wireless sensor networks. IEEE Transactions on Information Forensics and Security. 2008;3(2):273–289.

[30] Reboredo H, Xavier J, and Rodrigues MRD. Filter design with secrecy constraints: The MIMO Gaussian wiretap channel. IEEE Transactions on Signal Processing. 2013;61(15):3799–3814.

[31] Liao WC, Chang TH, Ma WK, *et al.* QoS based transmit beamforming in the presence of eavesdroppers: An optimized artificial-noise-aided approach. IEEE Transactions on Signal Processing. 2011;59(3):1202–1216.

[32] Goel S, and Negi R. Guaranteeing secrecy using artificial noise. IEEE Transactions on Wireless Communications. 2008;7(6):2180–2189.

[33] Dong L, Han Z, Petropulu AP, *et al.* Improving wireless physical layer security via cooperating relays. IEEE Transactions on Signal Processing. 2010;58(3):1875–1888.

[34] Gerbracht S, Wolf A, and Jorswieck EA. Beamforming for Fading Wiretap Channels with Partial Channel Information. In: International ITG Workshop on Smart Antennas (WSA). IEEE; 2010. p. 394–401.

[35] Guo D, Shamai S, and Verdú S. Mutual information and minimum mean-square error in Gaussian channels. IEEE Transactions on Information Theory. 2005;51(4):1261–1282.

[36] Feng X, Loparo KA, and Fang Y. Optimal state estimation for stochastic systems: An information theoretic approach. IEEE Transactions on Automatic Control. 1997;42(6):771–785.

[37] Kay SM. Fundamentals of statistical signal processing: Estimation theory. Upper Saddle River, NJ, USA: Prentice-Hall, Inc.; 1993.

[38] Boyd S, and Vandenberghe L. Convex optimization. New York, NY, USA: Cambridge University Press; 2004.

[39] Yoo T, and Goldsmith A. On the optimality of multiantenna broadcast scheduling using zero-forcing beamforming. IEEE Selected Areas in Communications. 2006;24(3):528–541.

[40] Guo X, Leong AS, and Dey S. Power Allocation for Distortion Minimization in Distributed Estimation with Security Constraints. In: Proc. IEEE Symposium on Signal Processing Advances in Wireless Communications (SPAWC). Toronto, Canada; 2014. p. 6 pages.

[41] Guo X, Leong AS, and Dey S. Estimation in wireless sensor networks with security constraints. IEEE Transactions on Aerospace and Electronic Systems. 2017;53(2):544–561.

[42] Goldsmith A. Wireless communications. Cambridge: Cambridge University Press; 2005.

[43] Guo X, Leong AS, and Dey S. Distortion outage minimization in distributed estimation with estimation secrecy outage constraints. IEEE Transactions on Signal and Information Processing over Networks. 2017;3(1):12–28.

[44] Luenberger DG. Optimization by vector space methods. John Wiley & Sons; 1968.

[45] Messerschmitt D. Stationary points of a real-valued function of a complex variable. Tech Rep UCB/EECS-2006-93. Berkeley: EECS Department, University of California; 2006.
[46] Sorber L, Barel MV, and Lathauwer LD. Unconstrained optimization of real functions in complex variables. SIAM Journal on Optimization. 2012;22(3): 879–898.
[47] Wiese M, Johansson KH, Oechtering TJ, *et al.* Secure Estimation for Unstable Systems. In: Proc. IEEE Conf. Decision and Control. Las Vegas, NV; 2016. p. 5059–5064.
[48] Tsiamis A, Gatsis K, and Pappas GJ. State Estimation with Secrecy Against Eavesdroppers. In: Proc. IFAC World Congress. Toulouse, France; 2017. p. 8715–8722.
[49] Leong AS, Quevedo DE, Dolz D, *et al.* On Remote State Estimation in the Presence of an Eavesdropper. In: Proc. IFAC World Congress. Toulouse, France; 2017. p. 7600–7605.
[50] Tsiamis A, Gatsis K, and Pappas GJ. State Estimation Codes for Perfect Secrecy. In: Proc. IEEE Conf. Decision and Control. Melbourne, Australia; 2017. p. 176–181.
[51] Leong AS, Redder A, Quevedo DE, *et al.* On the Use of Artificial Noise for Secure State Estimation in the Presence of Eavesdroppers. In: Proc. European Control Conf. Limassol, Cyprus; 2018.

Chapter 12

Robust fusion of unreliable data sources using error-correcting output codes

Aditya Vempaty[1], Bhavya Kailkhura[2], and Pramod K. Varshney[3]

The emergence of big and dirty data era demands new distributed learning and inference solutions to tackle the problem of inference with corrupted data. The central goal of this chapter is to discuss the presence of corrupted data in the context of distributed inference networks (DINs) and discuss coding-theoretic strategies to ensure reliable inference performance in several practical scenarios. It discusses a generalization of the classical *Byzantine Generals problem* in the context of distributed inference to different topologies. Over the last three decades, research community has extensively studied the impact of imperfect transmission channels or sensor faults on distributed inference systems. However, corrupted (Byzantine) data models, considered in this chapter, are philosophically different from the imperfect channels or faulty sensor cases. Byzantines are intentional and intelligent and therefore can optimize over the data corruption parameters. While learning their behavior and actively countering them is a viable approach, this chapter presents a new paradigm of mitigation strategies that use coding-theoretic results. The general approach of error-correcting output codes (ECOC) for data fusion is presented and its applicability for several inference problems in practice dealing with unreliable data including Byzantines is shown. This approach is then shown to be applicable to a wider range of inference problems such as classification using crowdsourced data.

12.1 Introduction

Inference involves drawing conclusions from observations. In the signal processing literature, typical inference problems are of two major kinds: detection (or more generally, classification) and estimation. In classification, the phenomenon of interest (POI) to be inferred is of finite possibilities, say M possible classes. This is represented

[1]Xio Research Inc., USA
[2]Lawrence Livermore National Laboratories, USA
[3]Department of Electrical Engineering and Computer Science, Syracuse University, USA

by M possible hypotheses: $H_0, \ldots, H_{M-1}$. Multiple observations $x = [x_1, \ldots, x_N]$ are made regarding the phenomenon and a data fusion rule such as the maximum *a posteriori* (MAP) rule and the simple majority rule, are used to make a final decision. When the number of hypotheses $M = 2$, it is referred to as the detection problem. On the other hand, in a typical estimation problem, the goal is to estimate the value of an unknown parameter θ (which could possibly be a vector). This value is continuous and can take any value in the set Θ. Similar to the classification problem, multiple observations are taken and an estimator is used to make an estimate. If the parameter θ is time varying, it is typically referred to as the tracking problem. We focus on DINs for the majority of the chapter. We show how unreliable data sources in such networks can be handled using coding-theoretic approaches. In Section 12.5, we show the applicability of the approach to an inference problem (classification) beyond DINs. Specifically, we consider classification using crowd-sourced data. In the remainder of this section, we present some basics on DINs and Byzantines.

12.1.1 Distributed inference networks

In a typical distributed inference (classification or estimation) network, a group of spatially distributed sensing entities are employed that provide these multiple observations and collaborate to make inferences about a given POI. These agents transmit their observations over (possibly) imperfect channels to a fusion center (FC) who then fuses the data using a data fusion rule to infer about the phenomenon. Several aspects of such a framework have been studied (see [1] and references therein): network topology, decision rules, effect of wireless channels, effect of spatiotemporal dependence, etc. DINs have attracted much recent attention due to a variety of applications in civilian and military domains. These include distributed spectrum sensing, traffic and environment monitoring, medical monitoring, power networks, localization and surveillance, etc. While there have been several studies on DINs, the aspect of DINs with corrupted data has only found interest in the recent past. This chapter focuses on data fusion in DINs with corrupted data, where the cause of corruption could be due to faults, imperfect channels, or Byzantine attacks.

12.1.2 Byzantine generals problem

In 1982, Lamport *et al.* presented the *Byzantine generals problem* as [2] "a group of generals of the Byzantine army camped with their troops around an enemy city. Communicating only by messengers, the generals must agree upon a common battle plan. However, one or more of them may be traitors who will try to mislead the others. The problem is to find an algorithm to ensure that the loyal generals will reach agreement." A sharp characterization was derived for the Byzantine generals power. It was determined that when less than 1/3 of them are Byzantine generals, it is possible for the loyal generals can reach a consensus agreement irrespective of the Byzantine generals' actions. However, if the fraction is above 1/3, consensus can no longer be guaranteed. This Byzantine generals problem can be applied to a variety of applications in security, where Byzantine generals play the role of internal adversaries. Such a setup is even possible in DINs, where few agents in the network are attacked who send falsified information to the FC for processing and inference.

The rest of the chapter is organized as follows: Section 12.2 presents the general approach of ECOC for data fusion. In Section 12.3, its applicability in dealing with unreliable data in parallel networks for inference is presented. Section 12.4 then extends this approach to tree-based networks. After completing the discussion on DINs, we then exemplify the broad applicability of the ECOC approach by discussing its use for classification using crowdsourced data. Finally, we conclude in Section 12.6 with some discussion.

12.2 Error-correcting output codes for data fusion

As discussed in the first part of the chapter, malicious or corrupted data affects the typical data fusion methods and produces erroneous final decisions. There are two major solution methodologies: active methods which are learning-based where the sources of corrupted data are identified via learning methods and appropriate action is taken; and passive methods where error-tolerant robust-methods are designed that handle errors in the data. In this chapter, the focus is on the second type of methods that use ECOC to design robust error-tolerant data fusion methods. ECOC [3] were originally proposed by Dietterich and Bakiri for multiclass learning problems. It is a way to solve multiclass learning problems by learning multiple binary classifiers. Let there be M classes. In this approach, each class is assigned a unique binary string of length N, which are referred to as *codewords* for the class. Now, N binary functions are learned, each of which helps in distinguishing between two distinct subsets of classes. Given such an $M \times N$ codematrix (where rows are the classes and columns are the binary classifiers), a new datapoint is classified by first evaluating each of the N binary classifiers to generate an N-bit string. This string is then compared to each of the M codewords, and the datapoint is assigned to the class whose codeword is closest, which is determined by a predecided distance measure.

As an example, consider the use of ECOCs for the digit-recognition task as described in [3]. In this problem, the binary classifiers have been chosen to be meaningful and interpretable. Figure 12.1 shows the $N = 6$ binary functions to classify a given digit into $M = 10$ different classes. Hence, the codematrix is of size 10×6. Observe that every row is distinct and represents the distinct codeword for the class. The binary functions in this case have been chosen to be interpretable and as follows: the first column represents the binary function f_{vl} which checks whether there is a vertical line (vl) in the digit. Similarly, $f_{hl}, f_{dl}, f_{cc}, f_{ol}$, and f_{or}, respectively, check the presence of a horizontal line (hl), diagonal line (dl), closed curve (cc), open curve to left (ol), and open curve to right (or). During learning, each binary function is learned specifically by providing samples with and without the characteristic. These

Class	Code word					
	v1	h1	d1	cc	o1	or
0	0	0	0	1	0	0
1	1	0	0	0	0	0
2	0	1	1	0	1	0
3	0	0	0	0	1	0
4	1	1	0	0	0	0
5	1	1	0	0	1	0
6	0	0	1	1	0	1
7	0	0	1	0	0	0
8	0	0	0	1	0	0
9	0	0	1	1	0	0

Figure 12.1 Error-correcting output code for ten digit multiclass learning problem [3]

binary classifiers are learnt to be high-accuracy classifiers for the property that they are trained on. They are also chosen to ensure that they are unique across columns to avoid redundancy and to ensure that they can be learned distinctly across columns and independent of the original classes.

When a new digit image arrives, it is run through each of the $N = 6$ binary classifiers to get a N-length bit sequence, such as 110001. The distance of this string with each of the $M = 10$ codewords is calculated using a predetermined distance measure, such as Hamming distance. The Hamming distance, which is the number of different bits between the codewords, is closest for the codeword 110000, which corresponds to class 4. Therefore, we predict the digit to be 4 using ECOC. Note that due to the nonzero error rate of the binary classifiers, we get erroneous response such as 110001 for an image of digit 4. In this case, the binary classifier corresponding to f_{or} which verifies if there is open curve to the right has provided a false alarm. In spite of such an error, due to the error-correcting nature of ECOC, we obtained a correct multiclass classification. The number of errors that an ECOC using minimum Hamming distance criterion can handle is a function of the minimum Hamming distance between any pair of codewords. An ECOC, with a minimum Hamming distance between codewords of d, can correct up to $\lfloor((d-1)/2)\rfloor$ errors [3], where $\lfloor\cdot\rfloor$ is the floor function.

Rest of the chapter discusses how the error-correcting capability of ECOCs can be helpful to handle corrupted data in DINs to ensure robust data fusion. In Section 12.3, the use of ECOCs for inference in parallel DINs is discussed. In Section 12.4, the setup is changed to inference in tree DINs. Finally, in Section 12.5, we show the benefit of using ECOC to address unreliable workers in the paradigm of classification using crowdsourced data.

12.3 Inference in parallel networks

In this section, we discuss inference in parallel networks by considering two different problems: distributed classification in wireless sensor networks (WSNs) in the presence of sensor faults and target localization in the presence of Byzantine attacks.

12.3.1 Distributed classification

As discussed before, classification is one of the major inference tasks in WSNs. Classification is performed using observations from the distributed sensors in the network which typically have constrained resources such as energy and/or bandwidth. Therefore, the local sensors send quantized observations to the FC rather than their raw observations. Based on this quantized data, the FC performs classification fusion. Clearly, classification performance depends on the data sent by these local sensors. When faulty sensors are present in the network, the performance of the network deteriorates. In [4], the use of ECOC was proposed for distributed multi-class classification for WSNs consisting of sensors sending binary decisions. These binary decisions from local sensors, possibly in the presence of faults, are forwarded to the FC that determines the final classification result. This approach was referred to as distributed classification fusion using error correcting codes (DCFECC).

12.3.1.1 Fault-tolerant distributed classification

The distributed classification problem considered in [4] is depicted in Figure 12.2. For such a setting consisting of M hypotheses to be distinguished by N local sensors, the expression for error probability was derived as a function of the code matrix $\mathbf{C}$. Let $A_{\mathbf{u}}^{j}$ be the cost that the received vector at the FC is $\mathbf{u}$ and the true hypothesis is H_j. These costs are determined by the decision regions of codewords. According to the fault-tolerant fusion rule used at the FC, the decision region of the codeword $\mathbf{r}_j$ is given as:

$$D(\mathbf{r}_j) = \{\mathbf{u} | d_H(\mathbf{u}, \mathbf{r}_j) \leq d_H(\mathbf{u}, \mathbf{r}_l) \ \forall \mathbf{r}_l \in \mathbf{C}\}. \tag{12.1}$$

In order to minimize the probability of misclassification, set $A_{\mathbf{u}}^{j} = 0$ if $\mathbf{u}$ is in the decision region of $\mathbf{r}_j$ corresponding to hypothesis H_j; otherwise, set $A_{\mathbf{u}}^{j} = 1$. Whenever the received vector $\mathbf{u}$ simultaneously belongs to decision regions of $\mathbf{r}_{l_0}, \ldots, \mathbf{r}_{l_{q-1}}$, where $q > 1$, for all $l = l_0, \ldots, l_{q-1}$, set $A_{\mathbf{u}}^{l} = 1 - 1/q$, i.e., the tie-break rule.

According to the above cost assignment, the probability of decision error can be written as [4]:

$$\begin{aligned} P_e = \sum_{i_1,\ldots,i_N,i_1^*,\ldots,i_N^*,l} \int_{y_1,\ldots,y_N} & P_l P(u_1 = i_1 | u_1^* = i_1^*) \\ & \times \cdots \times P(u_N = i_N | u_N^* = i_N^*) P(u_1^* = i_1^* | y_1) \\ & \times \cdots \times P(u_N^* = i_N^* | y_N) P(y_1, \ldots, y_N | H_l) A_{i_1,\ldots,i_N}^{l}, \end{aligned} \tag{12.2}$$

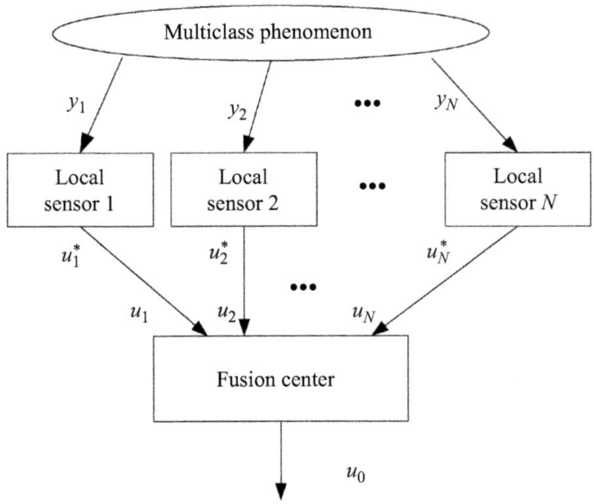

Figure 12.2 Distributed classification system

where, as shown in Figure 12.2, y_is are the local observations, u_i^*s are the local decisions, u_is are the received decisions at the FC, and P_l is the prior probability of hypothesis H_l. Note that u_i^* and u_i may be different because of permanent or transient faults or due to the malicious behavior of sensors or due to channel errors.

12.3.1.2 Numerical results

In [4], the proposed DCFECC approach was evaluated using numerical simulations. A system consisting of $N = 7$ sensors performing a 4-ary classification is considered. The four hypotheses ($M = 4$) are considered to be equally likely and all the sensor measurements are assumed to be independent and identically distributed (i.i.d.). The probability density function for each hypothesis is assumed to be a Gaussian with the same variance ($\sigma^2 = 1$) but with different means 0, s, $2s$, and $3s$, respectively. The signal-to-noise ratio (SNR) at each local sensor defined as $20 \log_{10} s$ is varied. Figure 12.3 shows the performance of DCFECC in the presence of stuck-at permanent faults at local sensors and data transmission over ideal channels. It is assumed that the sensors with a stuck-at fault always send a decision "1" to the FC. The code matrix designed by simulated annealing is chosen for the DCFECC approach and is given by:

$$\mathbf{C}_1 = [3, 8, 14, 12, 9, 12, 9]. \tag{12.3}$$

Here and in the sequel, code matrices are represented as a vector of M bit integers. Each integer r_j represents a column of the code matrix $\mathbf{A}$ and can be expressed as $r_j = \sum_{l=0}^{M-1} a_{lj} \times 2^l$. For instance, the integer $t_5 = 9$ in column 5 of code matrix $\mathbf{C}_1$ represents $c_{05} = 1$, $c_{15} = 0$, $c_{35} = 0$, and $c_{45} = 1$.

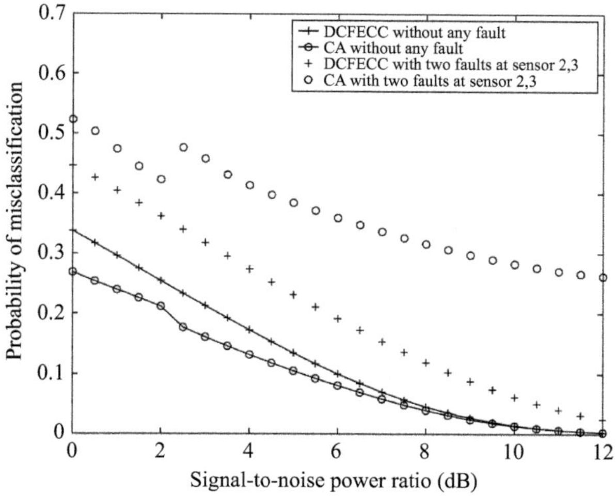

Figure 12.3 Performance comparison of DCFECC and the conventional approach (CA) when two faults are present [4]

The performance is also compared with the conventional approach (CA) where the optimal MAP-based fusion rule using the Chair–Varshney rule [5] is used. As can be seen from Figure 12.3, DCFECC is more robust to stuck-at faults than CA.

Several extensions to the above framework have also been considered. In the above set-up, it is assumed that the channels between the local sensors and the FC are ideal. In the presence of nonideal channels, one can use soft-decisions at the FC to reduce the errors due to channel uncertainties. This is the idea behind the distributed classification using soft-decision decoding (DCSD) approach proposed in [6]. Some other extensions include DCFECC using nonbinary codes [7] and suboptimal code design schemes based on error bounds [8].

12.3.2 Target localization

Another major use of WSNs is for estimation of a target location. In [9], a localization scheme was proposed based on hierarchical classification. This algorithm is iterative where M-ary classification is performed at each iteration. The region of interest (ROI) is split into M regions, and the M-ary classification at each algorithm decides the ROI for the next iteration. During each of the iterations, the M-ary classification is performed using the ECOC described in Section 12.2. The code matrix used in this process is designed as follows: each row represents a possible region and each column i represents ith sensor's binary decision rule, and the fusion rule at the FC is the ECOC rule of using minimum Hamming distance fusion rule. The optimal region split can be determined offline using k-means clustering [10]. For instance, when the sensors are deployed in a uniform grid, the optimal splitting is uniform as shown in Figure 12.4. The FC estimates the target location using the received data u. In this

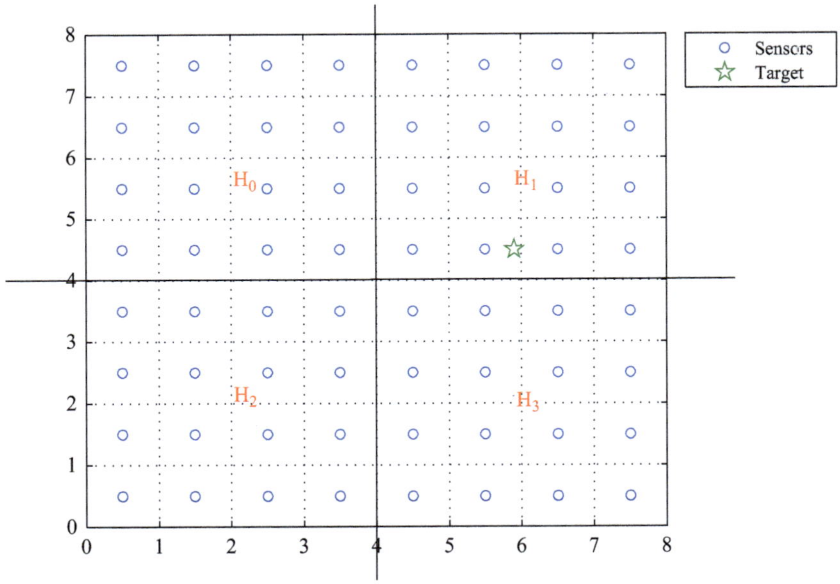

Figure 12.4 Equal region splitting of the ROI for localization as a 4-hypothesis test [9]

section, we show how ECOC can be used for target localization in the presence of Byzantines.

Since there are N sensors in the network which are split into M regions at every iteration, the number of sensors in the new ROI after every iteration is reduced by a factor of M. After k such iterations, ROI has (N/M^k) sensors. The code matrix for ECOC-based classification at the $(k+1)$th iteration would be of size $M \times (N/M^k)$. After k^{stop} iterations, there are only $(N/M^{k^{stop}})$ sensors present in the ROI and the final location estimate $\hat{\theta} = [\hat{\theta}_x, \hat{\theta}_y]$ can be obtained by taking an average of locations of the $(N/M^{k^{stop}})$ sensors present in the ROI:

$$\hat{\theta}_x = \frac{M^{k^{stop}}}{N} \sum_{i \in ROI_{k^{stop}}} x_i \tag{12.4}$$

and

$$\hat{\theta}_y = \frac{M^{k^{stop}}}{N} \sum_{i \in ROI_{k^{stop}}} y_i, \tag{12.5}$$

where $ROI_{k^{stop}}$ is the ROI at the last step.

The code matrix at each iteration can be designed as follows. Let R_j^k denote the region represented by the hypothesis H_j for $j = 0, 1, \ldots, M-1$ and let S_j^k be the set of sensors that lie in the region R_j^k. For the jth row of the code matrix, only those sensors that are in R_j^k have a "1" as their elements in the code matrix. In other words, the elements of the code matrix are given by:

$$c_{(j+1)i}^k = \begin{cases} 1 & \text{if } i \in \mathcal{S}_j^k \\ 0 & \text{otherwise} \end{cases}, \tag{12.6}$$

for $j = 0, 1, \ldots, M-1$ and $i \in ROI_k$.

Since the final ROI could be one of the M regions, a metric of interest is the probability of "zooming" into the correct region. The final region of the estimated target location is the same as the true target location, if and only if we "zoom" into the correct region at every iteration of the proposed scheme. If P_d^k denotes the detection probability at the $(k+1)$th iteration, the overall detection probability is given by:

$$P_D = \prod_{k=0}^{k^{stop}} P_d^k. \tag{12.7}$$

In [9], theoretical performance of the scheme was analyzed under some symmetry assumptions.

Theorem 12.1. *[9] Under symmetric network assumptions,* $\lim_{N\to\infty} P_D = 1$.

Figure 12.5 shows that the region detection probability increases as the number of sensors approaches infinity. Observe that for a fixed noise variance, the region detection probability increases with increase in the number of sensors. Also, for a fixed number of sensors, the region detection probability decreases with σ when the number of sensors are low. But when the number of sensors is large, the reduction in region detection probability with σ is negligible and as $N \to \infty$, the region detection probability converges to 1.

Another metric to analyze the quality of the approach is the robustness to the presence of malicious sensors (Byzantines) in the network. The use of ECOC provides robustness to the presence of Byzantines in the network due to the error-correction capability. For the basic scheme described so far, each column in C^k contains only one "1" and every row of C^k contains exactly (N/M^{k+1}) "1"s. Therefore, the minimum Hamming distance of C^k is $(2N/M^{k+1})$ and, at the $(k+1)$th iteration, it can tolerate a total of at most $(N/M^{k+1}) - 1$ faults (data falsification attacks) due to the presence of Byzantines in the network. This value is not very high and the basic scheme was therefore extended in [9] to a scheme which can handle more Byzantine faults.

The performance can be improved by using an exclusion method for decoding where the best two regions are kept for the next iteration and using weighted average

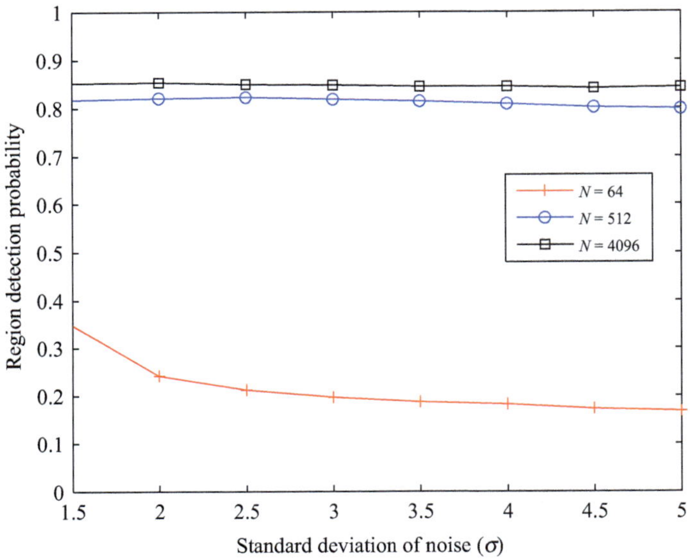

Figure 12.5 Region detection probability versus the standard deviation of noise with varying number of sensors [9]

to estimate the target location at the final step. Therefore, the target location estimate is given by:

$$\hat{\theta}_x = \frac{\sum_{i \in ROI_{k_{stop}}} u_i x_i}{\sum_{i \in ROI_{k_{stop}}} u_i} \tag{12.8}$$

and

$$\hat{\theta}_y = \frac{\sum_{i \in ROI_{k_{stop}}} u_i y_i}{\sum_{i \in ROI_{k_{stop}}} u_i}. \tag{12.9}$$

This exclusion method resulted in a better performance compared to the basic coding scheme since it keeps the two best regions after every iteration. The fault tolerance performance improves and a total of at most $(2^{k+1}N/M^{k+1}) - 1$ faults can be tolerated.

Proposition 12.1. *[9] The maximum fraction of Byzantines that can be handled at the $(k+1)$th iteration by the proposed exclusion method based coding scheme is limited by $\alpha_f^k = (2/M) - (M^k/2^k N)$.*

Theorem 12.2. *[9] Under symmetric network assumptions, when the fraction of Byzantines are in minority,* $\lim_{N \to \infty} P_D = 1$.

Figure 12.6 shows the performance of the proposed schemes in terms of the mean square error (MSE) of the estimated target location as a function of the fraction of

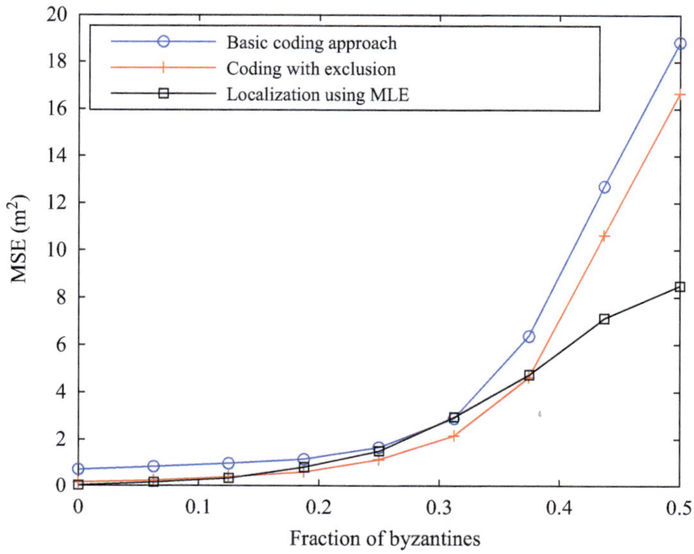

Figure 12.6 MSE comparison of the three localization schemes: basic coding scheme, coding with exclusion, and traditional ML-based localization [9]

Byzantines α in the network when compared with the traditional maximum likelihood estimation [11]. As can be seen from Figure 12.6, the performance of the exclusion method based coding scheme is better than the basic coding scheme and outperforms the traditional MLE-based scheme when $\alpha \leq 0.375$. When $\alpha > 0.375$, the traditional MLE-based scheme has the best performance.

The approach was also extended for the case of nonideal channels in [9], where soft-decoding approach was used to counter the effect of nonideal channels. The use of soft-decision decoding similar to DCSD [6] is discussed to compensate for the loss due to the presence of fading channels between the local sensors and the FC. The performance of the proposed schemes in terms of the Byzantine fault tolerance capability and probability of detection of the target region is derived. Performance bounds are also presented which can be used for system design.

12.4 Inference in tree networks

There are many practical situations where parallel topology cannot be implemented due to several factors, such as the FC being outside the communication range of the nodes and limited energy budget of the nodes [12]. In such cases, a multi-hop network is employed, where nodes are organized hierarchically into multiple levels (tree networks). In this section, we discuss results from [13] that considers classification and estimation problems in tree networks.

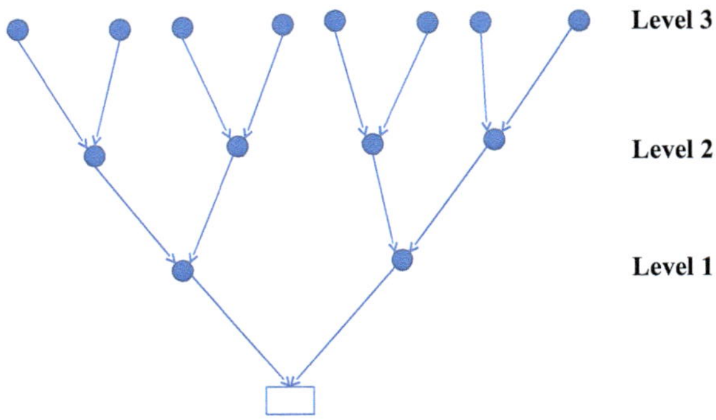

Figure 12.7 A distributed inference system organized as a perfect binary tree; T(3, 2) is shown as an example [13]

Consider a perfect tree, $T(K, N)$, rooted at the FC (see Figure 12.7). Nodes at level k, for $1 \leq k \leq K-1$, are referred to as intermediate nodes and nodes at the last level of the tree, i.e., $k = K$, are called the leaf nodes. In a perfect tree, all the intermediate nodes have an equal number of immediate successors, and all leaf nodes are at the same depth. The number of such immediate successors N is referred to as the degree of the tree. Each node j at level k performs two basic operations:

- Depending on the task, senses data regarding the phenomenon and/or collects data from its successors at level $k+1$, denoted by $S^{k+1}(j)$.
- Compresses the data available at node j about the phenomenon and transmits a 1-bit version to its predecessor at level $k-1$, denoted by $P^{k-1}(j)$.

Local observation of node j at level k is denoted as y_j^k. Received data vector at node j of level k from its successors $S^{k+1}(j)$ at level $k+1$ is denoted as $\mathbf{v_j^k} \in \{0, 1\}^N$. After processing the data at the node according to a processing model, every node j at level k sends its one-bit local decision $u_j^k \in \{0, 1\}$ to its immediate predecessor. This processing model was designed based on the inference problem considered, i.e., Figure 12.8 for classification or Figure 12.10 for estimation. Finally, the FC receives the inference vector $\mathbf{u^1} = (u_1^1, \ldots, u_N^1) \in \{0, 1\}^N$ and fuses this data to infer about the underlying phenomenon. Given a tree network, the objective is to find the appropriate processing scheme for nodes at all levels depending on the inference problem considered.

12.4.1 Distributed classification

Similar to parallel topology in [4], the classification problem was modeled as an M-ary hypotheses testing problem and ECOC was used for classification. Each

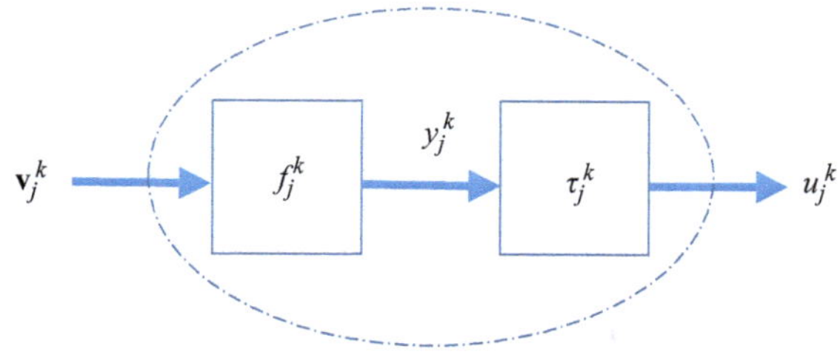

Figure 12.8 Data processing for distributed classification at node j at level $1 \le k \le K-1$. *Here* $\mathbf{v}_j^k \in \{0,1\}^N$, $y_j^k \in \{1, \ldots, M\}$, *and* $u_j^k \in \{0,1\}$. *Therefore, the mappings are* $f_j^k: \{0,1\}^N \to \{1, \ldots, M\}$ *and* $\tau_j^k: \{1, \ldots, M\} \to \{0,1\}$ *[13]*

node j at level k, for $1 \le k \le K-1$ in $T(K, N)$, performs two basic operations (see Figure 12.8):

- Collects data from its successors $S^{k+1}(j)$ and fuses their data using fusion rule $f_j^k(\cdot)$ to locally decide the hypothesis, denoted by y_j^k.
- Compresses the decision y_j^k at node j about the hypothesis and transmits a 1-bit version u_j^k to its predecessor $P^{k-1}(j)$ using the transmission mapping $\tau_j^k(\cdot)$.

For the leaf nodes (level K), there are no successors and, therefore, only the second operation needs to be performed. And, for the FC (level "0"), only the first operation needs to be performed. Each of the functions $f_j^k(\cdot)$ and $\tau_j^k(\cdot)$ depends on the code matrix used at level k. The transmission mapping of the leaf nodes can be obtained by employing a person-by-person approach as in [4]. For $1 \le k \le K-1$, the transmission mapping $\tau^k(\cdot)$ is straight-forward and is given as follows:

$$u_j^k = \tau_j^k(y_j^k) = c^k_{y_j^k j}, \quad \text{if } 1 \le k \le K-1. \tag{12.10}$$

For every intermediate node, the fusion rule $f_j^k(\cdot)$ is the minimum Hamming distance fusion rule. Therefore, the performance of the scheme depends on the minimum Hamming distance of the code matrices.

It was shown in [13] that there is a recursive structure, where the probability of misclassification at level k is dependent on the confusion matrix of level $k+1$. Therefore, the performance at the FC depends on all the code matrices in a recursive manner. The authors in [13] assumed that each node of the same level uses the same code matrix which is designed by optimizing on a person-by-person sequential basis. They recursively designed the code matrix at level $k-1$ to fuse data from level k using offline approaches such as simulated annealing or cyclic-column replacement.

It was shown that under certain conditions, the ECOC-based distributed classification scheme is asymptotically optimal (in N or K) as long as the probabilities of correct local classification for all hypotheses of the leaf nodes are greater than one half.

The performance of the ECOC-based distributed classification scheme was evaluated using simulations. Consider a tree network $T(3, 7)$ consisting of a total $N_{total} = 400$ nodes, including the FC. The leaf nodes sense the environment to identify

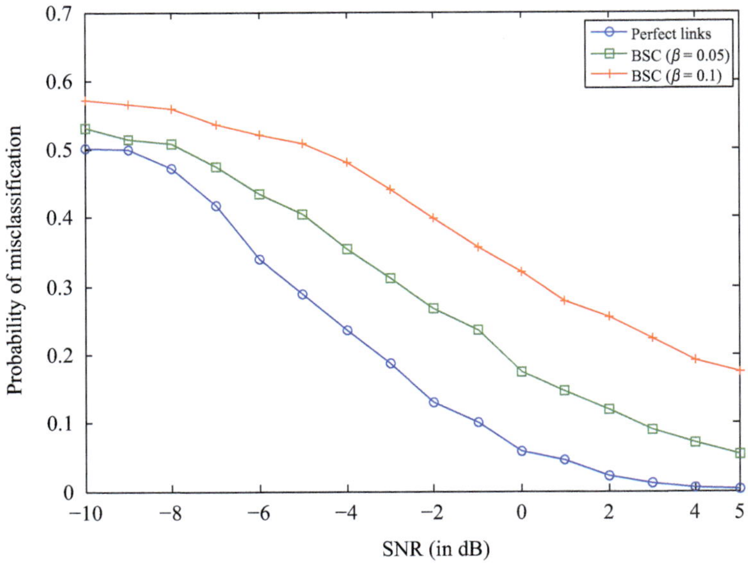

Figure 12.9 Probability of misclassification versus SNR [13]

among four ($M = 4$) equally likely hypotheses. Under each hypothesis, the probability density function is assumed to be Gaussian distribution with the same variance ($\sigma^2 = 1$) but with different means $0, s, 2s$, and $3s$, respectively. Figure 12.9 plots the final probability of misclassification at the FC with varying SNR values. Note that this probability of misclassification is empirically found by performing $N_{mc} = 5{,}000$ Monte Carlo runs. As we can observe, the performance of the scheme improves with increasing SNR and approaches 0 as early as 5 dB. In order to show the fault-tolerance capability of the scheme, consider the case when the links between the levels are binary symmetric channels with crossover probabilities $\beta = 0.05$ and $\beta = 0.1$. As shown in Figure 12.9, the ECOC-based scheme still performs reasonably well even in the presence of imperfect data due to nonideal channels modeled as binary symmetric channels.

12.4.2 Distributed estimation

Building on the classification results, the problem of estimation of a parameter θ was also considered in [13] using tree-based networks. The parameter estimation problem was broken into a sequence of M-ary decision making problems, and each of these M-ary decision-making problems was solved using a technique similar to the approach discussed in [9].

Each node j at level k, for $1 \leq k \leq K - 1$, performs four basic operations (see Figure 12.10):

- Collects data from its successors $S^{k+1}(j)$ and collaborates with other nodes at level k.

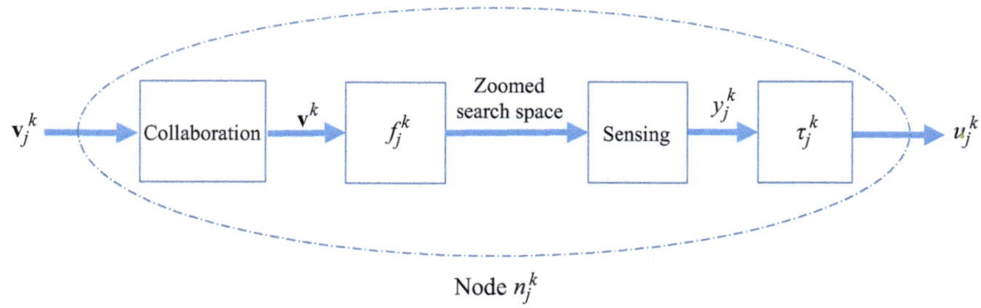

Figure 12.10 Data processing for distributed estimation at node j at level $1 \le k \le K-1$. Here, $\mathbf{v}_j^k \in \{0,1\}^N$, $\mathbf{v}^k \in \{0,1\}^{N^k}$, $y_j^k \in \mathbb{R}$, and $u_j^k \in \{0,1\}$. Therefore, the mappings are $f_j^k : \{0,1\}^{N^k}, \rightarrow \{1,\ldots,M\}$ and $\tau_j^k : \mathbb{R} \rightarrow \{0,1\}$ [13]

- Decides the new parameter space (hypotheses to test) by fusing data using fusion rule $f_j^k(\cdot)$.
- Acquires observation y_j^k and performs hypothesis testing to determine the new parameter space.
- Compresses the observation at node j about the hypothesis (new parameter space) and transmits a 1-bit version to the predecessor $P^{k-1}(j)$ using the transmission mapping $\tau_j^k(\cdot)$.

For the leaf nodes (level K), there are no successors and, therefore, only the third and fourth operations need to be performed. The FC (level "0") collects data from its successors and makes the final decision regarding the region where θ belongs. Transmission mappings and fusion rules for nodes at all levels were optimized (maximize the estimation performance at the FC) in a recursive manner similar to the previous section. It was shown in [13] that the proposed scheme is asymptotically optimal when both N and K tend to infinity.

Simulation results were provided to evaluate the performance of the proposed scheme. As before, consider a tree network $T(3,7)$ consisting of a total $N_{total} = 400$ nodes, including the FC. The observation at each node is Gaussian distributed with unknown mean θ and variance σ^2. This unknown parameter θ is uniformly distributed in $(0, \theta_{\max})$ where the region size is varied by varying the maximum value $\theta_{\max}$. At each level, the nodes perform an M-ary classification where $M = 4$.

Figure 12.11 shows the MSE between the true value of θ and its estimate $\hat{\theta}$ at the FC as a function of $\theta_{\max}$ when $\sigma^2 = 1$. This value of MSE was empirically found by performing $N_{mc} = 5{,}000$ Monte Carlo runs. As we can observe, the performance of the scheme gets worse with increasing region size. This is because, when the range of θ is increased while the total number of possible estimates remains fixed, the error due to quantization increases. The authors in [13] also simulated the case when the links between the levels are modeled as binary symmetric channels with crossover probability $\beta = 0.1$. As shown in Figure 12.11, the proposed scheme is quite robust to the presence of imperfect data arising due to nonideal channels modeled as binary

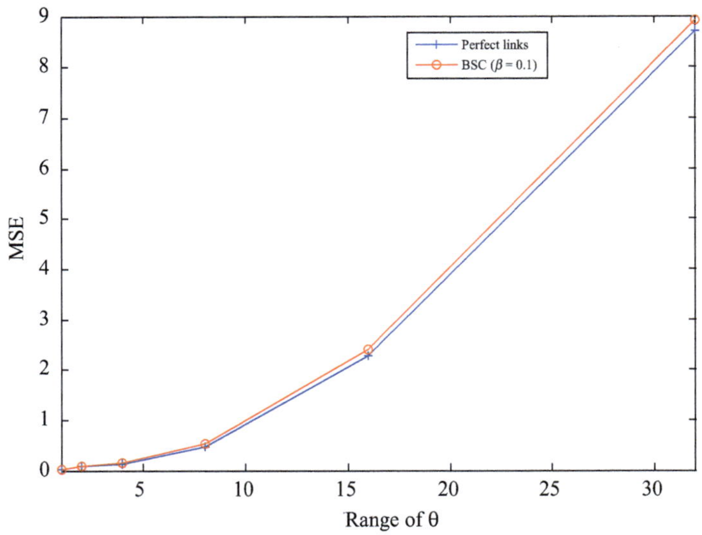

Figure 12.11 MSE as a function of the range of θ [13]

symmetric channels. As alluded to before, this robustness in performance is due to the use of ECOC.

12.5 Classification using crowdsourced data

In this section, we show that the basic concept of using ECOC can also be used in crowdsourcing systems to achieve reliable classification. In [14], ECOC was used to design the questions to be posed to the crowd workers performing an image classification problem. Consider the task of classifying a dog image into one of four breeds: Pekingese, Mastiff, Maltese, or Saluki [14]. Due to the nonexperts present in the crowd, the questions should be simple. As an example, the binary question of whether a dog has a snub nose or a long nose differentiates between {Pekingese, Mastiff} and {Maltese, Saluki}, whereas the binary question of whether the dog is small or large differentiates between {Pekingese, Maltese} and {Mastiff, Saluki}. Using ECOC and task taxonomy information, such questions can be derived that provide reliably inference even with unreliable workers.

The code matrix is designed as an optimization problem by maximizing the performance as a function of the code matrix. For this purpose, the theoretical performance is characterized as a function of the code matrix by developing a statistical model of the system. As part of modeling, worker j decides the true class (local decision y_j) with probability p_j and makes the wrong classification with uniform probability:

$$p(y_j|H_m) = \begin{cases} p_j & \text{if } y_j = m \\ \frac{1-p_j}{M-1} & \text{otherwise,} \end{cases} \tag{12.11}$$

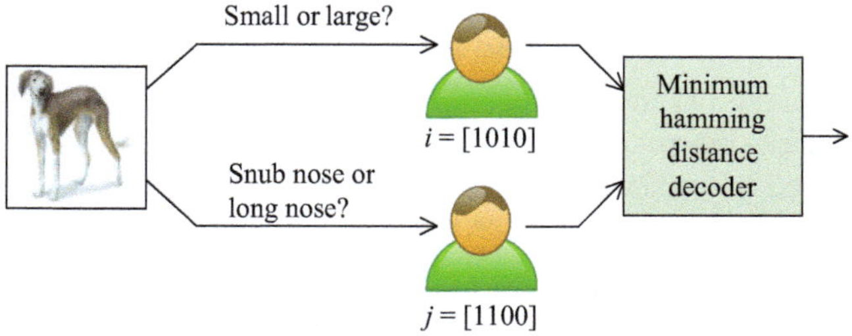

Figure 12.12 A schematic diagram showing binary questions posed to workers and the decoding rule used by the task manager [14]

This uniform noise model is the "worst case" in terms of information and would relate to an upper bound on the performance of the system. For every worker j, let a_j be the corresponding column of $\mathbf{A}$ and recall hypothesis $H_l \in \{H_0, H_1, \ldots, H_{M-1}\}$ is associated with row l in $\mathbf{A}$. The local workers send a binary answer u_j based on decision y_j and column a_j. An illustrative example is shown in Figure 12.12 for the dog breed classification task above. Let the columns corresponding to the ith and jth workers be $a_i = [1010]'$ and $a_j = [1100]'$, respectively. The ith worker is asked: "Is the dog small or large?" since the worker is to differentiate between the first (Pekingese) or third (Maltese) breed and the others. The jth worker is asked: "Does the dog have a snub nose or a long nose?" since the worker is to differentiate between the first two breeds (Pekingese, Mastiff) and the others. These questions can be designed using taxonomy and dichotomous keys [15]. Knowing that Pekingese and Maltese are small dogs while the other two breeds are large, we can design the appropriate question as "Is the dog small or large?" for ith worker whose corresponding column is $a_i = [1010]'$. The task manager makes the final classification as the hypothesis corresponding to the codeword (row) that is closest in Hamming distance to the received vector of decisions.

Due to the anonymity of crowds, the specific reliability of any specific worker cannot be estimated and hence the reliability p_j is modeled to be drawn from a common distribution that characterizes the crowd. In [14], three different crowd models are considered: individual and independent crowd workers; crowd workers governed by peer-dependent reward schemes; and crowd workers with common sources of information. We describe the individual and independent crowd workers in this section and refer the reader to [14] for further details.

The system consisting of individual and independent workers can be modeled as one where the workers' reliabilities are drawn i.i.d. from the specific distribution. Two crowd reliability models, namely, a spammer–hammer model and a beta model are typically considered. In a spammer–hammer model, the crowd consists of two kinds of workers: spammers and hammers. Spammers are unreliable workers that make a decision at random, whereas hammers are reliable workers that make a decision with high reliability. The quality of the crowd, Q, is governed by the fraction of hammers. In a beta model, the reliabilities of workers are drawn from a beta distribution with parameters α and β.

For the ECOC-based approach to reliable crowdsourcing, its performance is determined in terms of average misclassification probability for classification under minimum Hamming distance decoding. Suppose N workers take part in an M-ary classification task. Let $\mathbf{p}$ denote the reliabilities of these workers, such that p_j for $j = 1, \ldots, N$ are i.i.d. random variables with mean μ. Define this to be an (N, M, μ) crowdsourcing system.

Proposition 12.2. *[14] Consider an (N, M, μ) crowdsourcing system. The expected misclassification probability using code matrix $\mathbf{A}$ is:*

$$P_e(\mu) = \frac{1}{M} \sum_{i,l} \prod_{j=1}^{N} \Big[\left(\mu a_{lj} + \frac{(1-\mu)}{(M-1)} \sum_{k \neq l} a_{kj} \right) (2i_j - 1) + (1 - i_j) \Big] C_i^l, \tag{12.12}$$

where $\boldsymbol{i} = [i_1, \ldots, i_N] \in \{0, 1\}^N$ is the received codeword and C_i^l is the cost associated with a global decision H_l when the received vector is $\mathbf{i}$. This cost is:

$$C_i^l = \begin{cases} 1 - \frac{1}{\rho} & \text{if } \mathbf{i} \text{ is in decision region of } H_l \\ 1 & \text{otherwise.} \end{cases} \tag{12.13}$$

where ρ is the number of decision regions[1] $\boldsymbol{i}$ belongs to; ρ can be greater than one when there is a tie at the task-manager and the tie-breaking rule is to choose one of them randomly.

To get more insight, a crowdsourcing system with coding was simulated as follows: N workers take part in a classification task with $M = 8$ equiprobable classes. A good code matrix $\mathbf{A}$ for $N = 15$ and $M = 8$ is found by cyclic column replacement [4] to optimize the derived performance measures in [14]:

$$\mathbf{A} = [150, 150, 90, 240, 240, 153, 102, 204, 204, 204, 170, 170, 170, 170, 170]. \tag{12.14}$$

The code matrix for the system with $N = 90$ and $M = 8$ is formed suboptimally by concatenating the columns of (12.14) six times due to the computational cost to optimize for large system size.

Figure 12.13 shows the performance when $M = 8$ and N takes the two values: $N = 15$ and $N = 90$. These figures suggest that the gap in performance generally increases for larger system size. Similar observations have been made for the beta model of crowds, see [14] for details.

Similar observations can also be made for a system consisting of peer-dependent reward scheme with dependent observations. Figure 12.14 shows the performance of such a system using both majority- and coding-based approaches. The plots are for a system with N crowd workers performing an 8-ary classification task ($M = 8$).

[1]For each H_l, the set of $\mathbf{i}$ for which the decision H_l is taken is called the decision region of H_l.

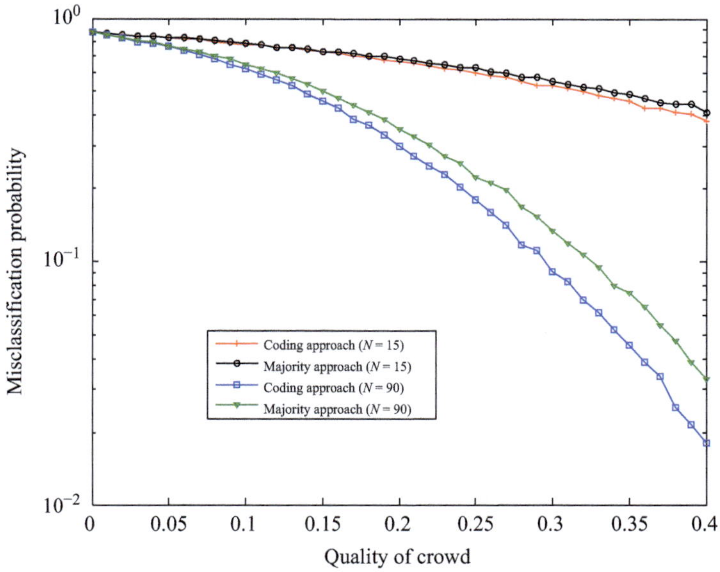

Figure 12.13 Misclassification probability as a function of crowd quality using ECOC- and majority-based approaches with the spammer–hammer model, (M = 8) [14]

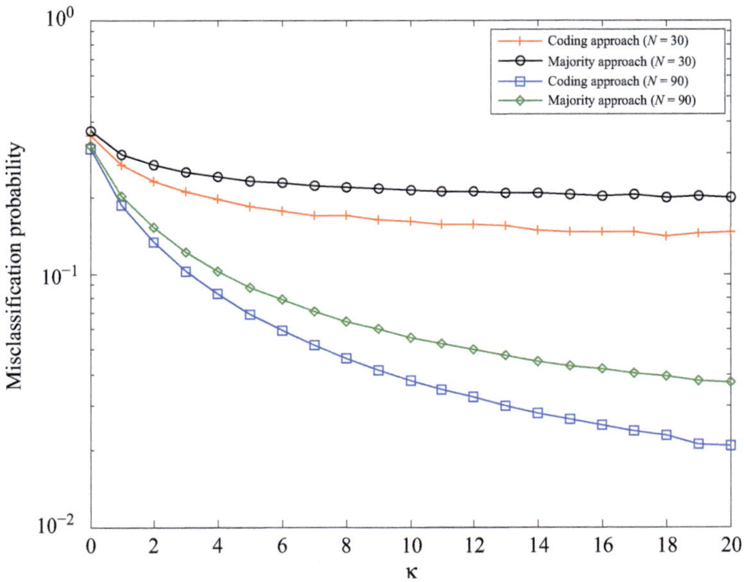

Figure 12.14 Misclassification probability as a function of κ (which denotes the amount of dependent information) using ECOC- and majority-based approaches with the beta ($\alpha = 0.5$, $\beta = 0.5$) model and $\rho = -0.5$, (M = 8) [14]

12.6 Discussion

In this chapter, we considered the general framework of distributed classification, localization, and estimation problems in both parallel and tree networks. We discussed an analytically tractable scheme to solve these problems which are asymptotic optimality. Since the discussed schemes use ECOC, they work well even in scenarios with unreliable data. It should be pointed out that the discussed schemes are not limited to wireless sensor networks, although the application of wireless sensor networks has been considered for most part in this chapter. To show the applicability beyond WSNs, a crowdsourcing setup was discussed in the final section. There are still many interesting problems that remain to be explored in the future work such as an analytical study of the sensitivity of the performance of the proposed scheme in the erroneous links case and the imperfect knowledge of the statistical distribution of the observations.

Acknowledgments

This work was performed in part under the auspices of the US Department of Energy by Lawrence Livermore National Laboratory under contract DE-AC52-07NA27344 (LLNL-BOOK-745286) and ARO under Grant W911NF-14-1-0339.

References

[1] Veeravalli VV, and Varshney PK. Distributed inference in wireless sensor networks. Philos Trans R Soc A. 2012;370(1958):100–117.
[2] Lamport L, Shostak R, and Pease M. The Byzantine generals problem. IEEE Trans Autom Control. 1982;4(3):382–401.
[3] Dietterich TG, and Bakiri G. Solving multiclass learning problems via error-correcting output codes. J Artif Intell Res. 1995;2:263–286.
[4] Wang TY, Han YS, Varshney PK, *et al.* Distributed fault-tolerant classification in wireless sensors networks. IEEE J Sel Areas Commun. 2005;23(4):724–734.
[5] Chair Z, and Varshney PK. Optimal data fusion in multiple sensor detection systems. IEEE Trans Aerosp Electron Syst. 1986;AES-22(1):98–101.
[6] Wang TY, Han YS, Chen B, *et al.* A combined decision fusion and channel coding scheme for distributed fault-tolerant classification in wireless sensors networks. IEEE Trans Wireless Commun. 2006;5(7):1695–1705.
[7] Wang TY, Han YS, and Varshney PK. Fault-tolerant distributed classification based on non-binary codes in wireless sensor networks. IEEE Commun Lett. 2005;9(9):808–810.
[8] Yao C, Chen PN, Wang TY, *et al.* Performance analysis and code design for minimum Hamming distance fusion in wireless sensor networks. IEEE Trans Inf Theory. 2007;53(5):1716–1734.

[9] Vempaty A, Han YS, and Varshney PK. Target localization in wireless sensor networks using error correcting codes. IEEE Trans Inf Theory. 2014;60(1): 697–712.
[10] Berkhin P. Survey of clustering data mining techniques. Accrue Software, Inc., San Jose, CA, USA; 2002.
[11] Niu R, and Varshney PK. Target location estimation in sensor networks with quantized data. IEEE Trans Signal Process. 2006;54(12):4519–4528.
[12] Lin Y, Chen B, and Varshney PK. Decision fusion rules in multi-hop wireless sensor networks. IEEE Trans Aerosp Electron Syst. 2005;41(2):475–488.
[13] Kailkhura B, Vempaty A, and Varshney PK. Distributed inference in tree networks using coding theory. IEEE Trans Signal Process. 2015;63(14): 3715–3726.
[14] Vempaty A, Varshney LR, and Varshney PK. Reliable Classification by Unreliable Crowds. In: Proc. IEEE Int. Conf. Acoust., Speech, Signal Process. (ICASSP 2013); 2013. p. 5558–5562.
[15] Rocker J, Yauch CM, Yenduri S, *et al.* Paper-based dichotomous key to computer based application for biological identification. J Comput Sci Coll. 2007;22(5):30–38.

Chapter 13
Conclusions and future perspectives

This edited book has dealt with data fusion in wireless sensor networks (WSNs) from a statistical signal-processing perspective. The effective use of data fusion in sensor networks is not new and has had extensive application to surveillance, security, traffic control, health care, environmental and industrial monitoring in the last decades. However, the rising paradigms of Internet-of-Things (IoT) and cyber-physical systems is fueling the huge, steadily increasing and pervasive (i.e., virtually in every spot) presence of wireless-connected network-enabled devices, in most cases inexpensive sensors, in everyday life (including most workplaces). On one hand, this represents an unprecedented opportunity toward the accomplishment of the innovative concepts of Smart Cities, Smart Factories, Industry 4.0 and Society 5.0. On the other hand, the stringent bandwidth and energy limitations imposed by the network coupled with sensor-fabrication (e.g., size and cost) constraints makes the design of inference systems, able to fuse sensors' information to obtain high-level information of a certain phenomenon of interest, very challenging. Such picture is worsened by the obvious necessity of dealing with measurements drawn from heterogeneous (multimodal) sensors, which provide information-diversity and enable the collection of different kinds of data (in big volumes) about the surrounding scenario.

The aim of this book was to provide a systematic overview of the relevant issues and corresponding methodological solutions to some of the key milestones within this (wide) research field. These include the **sensing** and **reporting** phases, the **architectural choice** of the WSN (with the corresponding implications) and some selected **cross-layer aspects**, imposed by both the context and novel technological advancements. We remark that, although there is (and there will be) a strong interplay among the above aspects in the design of "smart" WSNs (i.e., enabled by data fusion techniques), the book has been divided into four parts for reader's convenience. The reason has been to stress each aspect and the corresponding issues originating from each milestone. Finally, the aims of this book have been accomplished, according to our view, only by analyzing, studying and contributing to these milestones by means of solid methodological tools belonging to statistical signal processing.

The **first part** of the book focused on problems concerning the sensing phase, i.e., the whole WSN process which is to collect the information and prepare (viz., process) it before transmission to a central unit (i.e., the fusion center in the centralized case) or to peer sensor nodes (in the distributed case) to perform data fusion. The sensing phase is tightly coupled with and severely constrained by the bandwidth- and energy-budgets enforced on each sensor in the network. Therefore, advanced compression methodologies (such as those originated from compressive sensing theory) and aggressive quantization strategies have been shown to be the main tools for coping with such constraints, since complete signal reconstruction is not necessary. However, these measurement-coarsening techniques require a suitable design for inference management. Equally important, in most cases, due to a lack of knowledge of the parameters of the phenomenon of interest to be inferred, it is not possible to resort to theoretically optimal solutions. Accordingly, one alternative class of (suboptimal) approaches is represented, for detection problems, by composite-hypothesis testing tools, such as the generalized likelihood-ratio test (when its complexity is feasible for the problem under consideration) or by generalized score tests, which are usually less computationally demanding. Differently the second class of approaches, being data-driven in its nature, falls within the deep learning umbrella, which has been shown to be promising in acquiring signals to perform a variety of inference tasks (e.g., detection, classification and estimation).

Future perspectives in the context of sensing phase are represented by the design of more sophisticated fusion techniques for accurately modeled (inference-related) phenomena, jointly taking into account wise compression and rough quantization. While the latter two aspects are imposed by the technological constraints of increasingly cheaper sensor nodes operating in an increasingly crowded environment (e.g., the massive IoT paradigm), the sensing-model uncertainty aspect remains unavoidable in spite of huge design efforts and should be carefully taken into account by the WSNs researchers/engineers. The conception and design of fusion rules robust to uncertainties may be accomplished with either the use of model-based techniques (e.g., composite-hypothesis testing tools) or data-driven techniques (e.g., deep learning) or, more likely, a wise combination of both. Indeed, while the former class retains the power of phenomenon-explainability, it is severely challenged by the need for solving more and more complicated WSN-based inference tasks. On the other hand, while the latter class has the advantage of not requiring an explicit model definition, the huge-amount-of-data requirement (not even available at a single learning point, as in the case of WSNs) and the gray-box nature of such techniques still constitutes an existing gap toward practical WSN applications.

The **second** and **third parts** covered issues related to the reporting phase (i.e., the process of information exchange), in centralized (to the fusion center) and decentralized (among peer sensor nodes) architectures, respectively. The motivation for both architectural choices was to consider a complete spectrum of alternatives. Indeed, the former class usually requires a lower energy-consumption in the reporting phase, whereas the latter class avoids the single-point-of-failure problem represented by the fusion center. Equally important, reporting-phase issues are better understood in centralized architectures and any theoretical advancement (or even finding) in such a case naturally reflects on the distributed case. One of the main outcomes of this part is that the choice of a certain communication protocol (mostly induced by the WSN constraints) introduces peculiar challenges and issues in the design of effective fusion rules (and, more generally, WSN-based inference systems). Such outcome significantly highlights the interplay between communication theory and array processing in the design of WSNs. For example, this was clearly shown with reference to the use of (distributed) multiple-input multiple-output (MIMO) paradigm in the reporting phase, with its very appealing massive MIMO counterpart (i.e., the use of a large-size antenna array at the fusion center). On the other hand, in the case of distributed architectures, the design of fusion rules, whether for detection or estimation purposes, is very challenging and the convergence of such networks to the (hopefully true) desired inference results is not an easy task at all. This holds true even in the case of very simple reporting channel models. Therefore, part of current efforts has been devoted to the design of distributed counterpart to well-known fusion techniques and analyzing their convergence properties as a function of the network, including the nodes (time-varying) connectivity.

Future perspectives in this context will account for the (only conceptually-) straightforward integration of advanced communication protocols within distributed architectures, with the corresponding design of sophisticated fusion techniques and their convergence & stability analysis. On the other hand, a significant interest is also represented by the adoption of innovative communication protocols (such as millimeter-wave transmission) into WSNs which, as shown in this part of the book, cannot simply mimic results from communication theory.

Finally, the **last part** of the book emphasized the benefits and consequences originating from cross-layer aspects, such as those related to *energy harvesting* and *security* (both from passive and active viewpoints). Indeed, these two topics have obtained the raising interest of the scientific community due to the potential capabilities in extending WSNs lifetime and increasing their privacy and data-integrity level (in the face of the shared wireless medium and the growing presence of malicious parties apt to information leakage and deception). Clearly, the presence of energy-harvesting sensor nodes constitutes itself a true technological advancement to WSN lifetime extension. However, the path to the design of highly energy-efficiency sensor policies (related to sleeping cycles, energy-spent during a transmission slot, etc.) is still long and heavily depends on the accuracy of the predicted battery level, which is challenging due to the high-uncertainty and variability of the harvested source. On the other hand, the sensitivity of data-fusion techniques to passive or active malicious attacks and the consequent design of techniques which exhibit robustness to security issues is the object of current research and has been one of the objectives of the last part.

Future perspectives in the context of energy harvesting will account for the derivation of more realistic energy-harvesting models, allowing to predict very accurately the battery level of each sensor in the network and to jointly design sensor policies and fusion techniques making an optimal use of available energy. On the other hand, additional efforts toward the understanding of theoretical limits of secure inference need to be done, considering a varying degree of knowledge of the phenomenon to be inferred by malicious parties. Indeed, their presence and the corresponding design of privacy-robust sensor-policies (and fusion techniques) theoretically implies a performance degradation with respect to a nominal (i.e., "secure-free") WSN, which can be only mitigated with finer processing. In this respect, advanced channel-access techniques discussed in the second part of the book may represent one of the key enabling technologies.

Index